AF678504

Amphiphiles: Molecular Assembly and Applications

ACS SYMPOSIUM SERIES **1070**

Amphiphiles: Molecular Assembly and Applications

Ramanathan Nagarajan, Editor
Natick Soldier Research, Development & Engineering Center (NSRDEC)

Sponsored by the
ACS Division of Colloid and Surface Chemistry

American Chemical Society, Washington, DC

Distributed in print by Oxford University Press, Inc.

Library of Congress Cataloging-in-Publication Data

Amphiphiles : molecular assembly and applications / Ramanathan Nagarajan, editor.
p. cm. -- (ACS symposium series ; 1070)
Includes bibliographical references and index.
ISBN 978-0-8412-2650-0
1. Amphiphiles. 2. Hydrophile-lipophile balance. 3. Self-assembly--Chemistry. I. Nagarajan, R. (Ramanathan)
QD382.A47A47 2011
547.7--dc23

2011031439

The paper used in this publication meets the minimum requirements of American National Standard for Information Sciences—Permanence of Paper for Printed Library Materials, ANSI Z39.48n1984.

Copyright © 2011 American Chemical Society

Distributed in print by Oxford University Press, Inc.

All Rights Reserved. Reprographic copying beyond that permitted by Sections 107 or 108 of the U.S. Copyright Act is allowed for internal use only, provided that a per-chapter fee of $40.25 plus $0.75 per page is paid to the Copyright Clearance Center, Inc., 222 Rosewood Drive, Danvers, MA 01923, USA. Republication or reproduction for sale of pages in this book is permitted only under license from ACS. Direct these and other permission requests to ACS Copyright Office, Publications Division, 1155 16th Street, N.W., Washington, DC 20036.

The citation of trade names and/or names of manufacturers in this publication is not to be construed as an endorsement or as approval by ACS of the commercial products or services referenced herein; nor should the mere reference herein to any drawing, specification, chemical process, or other data be regarded as a license or as a conveyance of any right or permission to the holder, reader, or any other person or corporation, to manufacture, reproduce, use, or sell any patented invention or copyrighted work that may in any way be related thereto. Registered names, trademarks, etc., used in this publication, even without specific indication thereof, are not to be considered unprotected by law.

PRINTED IN THE UNITED STATES OF AMERICA

Foreword

The ACS Symposium Series was first published in 1974 to provide a mechanism for publishing symposia quickly in book form. The purpose of the series is to publish timely, comprehensive books developed from the ACS sponsored symposia based on current scientific research. Occasionally, books are developed from symposia sponsored by other organizations when the topic is of keen interest to the chemistry audience.

Before agreeing to publish a book, the proposed table of contents is reviewed for appropriate and comprehensive coverage and for interest to the audience. Some papers may be excluded to better focus the book; others may be added to provide comprehensiveness. When appropriate, overview or introductory chapters are added. Drafts of chapters are peer-reviewed prior to final acceptance or rejection, and manuscripts are prepared in camera-ready format.

As a rule, only original research papers and original review papers are included in the volumes. Verbatim reproductions of previous published papers are not accepted.

ACS Books Department

Contents

Indexes

Preface

This book results from the symposium "Surfactants and Amphiphilic Polymers: Fundamentals and Applications" held at the 239th ACS National Meeting in San Francisco, CA on March 21-25, 2010. The symposium was sponsored by the ACS Division of Colloid and Surface Chemistry as part of its continuing symposia series on surfactants and association colloids. Approximately 70 papers were presented at this symposium and the book chapters represent a sampling of problems addressed.

Amphiphilic molecules, whether they are low molecular weight conventional surfactants or high molecular weight amphiphilic polymers, display characteristic molecular self-assembly behavior in solutions, at interfaces and in bulk, generating nanoscale structures. These nanoscopic assemblies are being widely exploited for applications in materials, energy, pharmaceutical and biomedical technologies. There is great interest in the design of novel amphiphilic systems that will give rise to interesting structural morphologies following molecular assembly. This would allow the amphiphilic systems to be tailored for specific applications. Successful development of such amphiphilic systems requires (i) methods to synthesize novel amphiphilic molecules, (ii) understanding the molecular assembly behavior in a variety of environments as a function of the chemical structure of the amphiphiles, and (iii) finding ways to utilize the molecular assemblies in a robust way in practical applications. The proposed book addresses these essential themes in the study of amphiphilic systems.

The papers incorporated in the book cover a wide variety of amphiphiles including low molecular weight surfactants with novel ionic head groups, surfactants with chirality, peptoid amphiphiles, block copolymers, and block polyelectrolytes. In addition, other self assembling systems such as self-assembled monolayers of small molecules and polymer brushes are considered. Each of the papers includes one or more of the following aspects: synthesis of amphiphiles with novel chemical features, characterizing the assembly of amphiphiles in solutions, bulk or at interfaces, application of the amphiphilic system, or modeling the assembly behavior of the amphiphilic system.

The book starts with comparative molecular scale theories of self-assembly of low molecular weight surfactants and amphiphilic block copolymers. In this first Chapter, it is shown that analogous formulations of free energy models for aggregation coupled to molecular packing considerations can predict the self-assembly behavior for both classes of amphiphiles. Changes in the morphology of aggregates resulting from variations in the chemical structures of the amphiphile or in the solution conditions are satisfactorily predicted by the phenomenological theory.

The molecular structural features of amphiphiles and their impact on self-assembly and/or potential applications are considered in many chapters. Ilies and coworkers have presented a comprehensive discussion of design of amphiphiles with pyridinium polar head for nucleic acid delivery in Chapter 2. The impact of various structural characteristics of the amphiphile (hydrophobic chain type, linkage type, counterion) on the transfection activity and cytotoxicity is discussed for various classes of pyridinium amphiphiles (detergents, lipids, gemini surfactants, lipophilic polycations). In Chapter 3, Robinson et al discuss their synthesis of a family of novel amphiphile, helical peptoids, with structural motifs similar to the m13 bacteriophage. They show that is possible to tune the interaction of peptoids with double-stranded DNA through variation of the side group sequence, thereby suggesting a potential use of such peptoids in gene delivery and structural DNA applications. Carbohydrate surfactants based on uronic acid are described in Chapter 4. In this chapter, experimental measurements of both dynamic and equilibrium properties at air-liquid and liquid-liquid interfaces for the uronic amphiphiles relevant to their use in emulsions and foams are reported.

Lo Nostro et al, discuss the self-assembly and interfacial properties of amphiphiles with chiral head groups, L-ascorbyl-alkanoates (L-ASC) and D-isoascorbyl-alkanoates (D-ASC), in Chapter 5, in terms of the differing stereochemistries of the surfactant head groups that determine the inter- or intramolecular interactions. Using differential scanning calorimetry, infrared spectra, and X-ray diffraction experiments in the solid and in the coagel states, they reveal chirality dependent phase behavior for mixtures of L-ASC12 and D-ASC12. In Chapter 6, Klapper et al show that oil-in-oil type of emulsions can be effectively stabilized by amphiphilic polymers. Using polystyrene-b-polypentafluro styrene and polyisoprene-b-polymethyl methacrylate, they formulated stable emulsions of *n*-hexane / DMF and toluene / perfluoro methyl cyclohexane which were then used as aprotic reaction media for emulsion polymerization applications. The molecular assembly between fluorinated polyelectrolytes such as perfluoro sulfonic acid PTFE copolymer (i.e. Nafion®) and poly(vinyl alcohol) (PVA) or poly(allyl amine) (PAH) is shown to involve fluorinated hydrogen bonds by Farhat. In this chapter, the formation of hydrogels from these systems is described and it is shown that these hydrogels can effectively replace liquid electrolyte solutions for the purpose of making miniature solid-state electrochemical systems.

Novel characterization methods have been employed to look at the structural and morphological properties of self-assembled nanostructures. Electrokinetic sonic amplitude technique has been used by Eisenbach and coworkers to determine the nature of adsorbed layers of amphiphilic polymers on hydrophilic titanium dioxide and hydrophobic copper phthalocyanine surfaces, in Chapter 8. The contrasting impact of the surface-polymer interactions on the morphology of the polymer adsorbed layer is discussed. Atomic force microscopy and neutron reflectivity measurements are used by Takahar et al in Chapter 9 to study the thickness of polymer brush under different salt conditions. Existence of attractive electrostatic interactions at low salt concentrations and their screening at high salt concentrations are shown to dramatically change the structure and dimensions of the polymer at the interface. The application of quartz crystal microbalance with dissipation monitoring (QCM-D) for determining the viscoelastic properties of soft materials, such as vesicles and microemulsions near surfaces is reviewed by Lane in Chapter 10. Knock and Sanii use QCM-D as a mass sensing device to examine the adsorption of single and double chain ionic surfactants on hydrophilic and hydrophobic surfaces in Chapter 11. Atomic force microscopy, contact angle goniometry and Diffuse Reflectance Infrared Fourier Transform Spectroscopy were used to study the stability of molecular films assembled on steel surfaces In Chapter 12. Here, Lim et al assembled a molecular layer of amphiphilic alkyl phosphonic acid on the steel surface and thermally reacted to create a surface resistant to biofouling for application in medical devices.

A new class of amphiphiles termed amphiphilic inverted polymers with hydrophobic and hydrophilic moieties, alternately distributed along the main polymer backbone are described in Chapter 13. Here, Voronov and coworkers report the micelle formation properties of these polymers including critical micelle concentration, micelle size and solubilization and also their ability to invert their morphologies when the solvent polarity is modified. Biesalski et al have synthesized a new class of amphiphiles by modifying di-acetylenic fatty acids at the carboxylic acid head-group with either cell-adhesive or mussel-adhesive peptide sequences. In Chapter 14, they show that the resulting amphiphiles can be stabilized by UV-light induced polymerization of the hydrophobic tail into well-defined structures that allow the presentation of the adhesive groups at the vesicular surface.

Applications of self-assmbled amphihliar structures to material synthesis are discussed in many chapters. Mandal and Kruk manipulate block copolymer liquid crystal structures by controlling the amount of hydrocarbon solubilized and have used them as templates to produce large-pore organosilicas. In Chapter 15 they describe the preparation of organosilica materials with nominal pore sizes of 18 and 10 nm employing a synthesis procedure carried out at room temperature. In contrast, a template free synthesis of nanocapsules by polymerization of the macrocyclic compound, resorcinarenes, is described by Balasubramanian and Kalaitzis in Chapter 16. The photopolymerization is found to yield nanostructures of different morphologies besides nanocapsules, depending on the choice of the solvent. In Chapter 17, the preparation of polymer capsules with oil cores is reported by Adalsteinsson et al using a kinetically stable, miniemulsion as the template

The book also includes theoretical and computer simulation studies of new problems in amphiphile self-assembly. Molecular dynamic simulations in combination with statistical mechanics have been applied by Nielsen group to predict assembly of nanoparticles and surfactants at the oil-water interface for systems consisting only of nanoparticles, competing or cooperating mixtures of nanoparticles and nonionic surfactants and for nanoparticles chemically functionalized with surfactants on their surfaces in Chapter 18. The simulation results for molecular organization behavior as well as for the reduction of oil-water interfacial tensions are presented.

The contributing authors for thus book come from a wide cross-section of US universities and national laboratories and a third of the contributors are from abroad. The affiliations of the authors are representative of the broad theoretical and practical interest in amphiphile systems and their applications.

The editor is thankful to the numerous reviewers. The reviewers of individual chapters contributed prompt critical reviews that have helped improve the quality of the manuscripts. The Editor acknowledges support from ACS Books, especially Tim Marney, with the use of the newly introduced on-line manuscript submissions process.

The Editor acknowledges the support from Natick Soldier Research, Development & Engineering Center, that allowed him to organize the symposium and develop this book based on selected contributions.

R. Nagarajan

Molecular Sciences and Engineering Team
Natick Soldier Research, Development & Engineering Center
Natick, MA

Chapter 1

Amphiphilic Surfactants and Amphiphilic Polymers: Principles of Molecular Assembly

R. Nagarajan*

Molecular Sciences and Engineering Team, Natick Soldier Research, Development & Engineering Center, Natick MA 01760
***Ramanathan.Nagarajan@us.army.mil**

Amphiphilic surfactants and polymers display characteristic molecular self-assembly behavior in solutions, at interfaces and in bulk, generating nanoscale structures of different shapes. These nanoscale features determine many characteristics of these amphiphiles, relevant for their practical applications in materials, pharmaceutical and biomedical technologies. The ability to generate desired nanoscale morphologies by synthesizing novel amphiphiles would allow the amphiphilic systems to be tailored for specific applications. Critical to achieving this goal is an understanding of the link between the molecular structure of the amphiphiles and their self-assembly behavior. In this chapter, we review the principles of self-assembly for both conventional low molecular weight surfactants and amphiphilic block copolymers. The main emphasis is on demonstrating how general principles of thermodynamics and considerations of molecular packing together help predict the self-assembled morphologies given the amphiphile molecular structure. The similarity between the behavior of classical surfactants and amphiphilic block copolymers are identified. The self-assembly behavior of novel amphiphiles can be extrapolated from these fundamental principles.

Not subject to U.S. Copyright. Published 2011 by American Chemical Society

Introduction

The self-assembly of surfactants in solution has been widely investigated both experimentally and theoretically, because numerous practical applications take advantage of the resulting multimolecular aggregates. The structure of these aggregates influences the properties of surfactant solutions, such as for example, their solubilization capacity for hydrophobic substances or their viscous and viscoelastic properties, and consequently, the performance of surfactants in various applications. To select molecules that would yield desired structures such as spherical, globular or rodlike micelles, or spherical bilayer vesicles (Figure 1), or to custom-design novel amphiphiles to generate desired aggregate morphologies, it is necessary to know how the molecular structure of the surfactant controls the shape and size of the resulting aggregate.

Tanford (*1*) and Israelachvili, Mitchell and Ninham (*2*) pioneered two of the most important ideas to answer the above question and these ideas currently dominate our understanding of self-assembling systems. Tanford proposed the concept of opposing forces to formulate a quantitative expression for the standard free energy change on aggregation, which is the difference in the standard state free energy between a singly dispersed surfactant molecule present in solution and a surfactant molecule present in an isolated aggregate of any shape in solution. Using this free energy expression, he was able to explain why surfactant aggregates form in aqueous solutions, why they grow, and why they do not keep growing but are finite in size. Israelachvili, Mitchell and Ninham proposed the concept of a molecular packing parameter and demonstrated how the size and the shape of the aggregate at equilibrium can be predicted from a combination of molecular packing considerations and general thermodynamic principles. We can understand conceptually the molecular self-assembly principles governing surfactant behavior based on these two pioneering contributions.

More detailed free energy models for surfactant aggregation have been formulated (*3*, *4*) since the appearance of Tanford's model to predict accurately and quantitatively the aggregation behavior of surfactants. However the simple phenomenological framework provided by Tanford is more than adequate for understanding the self-assembly principles based on which the surfactants act. Therefore, we will use Tanford's free energy model for our analysis, incorporating in it a modification suggested by later theories.

Free Energy Model for Surfactants

Tanford (*1*) proposed a phenomenological model for the standard free energy change associated with the transfer of a surfactant molecule from its singly dispersed state in water to an aggregate of size g (containing g number of surfactant molecules, the aggregate being of any shape), composed of three principal contributions:

$$\left(\frac{\Delta\mu_g^o}{kT}\right) = \left(\frac{\Delta\mu_g^o}{kT}\right)_{\text{Transfer}} + \left(\frac{\Delta\mu_g^o}{kT}\right)_{\text{Interface}} + \left(\frac{\Delta\mu_g^o}{kT}\right)_{\text{Head}} \quad (1)$$

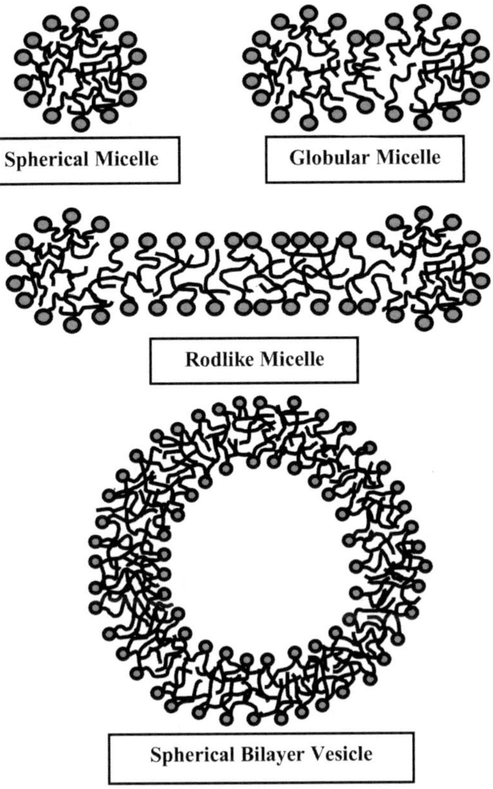

Figure 1. Schematic representation of surfactant aggregates in dilute aqueous solutions.

The first term $(\Delta\mu_g^o / kT)_{Transfer}$ is a negative free energy contribution arising from the transfer of the hydrophobic surfactant tail from its unfavorable contact with water to the hydrocarbon-like environment of the aggregate core. The transfer free energy contribution depends on the surfactant tail but not on the aggregate shape or size. The second term $(\Delta\mu_g^o / kT)_{Interface}$ provides a positive contribution to account for the fact that the entire surface area of the tail is not removed from water but there is still residual contact with water at the surface of the aggregate core. This is represented as the product of a contact free energy per unit area σ (or the water-hydrocarbon interfacial free energy) and the surface area per molecule of the aggregate core, a (namely, the surface area of the core of the aggregate divided by the aggregation number). The third term $(\Delta\mu_g^o / kT)_{Head}$ provides another positive contribution representing the repulsive interactions between the head groups that crowd at the aggregate surface. The repulsions may be due to steric interactions (for all types of head groups) and also electrostatic interactions (dipole–dipole interactions for zwitterionic head groups and ion–ion repulsions for ionic head groups). Since the repulsion would increase if the head groups come close to one another, Tanford proposed an expression with an inverse dependence on a. Thus, the standard free energy change per molecule on aggregation, proposed by Tanford has the form:

$$\left(\frac{\Delta\mu_g^o}{kT}\right) = \left(\frac{\Delta\mu_g^o}{kT}\right)_{Transfer} + \left(\frac{\sigma}{kT}\right)a + \left(\frac{\alpha}{kT}\right)\frac{1}{a} \qquad (2)$$

where α is the head group repulsion parameter, k the Boltzmann constant and T the temperature.

Starting from the free energy model of Tanford, the equilibrium aggregation behavior can be examined either by treating the surfactant solution as consisting of aggregates with a distribution of sizes or by treating the aggregate as constituting a pseudophase (5). If the aggregate is viewed as a pseudophase, in the sense of small systems thermodynamics, the equilibrium condition corresponds to a minimum in the standard free energy change on aggregation, $(\Delta\mu_g^o / kT)$. The minimization can be done with respect to either the aggregation number g or the area per molecule a, since they are dependent on one another through the geometrical relations given in Table 1. One obtains in this manner, the equilibrium condition:

$$\frac{\partial}{\partial a}\left(\frac{\Delta\mu_g^o}{kT}\right) = 0 \;\Rightarrow\; \left(\frac{\sigma}{kT}\right) - \left(\frac{\alpha}{kT}\right)\frac{1}{a^2} = 0 \;\text{ at } a = a_e \Rightarrow$$

$$a_e = \left(\frac{\alpha}{\sigma}\right)^{1/2} \qquad (3)$$

Here, a_e is the equilibrium area per molecule in the aggregate. The equilibrium area per molecule is directly controlled by the repulsions between the surfactant head groups (denoted by the parameter α). The critical micelle concentration (CMC, denoted as X_{CMC} in mole fraction units), in the pseudophase approximation, is obtained from the relation,

$$\ln X_{CMC} = \left(\frac{\Delta\mu_g^o}{kT}\right) = \left(\frac{\Delta\mu_g^o}{kT}\right)_{Transfer} + \left(\frac{\sigma}{kT}\right)a_e + \left(\frac{\alpha}{kT}\right)\frac{1}{a_e} \qquad (4)$$

In Tanford's free energy expression (eq.2), the first contribution, the tail transfer free energy, is negative. Hence, this contribution is responsible for the aggregation to occur. It affects only the CMC (as shown by eq.4) but not the equilibrium area per molecule, a_e (as shown by eq.3) and hence, the size and shape of the aggregate. The second contribution, the free energy of residual contact between the aggregate core and water, is positive and decreases in magnitude as the area per molecule, a, decreases. A decrease in the area per molecule a, corresponds to an increase in the aggregation number g, for all aggregate shapes, as shown on Table 1. Hence, this contribution promotes the growth of the aggregate. The third contribution, the free energy due to head group repulsions, is also positive and increases in magnitude if the area per molecule, a, decreases or the aggregation number, g, increases. Hence, this contribution is responsible for limiting the growth of aggregates to a finite size. Thus, Tanford's model

Table 1. Geometrical Relations for Spherical and Cylindrical Micelles and Bilayers*

Variable	*Sphere*	*Cylinder*	*Bilayer*
Volume of core $V = g\, v_o$	$4\pi R^3/3$	πR^2	$2R$
Surface area of core $A = g\, a$	$4\pi R^2$	$2\pi R$	2
Area per molecule a	$3v_o/R$	$2v_o/R$	v_o/R
Shape parameter v_o/aR	$1/3$	$1/2$	1
Largest aggregation number g_{max}	$4\pi \ell_o^3/3v_o$	$\pi \ell_o^2/v_o$	$2\ell_o/v_o$
Aggregation number g	$g_{max}\,(3v_o/a\ell_o)^3$	$g_{max}\,(2v_o/a\ell_o)^2$	$g_{max}\,(v_o/a\ell_o)$

* Variables V, A, g and g_{max} refer to the entire spherical aggregate, unit length of a cylinder or unit area of a bilayer. R is the radius of spherical or cylindrical micelle or the half-bilayer thickness. v_o and ℓ_o are the volume and extended length of the surfactant tail. g_{max} is the largest aggregation number possible for the given geometry based on the constraint that the aggregate core is filled and the tail cannot stretch beyond its extended length.

clearly identifies why aggregates form, why they grow and why they do not keep growing but remain finite in size.

In formulating the above phenomenological model, Tanford had recognized that the state of the hydrocarbon tail of the surfactant within an aggregate is different from that of a similar hydrocarbon in bulk liquid state, because one end of the surfactant tail is constrained to remain at the aggregate-water interface while the entire tail has to pack within a given aggregate shape maintaining liquid like density in the aggregate core. He accounted for this empirically by making a correction to the estimate for the transfer free energy contribution. Since this correction was taken to be independent of the aggregate shape and size, it cannot accurately account for the subtle molecular packing constraints which must depend on the size and shape of the aggregates. Detailed chain packing models to estimate the packing free energy, $(\Delta\mu_g^o/kT)_{Packing}$ as a function of the aggregate shape and size were developed following different approaches, by Gruen (*6*), Dill and Flory (*7*), Ben-Shaul and Gelbart (*8*), Nagarajan and Ruckenstein (*3*) and Puvvada and Blankschtein (*4*). This contribution has been shown to be important for quantitative predictions of a wide range of self-assembly phenomena (*3, 5, 9*). We will adopt the result from our previous work (*3*) here, because of the analytical nature and simplicity of the free energy expression formulated. This expression takes into account the fact that the surfactant tail has to deform non-uniformly along its length to fill the aggregate core with uniform density. The packing free energy contribution has the form

$$\left(\frac{\Delta\mu_g^o}{kT}\right)_{Packing} = \left(\frac{3\pi^2}{80}\right)\frac{R^2}{NL^2},\quad \left(\frac{5\pi^2}{80}\right)\frac{R^2}{NL^2},\quad \left(\frac{10\pi^2}{80}\right)\frac{R^2}{NL^2} \qquad (5)$$

for spheres, cylinders and bilayers. In the above equation L is a characteristic segment length that is taken to be 4.6 Å (see ref. (*3*) for details) and N is the number of segments in a tail such that $N L^3 = v_o$. Since $R = 3\, v_o/a$, $2\, v_o/a$, and v_o/a for the three geometries (Table 1), the packing free energy contribution can be rewritten as

$$\left(\frac{\Delta\mu_g^o}{kT}\right)_{\text{Packing}} = \frac{Q}{a^2},$$

$$Q_{\text{sph}} = \left(\frac{27}{8}\right) v_o L\,,\; Q_{\text{cyl}} = \left(\frac{20}{8}\right) v_o L\,,\; Q_{\text{bilayer}} = \left(\frac{10}{8}\right) v_o L \qquad (6)$$

where the symbol Q is used to denote the coefficient of $1/a^2$ in the free energy expression and it stands for Q_{sph} , Q_{cyl} or $Q_{bilayer}$ depending upon the aggregate shape. The equilibrium area a_e given before by eq.(3) is now obtained from

$$\frac{\partial}{\partial a}\left(\frac{\Delta\mu_g^o}{kT}\right) = 0 \Rightarrow \left(\frac{\sigma}{kT}\right) - \left(\frac{\alpha}{kT}\right)\frac{1}{a^2} - \frac{2\,Q}{a^3} = 0 \quad \text{at } a = a_e \Rightarrow$$

$$a_e = \left(\frac{\alpha}{\sigma} + \frac{2\,Q/a_e}{\sigma/kT}\right)^{1/2} \qquad (7)$$

Since the variable Q is dependent on the tail, the tail has direct influence over the equilibrium area a_e, when compared to eq.(3) based on the original free energy model of Tanford.

Molecular Packing in Surfactant Aggregates

The concept of molecular packing was introduced by Israelachvili, Mitchell and Ninham (*2*) to relate the equilibrium area per molecule calculated from the free energy considerations to the shape of the equilibrium aggregate. A molecular packing parameter is defined as $v_o/a_e\ell_o$, where v_o and ℓ_o are the volume and the extended length of the surfactant tail and a_e is the surface area of the hydrophobic core of the equilibrium aggregate expressed per molecule in the aggregate. If we consider a spherical micelle with a core radius R, made up of g molecules, then the volume of the core $V = g\, v_o = 4\pi\, R^3/3$, the surface area of the core $A = g\, a_e = 4\pi\, R^2$, and hence $R = 3\, v_o/a_e$, from simple geometrical relations (Table 1). If the micelle core is packed with surfactant tails without any empty space, then the radius R cannot exceed the extended length ℓ_o of the tail. Introducing this constraint in the expression for R, one obtains, $0 \le v_o/a_e\ell_o \le 1/3$, for spherical micelles.

For spherical, cylindrical or bilayer aggregates made up of g surfactant molecules, the geometrical relations for the volume V and the surface area A are given in Table 1. These geometrical relations, together with the constraint that at least one dimension of the aggregate (the radius of the sphere or the cylinder, or the half-bilayer thickness, all denoted by R) cannot exceed ℓ_o, lead to the following well-known (*2*) connection between the molecular packing parameter and the aggregate shape: $0 \le v_o/a_e\ell_o \le 1/3$ for sphere, $1/3 \le v_o/a_e\ell_o \le 1/2$ for cylinder, and $1/2 \le v_o/a_e\ell_o \le 1$ for bilayer.

For common surfactants, the ratio v_o/ℓ_o is a constant independent of tail length, equal to 21 Å^2 for single tail and 42 Å^2 for double tail (*1*). Consequently, only the area a_e reflects the specificity of the surfactant in the packing parameter $v_o/a_e\ell_o$.. Eq.(3) shows that the area a_e is influenced directly by the head group interactions in Tanford's free energy model while eq.(7) shows that the equilibrium area is also influenced by the surfactant tail packing in our extended free energy model. Therefore, if we know the equilibrium area per molecule and hence the molecular packing parameter, the shape and size of the equilibrium aggregate can be readily identified as shown above. This is the predictive sense in which the molecular packing parameter has found significant use in the literature.

Predicting Self-Assembly Behavior of Surfactants

The packing parameter $v_o/a_e\ell_o$ can be estimated using the area a_e obtained from eq.(7). One can observe that a_e will be small and the packing parameter will be large if the head group interaction parameter α is small. The area a_e will increase and the packing parameter will decrease, if the interfacial free energy per unit area σ decreases. If the chain length increases, then Q increases and this will cause an increase in the area a_e. Of the two terms contributing to the area, the head group repulsion term is usually larger than the chain packing term. These simple considerations allow one to predict many features of surfactant self-assembly as summarized below.

Nonionic Surfactants

For nonionic surfactants with oligo ethylene oxide as the head group, the steric repulsions between the head groups can be expected to increase if the number of ethylene oxide units in a head group increases. Correspondingly, the head group parameter α will increase in magnitude if the number of ethylene oxide units in the head group increases. Therefore, when the number of ethylene oxide units is small, α is small, a_e is small, $v_o/a_e\ell_o$ is large and bilayer aggregates (lamellae) are favored. For larger number of ethylene oxide units, α increases, a_e increases, $v_o/a_e\ell_o$ decreases and cylindrical micelles become possible. When the number of ethylene oxide units is further increased, $v_o/a_e\ell_o$ becomes small enough so that spherical micelles will form with their aggregation number g decreasing with increasing ethylene oxide chain length.

Further, the increase in α and a_e for the nonionic surfactants with increasing ethylene oxide chain length gives rise to an increase in the CMC as the head group size increases.

For a given ethylene oxide chain length and an equilibrium packing parameter corresponding to spherical micelles, increasing the chain length of the surfactant tail will cause an increase in the aggregation number of the micelles as predicted by the geometrical relation in Table 1.

Ionic and Zwitterionic Surfactants

Comparing nonionic and ionic surfactants, the head group interaction parameter α will be larger for the ionic surfactants than for the nonionic surfactants, because onc has to also consider ionic repulsions between the head groups in the former case. Therefore, the equilibrium area per molecule, a_e will be larger and $v_o/a_e\ell_o$ will be smaller for the ionics compared to the nonionics. As a result, ionic surfactants would form aggregates of smaller aggregation number compared to nonionic surfactants of the same tail length. Further, the increase in α and a_e for the ionic surfactants also gives rise to an increase in the CMC when ionic and nonionic surfactants of equal tail lengths are compared.

The head group repulsions for zwitterionic surfactants are intermediate between those for ionic and nonionic surfactants. Therefore, the CMC for zwitterionic surfactants and their aggregation numbers will be intermediate between those for ionic and nonionic surfactants of the same tail length.

Solution Conditions

For a given surfactant molecule, the head group repulsion can be modified, by a change in the solution conditions.. For example, adding salt to an ionic surfactant solution decreases electrostatic repulsions between ionic head groups; increasing the temperature for a nonionic surfactant molecule with ethylene oxide head group decreases steric repulsions between the nonionic head groups. Because the head group repulsion parameter α decreases, the equilibrium area per molecule, a_e will decrease and the packing parameter $v_o/a_e\ell_o$ will increase. Thus, one can achieve a transition from spherical micelles to rodlike micelles and possibly to bilayer aggregates, by modifying solution conditions (adding salt to ionic surfactants and increasing the temperature for the nonionic surfactants) that control head group repulsions.

For a given surfactant, one can decrease the aggregate-solvent interfacial tension σ by adding a small molecular weight alcohol such as ethanol or isopropanol to water. Because σ decreases, the equilibrium area per molecule, a_e will increase and the packing parameter $v_o/a_e\ell_o$ will decrease. Thus, one can achieve a transition from bilayer aggregates, to cylindrical micelles and spherical micelles and eventually to absence of micellization by the gradual addition of a low molecular weight alcohol to water.

Double Tail Surfactants

If single tail and double tail surfactant molecules are compared, for the same equilibrium area per molecule, a_e, the double tail molecule will have a packing parameter $v_o/a_e\ell_o$ twice as large as that of the single tail molecule. Therefore, the double tail molecule can self-assemble to form bilayer vesicles while the corresponding single tail molecule aggregates into only spherical or globular micelles.

Surfactant Mixtures

If we consider mixtures of ionic and nonionic surfactants, then the repulsive head group interaction parameter α decreases compared to that of the ionic surfactant. This would cause a decrease in the equilibrium area per molecule, a_e and an increase in the molecular packing parameter $v_o/a_e\ell_o$. As a result, the mixed ionic-nonionic micelles will be formed at a lower CMC and the micelle size would be larger, compared to that of the ionic surfactant.

If mixtures of single and double tail surfactants with similar head group repulsion parameter and equilibrium areas are combined, the molecular packing parameter $v_o/a_e\ell_o$ will lie between those for the single chain and double chain surfactants. Therefore, it would be possible to transform a bilayer vesicle formed by a double tail surfactant to a cylindrical micelle and then to a spherical micelle by the addition of a single tail surfactant.

Polar Organic Solvents

If the solvent is changed from water to a mixed aqueous-organic solvent, then the interfacial tension parameter σ decreases. For a given surfactant, this would lead to an increase in the equilibrium area per molecule a_e, and hence, a decrease in $v_o/a_e\ell_o$. Therefore, on the addition of a polar organic solvent to an aqueous surfactant solution, bilayers will transform into micelles, rodlike micelles into spherical micelles, and spherical micelles into those of smaller aggregation numbers including only small molecular clusters. Starting from phospholipid surfactants that have two hydrophobic tails per molecule, we can form bilayer vesicles, cylindrical micelles or spherical micelles, respectively, if we use an aqueous – organic solvent mixture with increasing amount of polar organic solvent in the solvent mixture.

Non-Polar Solvents

If the solvent is non-polar, the transfer free energy of the surfactant tail representing the change in state from a singly dispersed surfactant to one in an aggregate reduces in magnitude and can even become positive, thereby ceasing to be the driving force for aggregation. This is because of the essential similarity in the interactions between surfactant tails within an aggregate and that between the singly dispersed surfactant tail and the non-polar solvent. In this case, the nature of micellar aggregates changes and so-called reverse micelles with head groups in the aggregate interior and the tail groups in the aggregate exterior are formed. An analysis of this mode of self-assembly has shown that in binary surfactant-non-polar solvent systems, typically very small aggregates are formed and often there is no discernible critical micelle concentration accompanying the process.

All of the above predictions are in agreement with numerous experiments and are by now well established in the literature (see papers cited in ref. (*3–5, 9*)). One can thus see the evidence of the predictive power of the phenomenological free energy model and the molecular packing considerations, despite their remarkable simplicity. These concepts can be used to predict the self-assembly behavior of other novel amphiphiles and one example is the predictions made for bola amphiphiles that have a single hydrophobic chain separating two terminal polar head groups in each molecule (*10*).

Self-Assembly of Amphiphilic Block Copolymers

Amphiphilic block copolymer molecules exhibit self-assembly behavior similar to conventional low molecular weight surfactants. The different blocks of the copolymer (i.e. blocks A and B in an AB diblock or ABA or BAB triblock copolymers) can have differing affinities for the solvent, one being solvophobic (say, A block) and the other, solvophilic (say, B block). The two blocks thus resemble the tail and the head of a conventional low molecular weight surfactant. As a result, the block copolymer forms micelles in a variety of solvents that are selective to one block and non-selective to the other block. In these micelles, the solvophobic blocks constitute the core and the solvophilic blocks constitute the corona or the shell region of the micelle (Figure 2). For BAB triblock copolymers, the A block has to bend back and form a loop so as to ensure that the two B blocks are in the corona; similarly, in the case of a ABA copolymer, the B block has to form a loop and bend back so that the two A blocks can be inside the core.

Theoretical treatments of block copolymer micelles have been pioneered by de Gennes (*11*), Leibler, Orland and Wheeler (*12*), Noolandi and Hong (*13*) and Whitmore and Noolandi (*14*). de Gennes (*11*) analyzed the formation of a diblock copolymer micelle in selective solvents by minimizing the free energy per molecule of an isolated micelle with respect to the aggregation number or core radius. The micelle core was assumed fully segregated and devoid of any solvent. The free energy of formation of the core-corona interface and the elastic free energy of stretching of the core blocks control the micellization behavior. Leibler et al (*12*) treated the problem of micelle formation of a symmetric diblock copolymer in a homopolymer solvent. In their study and in de Gennes' work, the interface was taken to be sharp. Noolandi and Hong (*13*) and Whitmore and Noolandi (*14*) formulated mean field models taking into account the possibility of a diffuse interface between the core and corona regions. In none of these theories, the free energy contributions arising from the interactions between the solvophilic block of the block copolymer and the solvent played a significant role. Recognizing the similarity between this contribution and the head group repulsion in surfactant systems, Nagarajan and Ganesh (*15*) had proposed a free energy model for block copolymer self-assembly explicitly incorporating this important contribution. The present treatment and calculated results follow from this free energy model. The thermodynamic formulation of the free energy model is presented here for AB diblock and BAB triblock copolymers in such a way as to observe the close similarity between the model for surfactants and that for the block copolymers.

The free energy model developed here is applicable to both diblock and symmetric triblock copolymers and to block copolymer aggregates having spherical, cylindrical and lamellar (or bilayer) morphologies. The symbols A

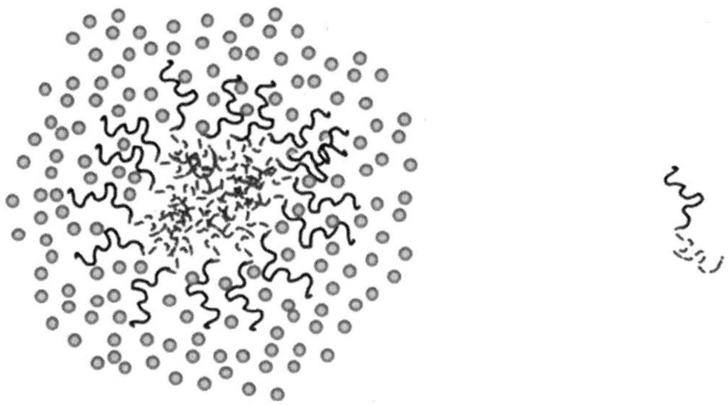

Figure 2. Schematic of diblock copolymer micelle. Thick line and broken line represent solvophilic and solvophobic blocks. Circles denote solvent selective to the solvophilic block and non-selective to the solvophobic block.

and B represent the solvophobic core and solvophilic corona blocks. We use the variable R to denote the solvophobic core dimension (radius for sphere or cylinder and half-bilayer thickness for lamella), D for the corona thickness, and a to denote the surface area of the aggregate core per constituent block copolymer molecule. The number of block copolymer molecules g in an aggregate, the aggregate core volume V_C, and the corona volume V_S all refer to the total quantities in the case of spherical aggregates, quantities per unit length in the case of cylindrical aggregates and quantities per unit area in the case of lamellar aggregates. The concentrations of polymer segments are assumed to be uniform in the core as well as in the corona, with φ_A standing for the volume fraction of the A segments in the core ($\varphi_A = 1$), and φ_B for the volume fraction of the B segments in the corona. The geometrical relations describing spherical, cylindrical and lamellar block copolymer aggregates are summarized in Table 2. If any two structural variables such as R and D are specified, all the remaining geometrical variables can be calculated through the relations given in Table 2.

In Table 2, v_A, v_B, and v_S, refer to the molecular volumes of A and B segments, and the selective solvent. N_A and N_B are the number of segments of blocks A and B for both AB diblock and for the symmetric BAB triblock copolymers.

Free Energy Model for Amphiphilic Block Copolymers

An expression for the free energy change on aggregation $\Delta\mu_g^o$ for block copolymers can be formulated by considering all the physicochemical changes accompanying the transfer of a singly dispersed copolymer molecule from the infinitely dilute solution state to an isolated micelle in the infinitely dilute solution state. Firstly, the transfer of the singly dispersed copolymer to the micellar core is associated with changes in the state of dilution and in the state of elastic deformation (or stretching) of the A block. Secondly, the B block of the singly dispersed copolymer is transferred to the corona region of the micelle and this transfer process also involves changes in the states of dilution and deformation of the B block. Thirdly, the formation of the micelle localizes the copolymer

Table 2. Geometrical Relations for Block Copolymer Aggregates

Property	*Sphere*	*Cylinder*	*Lamella*
Core volume $V_C = g\, N_A\, v_A$	$4\pi R^3/3$	πR^2	$2R$
Corona volume V_S	$V_C\,[(1+D/R)^3-1]$	$V_C\,[(1+D/R)^2-1]$	$V_C\,[(1+D/R)-1]$
Aggregation number g	$V_C\,/v_A$	$V_C\,/v_A$	$V_C\,/v_A$
Core surface area per molecule a	$3\,v_A/\,R$	$2\,v_A/\,R$	$v_A/\,R$
Volume fraction of B in corona φ_B	$(v_B/v_A)(V_C/V_S)$	$(v_B/v_A)\,(V_C/V_S)$	$(v_B/v_A)\,(V_C/V_S)$

such that the A block is confined to the core while the B block is confined to the corona. Fourthly, the formation of the micelle is associated with the generation of an interface between the micelle core made up of A blocks and the micelle corona consisting of solvent S and B blocks. Further, in the case of a BAB triblock copolymer, folding or loop formation of the A block occurs ensuring that the B blocks at the two ends are in the solvent domain while the folded A block is within the solvophobic core of the micelle. This provides an additional free energy contribution. The overall free energy of micellization can be obtained as the sum of the above individual contributions:

$$\begin{aligned}(\Delta\mu_g^o) = (\Delta\mu_g^o)_{A,dil} + (\Delta\mu_g^o)_{A,def} + (\Delta\mu_g^o)_{B,dil} + (\Delta\mu_g^o)_{B,def} \\ + (\Delta\mu_g^o)_{loc} + (\Delta\mu_g^o)_{int} + (\Delta\mu_g^o)_{loop}\end{aligned} \qquad (8)$$

One may observe that many of the above free energy contributions are analogous to those considered in eq.(6) for conventional surfactants. The A block dilution contribution is equivalent to the tail transfer free energy for the surfactant. The A block deformation free energy is identical to the tail packing energy for the surfactant. The B block dilution and deformation contributions are equivalent to head group repulsions for surfactant. The interfacial free energy term has the same meaning for both the conventional surfactant and the block copolymer. Only the localization and looping free energy contributions are additions for block copolymer that had not been considered for surfactants. Indeed for bola surfactants that have two head groups separated by a hydrophobic chain, one would have to consider the looping contribution just as for the analogous triblock copolymer. These two contributions are found to be less important relative to the other contributions with respect to the determination of equilibrium aggregate shape and size. In essence, the free energy model for the block copolymer remains essentially the same as that for the conventional surfactant molecules, from a phenomenological point of view.

Expressions for each of the contributions appearing in eq.(8) are formulated below. First, the transfer of the singly dispersed copolymer to the micellar core is associated with changes in the state of dilution and the state of deformation of A block. The corresponding free energy change is calculated using Flory theory for polymer solutions (*16*), Semenov theory for elastic deformations in constrained systems (*17*) and de Gennes theory for the conformation of a collapsed polymer (*18*). Second, the formation of micelle is associated with the generation of an interface between the micelle core made up of A blocks and the micelle corona consisting of B blocks and the solvent S. The characteristic interfacial tension σ_{agg} between the core and corona regions can be determined in many approximate ways including based on the Prigogine theory for interfacial tension against solutions (*19*). Third, the B block of the singly dispersed copolymer is transferred to the corona region of the micelle and this transfer process also involves changes in the states of dilution and deformation of the B block. This is calculated using Flory theory for polymer solutions (*16*), Semenov theory for elastic deformations in constrained systems (*17*) and the Flory theory for swollen isolated polymer (*16*) with modification suggested by Stockmayer (*20*). Fourth, the formation of the micelle localizes the copolymer such that the A block is confined to the core while the B block is confined to the corona. This is calculated using a configurational volume restriction model (*15*). Fifth, in the case of a BAB triblock copolymer, there is a free energy contribution due to the folding or loop formation of the A block (which occurs to ensure that the B blocks at the two ends are in the corona while the folded A block is within the micelle core). This is calculated using the Jacobsen-Stockmayer model (*21*). All of the free energy contributions are briefly listed below and more detailed descriptions with explanations can be found in ref. (*22*).

Change in state of dilution of block A

$$\frac{(\Delta\mu_g^o)_{A,dil}}{kT} = -N_A\left[\frac{v_A}{v_S}\frac{1-\phi_{A1}}{\phi_{A1}}\ln(1-\phi_{A1}) + \frac{v_A}{v_S}(1-\phi_{A1})\chi_{AS}\right] - \left(\frac{\sigma_{AS}L_A^2}{kT}\right)\frac{6N_A^{1/2}}{\alpha_A} \tag{9}$$

Here, $L_K = (v_K)^{1/3}$ for K = A or B. The chain expansion parameter α_A and for the segment volume fraction φ_{A1} within the singly dispersed copolymer molecule in its standard state are calculated using the de Gennes theory for collapsed polymers (*18*) from the expressions:

$$\ln(1-\phi_{A1}) + \phi_{A1} + \chi_{AW}\phi_{A1}^2 = 0, \quad \alpha_A = (6/\pi)^{1/3}N_A^{-1/6}\phi_{A1}^{-1/3} \tag{10}$$

The interfacial tension σ_{AS} between block A and the solvent is calculated from

$$\sigma_{AS} = \left(\frac{\chi_{AS}}{6}\right)^{1/2}\frac{kT}{L_S^2}, \quad L_S = v_S^{1/3} \tag{11}$$

Change in state of deformation of block A

$$\frac{(\Delta\mu_g^o)_{A,def}}{kT} = \left[q \left(\frac{p\,\pi^2}{80} \right) \frac{R^2}{(N_A/q)\,L_A^2} \right] - \left[\frac{3}{2} \left(\alpha_A^2 - 1 \right) - \ln \alpha_A^3 \right] \tag{12}$$

Here, $q = 1$ for AB diblock and 2 for symmetric BAB triblock copolymers and $p = 3$ for spheres, 5 for cylinders, and 10 for lamella.

Formation of core-solvent interface

$$\frac{(\Delta\mu_g^o)_{int}}{kT} = \frac{\sigma_{agg}}{kT}\,a\,, \quad \sigma_{agg} = \sigma_{AS}\,(1-\phi_B) + \sigma_{AB}\,\phi_B \approx \sigma_{AS} \tag{13}$$

Here, σ_{agg} is calculated as that representing an interface between the core domain of A and the corona domain of B and solvent S. Anticipating the relative dilution of polymer B segments in the corona region, we approximate the interfacial tension σ_{agg} by σ_{AS}. Other approximations have been discussed in ref. (*22*).

Change in state of dilution of block B

$$\frac{(\Delta\mu_g^o)_{B,dil}}{kT} = N_B \left[\frac{v_B}{v_S} \frac{1-\phi_B}{\phi_B} \ln(1-\phi_B) + \frac{v_B}{v_S} (1-\phi_B)\,\chi_{BS} \right] - N_B \left[\frac{v_B}{v_S} \frac{1-\phi_{B1}}{\phi_{B1}} \ln(1-\phi_{B1}) + \frac{v_B}{v_S} (1-\phi_{B1})\,\chi_{BS} \right] \tag{14}$$

The chain expansion parameter α_B and the segment volume fraction φ_{B1} within the singly dispersed lock copolymer in its standard state are calculated using the Flory theory for swollen isolated polymers (*16*) with the modification suggested by Stockmayer (*20*)

$$\alpha_B^5 - \alpha_B^3 = 0.88\,(1/2 - \chi_{BS})\,(N_B/q)^{1/2}$$
$$\phi_{B1} = (6/\pi)\,(N_B/q)^{-1/2}\,\alpha_B^{-3} \tag{15}$$

Change in state of deformation of block B

$$\frac{(\Delta\mu_g^o)_{B,def}}{kT} = q \left[\frac{3}{2} \frac{L_B\,R}{(a/q)\,\phi_B} P \right] - q \left[\frac{3}{2} \left(\alpha_B^2 - 1 \right) - \ln \alpha_B^3 \right] \tag{16}$$

where P is a shape-dependent function given by $P = (D/R)/[1+(D/R)]$ for spheres, $P = \ln[1+(D/R)]$ for cylinders and $P = (D/R)$ for lamellae.

Localization of block copolymer

$$\frac{(\Delta\mu_g^o)_{loc}}{kT} = -q \ln\left[\frac{d\,L_B}{R\,(1+D/R)^d}\right] \tag{17}$$

Here, $d = 3$ for spheres, 2 for cylinders, and 1 for lamella.

Backfolding for a BAB block copolymer

$$\frac{(\Delta\mu_g^o)_{loop}}{kT} = \frac{3}{2}\beta \ln[\,N_A\,] \tag{18}$$

Here, the excluded volume parameter β is taken to be unity (*21*).

Molecular constants appearing in the free energy expressions

Illustrative calculations have been carried out for the diblock copolymers polystyrene-polyisoprene (PS-PI), polyethylene oxide-polypropylene oxide (PEO-PPO, denoted E_XP_Y) and the triblock copolymer PEO-PPO-PEO (denoted $E_XP_YE_X$). One may note that for PS-PB block copolymer, the micellization is examined in n-heptane which is a selective solvent for the PB block and is a non-solvent for the PS block. For the other block copolymers, the calculations have been done in water as the solvent with PEO being hydrophilic and PPO being hydrophobic.

The molecular volumes of the repeating units are estimated to be 0.1683 nm^3 for styrene, 0.0126 nm^3 for isoprene, 0.0965 nm^3 for propylene oxide and 0.0646 nm^3 for ethylene oxide (*15*). The molecular volume of n-heptane is 0.2447 nm^3 while that of water is 0.030 nm^3. For polyisoprene-heptane, the solvent is practically a theta solvent at 25°C (*13*) and correspondingly, the Flory interaction parameter χ_{AS} is taken to be 0.5. For polystyrene-heptane, the solvent is a very poor solvent and the interaction parameter χ_{BS} is taken to be 1.9 (*13*). The corresponding value for the PS-heptane interfacial tension is estimated using eq.(6) to be $\sigma_{AW} = 5.92$ mN/m.

The Flory interaction parameters χ_{AW} between PPO and water and χ_{BW} between PEO and water are estimated from available experimental activity data for polymer solutions (*23*). Since the hydrophilic domain of the aggregate is a dilute solution of PEO in water, χ_{BW} is taken from the dilute region of the activity data as equal to 0.2. The hydrophobic region of the aggregate is made of pure PPO and hence, χ_{AW} is taken from the concentrated region of the activity data as equal to 2.1. The corresponding value for the PPO-water interfacial tension is estimated using eq.(6) to be $\sigma_{AW} = 25.9$ mN/m.

Predicting Self-Assembly of Amphiphilic Block Copolymers

Influence of Free Energy Contributions

The role of various free energy contributions in influencing block copolymer micelle formation can be understood from their dependence on the aggregation number. The formation of micelles in preference to the singly dispersed state of the copolymer occurs because of the large negative free energy contribution arising from a change in the state of dilution of the solvent incompatible A block. This free energy contribution is a constant independent of the size of the micelle and hence does not govern the aggregation number of the micelle. The contribution favorable to the growth of the aggregates is provided by the interfacial energy. The geometrical relations dictate that surface area per molecule of the micelle decrease with an increase in the aggregation number. Consequently, the positive free energy associated with the formation of the interface between the micellar core and the corona decreases with increasing aggregation number of the micelle and thus this contribution promotes the growth of the micelle. The changes in the state of deformation of the A and the B blocks and the change in the state of dilution of the B block provide positive free energy contributions that increase with increasing aggregation number of the micelle. Therefore, these factors are responsible for limiting the growth of the micelle. The free energy of localization and the free energy change on looping or backfolding (in the case of the BAB triblock copolymer) are practically independent of g and thus have little influence over the determination of the equilibrium size and shape of the micelle.

Predicted Aggregation Behavior

For the polystyrene-polyisoprene (PS-PI) copolymer, the micelle formation in a hydrophobic solvent n-heptane which is selective to polyisoprene has been examined and the predicted results are compared against the experimental measurements (*24*) in Table 3. Results from photon correlation spectroscopy and viscosity measurements were interpreted assuming that the micelle core is completely devoid of any solvent. The values for the aggregation number g, the core radius R and the corona thickness D estimated from the measurements in this manner are compared against the model predictions and show reasonable agreement. Experimental observations (*24*) using electron microscopy indicated that the micelles are practically monodispersed, which is also in agreement with the predictions.

The equilibrium aggregate morphologies predicted for the aqueous solutions of commercially available PEO-PPO-PEO triblock copolymers (Pluronics) of various block compositions and molecular weights are summarized in Table 4. In this case, water is the selective solvent for the PEO lock and non-selective to the PPO block. The calculated morphologies correspond to concentrations above the critical micelle concentration and for dilute solutions. The critical micelle concentrations for most of the block copolymers shown on Table 4 are well below experimentally measurable values and only for the block copolymers that are rich in polyethylene oxide, the critical micelle concentrations become measurable.

Table 3. Micellization of PS-PI in n-heptane at 25°C

N_B	N_A	*R (nm)*	*D/R*	*g*	*R (nm)*	*D/R*	*g*
		Predictions			*Experimental*		
295	86	85	1.77	187	94	1.77	248
295	156	128	1.24	352	119	1.31	278
295	184	144	1.12	422	141	1.13	386
295	278	193	0.87	672	194	0.96	666
295	316	211	0.80	778	240	0.92	1113

The calculated results predict that at room temperature, the triblock copolymers with 20 weight % PEO (L62, L72, L92, L122) form lamellar aggregates, those with 30 weight % PEO form either lamellae (L63) or cylinders (P103, P123) while molecules containing 40 weight % or more of PEO form spherical aggregates. Lamellar aggregates are favored when the ratio of PEO to PPO is small whereas spherical aggregates are favored when the PEO to PPO ratio is large.

From the predicted hydrophobic and hydrophilic domain dimensions we observe that for a given size of the hydrophobic block, the hydrophobic domain size R increases from lamellar to cylindrical to spherical aggregate. In contrast, the hydrophilic domain size D is mainly controlled by the size of the B block and less dependent on the shape of the aggregate. The equilibrium area per molecule a increases from lamellar to cylindrical to spherical aggregate.

For comparison, the equilibrium aggregate morphologies predicted for the PEO-PPO diblock copolymers of various compositions and molecular weights are summarized in Table 5. The block compositions and molecular weights are kept identical to those of the commercially available Pluronic triblock copolymer and the entries in Table 5 are identified with the same trade names but with a prefix Di to indicate they are diblock copolymers.

The general pattern of aggregation for the diblock copolymers follows the same trend as for the triblock copolymers. At room temperature, the block copolymers with 20 weight % PEO (Di-L62, Di-L72, Di-L92, Di-L122) form lamellar aggregates, Di-L63 with 30 weight % PEO forms cylinders, while the other diblocks with 30 or larger weight% PEO form spheres. As for the triblocks, lamellar aggregates are favored when the ratio of PEO to PPO is small whereas spherical aggregates are favored when the PEO to PPO ratio is large.

The importance of solvophobic block is seen by comparing the results for PEO-PPO in water where water is a very good solvent for the PEO block compared to PS-PI in n-heptane where n-heptane is a theta solvent for the PB block. When S is a very good solvent for the B block, the positive free energy contributions resulting from changes in the states of dilution and of deformation of the B block (analog of head group repulsions) outweigh the contribution resulting from the changes in the state of deformation of the A block (analog of tail packing free energy). Under such conditions, the B block related free energy contributions strongly influence the structural properties of the equilibrium micelles.

Table 4. Predicted Aggregation Behavior of $E_xP_{2y}E_x$*

Trade Name	*Block copolymer description*	*R (nm)*	*D (nm)*	*a_e (nm²)*	*g*	*Aggregate Shape*
L62	$E_6P_{35}E_6$	1.75	0.96	1.93	1.04	Lamella
L72	$E_6P_{38}E_6$	1.87	0.96	1.96	1.02	Lamella
L92	$E_8P_{50}E_8$	2.19	1.22	2.20	0.91	Lamella
L122	$E_{11}P_{69}E_{11}$	2.65	1.60	2.52	0.80	Lamella
L63	$E_9P_{32}E_9$	1.45	1.39	2.13	0.94	Lamella
P103	$E_{17}P_{60}E_{17}$	3.87	2.17	2.99	8.1	Cylinder
P123	$E_{20}P_{70}E_{20}$	4.21	2.48	3.21	8.3	Cylinder
L64	$E_{13}P_{30}E_{13}$	3.42	1.64	2.54	58	Sphere
P84	$E_{19}P_{43}E_{19}$	4.19	2.24	2.97	74	Sphere
P104	$E_{27}P_{61}E_{27}$	5.09	2.99	3.47	94	Sphere
P65	$E_{19}P_{29}E_{19}$	3.06	2.18	2.74	43	Sphere
P75	$E_{24}P_{35}E_{24}$	3.36	2.63	3.02	47	Sphere
P85	$E_{26}P_{40}E_{26}$	3.66	2.82	3.16	53	Sphere
P105	$E_{37}P_{56}E_{37}$	4.38	3.73	3.70	65	Sphere
F77	$E_{52}P_{35}E_{52}$	2.75	4.31	3.69	26	Sphere
F87	$E_{61}P_{40}E_{61}$	2.92	4.84	3.96	27	Sphere
F127	$E_{100}P_{64}E_{100}$	3.70	6.98	5.00	35	Sphere
F68	$E_{77}P_{29}E_{77}$	2.13	5.13	3.94	15	Sphere
F88	$E_{104}P_{39}E_{104}$	2.48	6.39	4.55	17	Sphere
F98	$E_{118}P_{45}E_{118}$	2.67	7.03	4.86	18	Sphere
F108	$E_{133}P_{50}E_{133}$	2.83	7.66	5.14	20	Sphere

* *R* is the size of the hydrophobic A domain, *D* is the size of the hydrophilic B domain, a_e is the equilibrium area per molecule of the domain A-domain B interface, *g* is the aggregation number defined as the total number of molecules in a micelle in the case of sphere, the number of molecules per nm length in the case of cylinder and the number of molecules per nm² area in the case of lamella.

Comparison to Surfactant Behavior

The predicted aggregation pattern as a function of the block copolymer composition follows the geometrical packing concepts discussed earlier for small surfactant aggregates based on the interplay between the interfacial free energy for the formation of the aggregate core - water interface and the free energy of surfactant head group repulsions. For conventional surfactants, we observed that when the head group repulsions are weak, lamellar aggregates are favored while for strong head group repulsions, spherical aggregates are formed. Cylindrical aggregates result for intermediate values of head group repulsions. For the block copolymer aggregates, the free energy contributions associated with the hydrophilic B domain are analogous to the head group repulsions in classical surfactant micelles. Therefore the pattern of aggregation observed for increasing head group repulsions in surfactants is exactly reproduced in the case of block copolymers with increasing size of the hydrophilic block.

Table 5. Predicted Aggregation Behavior of $E_{2x}P_{2y}$*

Trade Name	*Block copolymer description*	*R (nm)*	*D (nm)*	*a_e (nm²)*	*g*	*Aggregate Shape*
Di-L62	$E_{12}P_{35}$	2.62	1.73	1.29	1.55	Lamella
Di-L72	$E_{12}P_{38}$	2.80	1.72	1.31	1.53	Lamella
Di-L92	$E_{16}P_{50}$	3.26	2.19	1.48	1.35	Lamella
Di-L122	$E_{22}P_{69}$	3.90	2.85	1.71	1.17	Lamella
Di-L63	$E_{18}P_{32}$	4.02	2.27	1.54	16.5	Cylinder
Di-P103	$E_{34}P_{60}$	8.29	3.69	2.09	413	Sphere
Di-P123	$E_{40}P_{70}$	9.02	4.21	2.25	455	Sphere
Di-L64	$E_{26}P_{30}$	5.07	2.91	1.71	189	Sphere
Di-P84	$E_{38}P_{43}$	6.16	3.95	2.02	235	Sphere
Di-P104	$E_{54}P_{61}$	7.43	5.24	2.38	292	Sphere
Di-P65	$E_{38}P_{29}$	4.48	3.81	1.88	134	Sphere
Di-P75	$E_{48}P_{35}$	4.89	4.56	2.07	145	Sphere
Di-P85	$E_{52}P_{40}$	5.33	4.89	2.17	164	Sphere
Di-P105	$E_{74}P_{56}$	6.33	6.43	2.56	196	Sphere
Di-F77	$E_{104}P_{35}$	3.93	7.27	2.58	75	Sphere
Di-F87	$E_{122}P_{40}$	4.18	8.14	2.77	79	Sphere
Di-F127	$E_{200}P_{64}$	5.27	11.6	3.52	99	Sphere
Di-F68	$E_{154}P_{29}$	3.04	8.51	2.76	42	Sphere
Di-F88	$E_{208}P_{39}$	3.53	10.6	3.20	49	Sphere
Di-F98	$E_{236}P_{45}$	3.81	11.6	3.42	53	Sphere
Di-F108	$E_{266}P_{50}$	4.00	12.6	3.62	56	Sphere

* R is the size of the hydrophobic A domain, D is the size of the hydrophilic B domain, a_e is the equilibrium area per molecule of the domain A-domain B interface, g is the aggregation number defined as the total number of molecules in a micelle in the case of sphere, the number of molecules per nm length in the case of cylinder and the number of molecules per nm² area in the case of lamella.

Some important differences arise between the behaviors of surfactants and block copolymers because of their differing molecular weights. For surfactants the main opposing forces controlling aggregation are the interfacial energy at the core-solvent interface and the head group repulsions. The tail packing plays a minor role because of the small molecular weight of surfactant tails. In the case of block copolymers, the tail packing can be very significant because of the high molecular weight. Consequently, we can have the interfacial free energy balanced either by the head group repulsions as for surfactants, or by the tail packing, or by a combination of both. This would differentiate from a quantitative point of view, how the aggregation patterns and aggregate sizes are dependent on the molecular weight, for the surfactant compared to block copolymers.

The tail transfer free energy is the dominant contribution to the net free energy change on micellization and thus affects the magnitude of the critical micelle concentration. Because the surfactant tails are usually shorter than the solvophobic block of the block copolymer, the magnitude of the tail transfer fee energy for the block copolymer is usually much larger than that for the surfactants. As a result, the cmc for the block copolymers can be very small and often not measurable.

The tail transfer free energy is the driving force for the aggregation for both surfactants and block copolymers. In the case of surfactants, the negative free energy change associated with the transfer of the tail is significant when water is the solvent and it diminishes in magnitude when water is replaced by polar organic solvents such as alcohols or non-polar solvents. Consequently, the aggregation tendency of surfactants is significantly diminished or eliminated when water is replaced by other solvents. In contrast, for block copolymers, the core forming block can be incompatible with a variety of solvents and therefore the tail transfer free energy can remain negative with large magnitude. This allows significant aggregation of the same block copolymer in a range of solvents. The aggregation phenomenon where entropy changes (hydrophobic effect) are important in the case of conventional surfactants is replaced by an aggregation process where enthalpy effects are also important in the case of block copolymers.

Another consequence of the negative tail transfer free energy is connected to the kinetics of the aggregation process. When the tail transfer free energy is large, it implies great solvophibicity for the tail and that it will be kinetically difficult to dissolve the amphiphilic molecule in the solvent. The concept of equilibrium self-assembly assumes that amphiphilic molecules are dynamically exchanged between the bulk solvent and the aggregate. However such dynamic exchange would be hindered if the amphiphilic molecule is not easily dissolved in the bulk solvent. In such cases, the generation of equilibrium aggregates spontaneously is not possible due to kinetic reasons. For most conventional surfactants this situation does not occur, with the exception of phospholipids. For phospholipids with two hydrophobic chains in their tails, the dissolution of the amphiphile in water poses a problem and often the formation of bilayers from phospholipids requires approaches that manipulate the kinetics, such as the use of a solvent to get the surfactant into solution before allowing the formation of bilayer vesicles. For many block copolymers, it is indeed very difficult to dissolve the amphiphilic polymer in the selective solvent directly and spontaneous aggregation is not possible (Pluronic block copolymers are an exception because of their molecular weights, triblock structure and the nature of polypropyleneoxide-solvent interactions). Kinetic procedures such as the use of a solvent that can dissolve both blocks of the block copolymer are employed as part of the pathway to generate aggregates in a selective solvent. Indeed, this has given rise to the concept of non-equilibrium aggregates in the case of block copolymers.

Additional differences exist between surfactants and block copolymers with respect to their ability to solubilize other small molecules within aggregate cores and the nature of solubilizate-induced aggregate shape transitions. These are discussed in our earlier studies focused on solubilization (*25*) by block copolymer aggregates.

Conclusions

The general principles of molecular self-assembly exhibited by amphiphilic molecules is examined in the framework of the free energy model governing the self-assembly process. The driving force for self-assembly arises from the solvophobic part of the amphiphilic molecule. The key variable that describes the structure of the self-assembled multimolecular aggregate is the equilibrium surface area of the aggregate per constituent amphiphilic molecule. This equilibrium area is determined by a balance of competing forces: repulsions between solvophilic parts of the amphiphile within the aggregate, the formation of the aggregate-solvent interface, and the molecular packing constraints imposed on the solvophobic part of the amphiphile. Since these factors depend on the molecular structure of the amphiphile and the solution conditions, they entirely determine the equilibrium area. The equilibrium area completely determines the shape of the equilibrium aggregate because of the geometrical rules governing molecular packing within different shapes.

The aggregation tendency of block copolymers is more general compared to that of the surfactants in the sense that aggregation of block copolymers is achievable in almost any solvent by selecting the appropriate block copolymer, whereas for low molecular weight surfactants, water constitutes the best solvent for promoting aggregation. Further, in the case of block copolymers, because of the large size of the solvophobic block, the CMC is often too small to be measurable when compared to the CMC of low molecular weight surfactants.

For surfactants, the equilibrium area is principally determined by the balance between head group repulsions and interface formation, with chain packing playing a minor role. In contrast, for block copolymers, we can have different situations with either or both of head group repulsions and chain packing balancing the free energy of formation of the interface. Most importantly, the equilibrium shapes of amphiphilic molecules, whether block copolymer or surfactants, are predictable by the same considerations of principle of opposing forces and molecular packing requirements.

Acknowledgments

Work supported by Department of Army In-House Laboratory Independent Research (DA-ILIR) Program, Natick Soldier Research, Development & Engineering Center.

References

1. Tanford, C. *The Hydrophobic Effect*; Wiley-Interscience: New York, 1973.
2. Israelachvili, J.; Mitchell, D. J.; Ninham, B. W. *J. Chem. Soc., Faraday Trans. 2* **1976**, *72*, 1525.
3. Nagarajan, R.; Ruckenstein, E. *Langmuir* **1991**, *7*, 2934.
4. Puvvada, S.; Blankschtein, D. *J. Chem. Phys.* **1990**, *92*, 3710; *J. Phys. Chem.* **1992**, *96*, 5567.
5. Nagarajan, R. In *Structure-Performance Relationships in Surfactants*; Esumi, K., Ueno, M., Eds.; Marcel Dekker: New York, 1997; Chapter 1, pp 1–89.
6. Gruen, D. W. R. *Biochim. Biophys. Acta* **1980**, *595*, 161; *J. Colloid Interface Sci.* **1981**, *84*, 281; *J. Phys. Chem.* **1985**, *89*, 146.
7. Dill, K. A.; Flory, P. J. *Proc. Natl. Acad. Sci. U.S.A.* **1980**, *77*, 3115; *Proc. Natl. Acad. Sci. USA* 1980, *78*, 676.
8. Ben Shaul. A; Szleifer, I.; Gelbart, W. M. *Proc. Natl. Acad. Sci. U.S.A.* **1984**, *81*, 4601; *J. Chem. Phys.* **1985**, *3*, 3597; Ben-Shaul, A.; Gelbart, W. M. *Rev. Phys. Chem.* **1985**, *36*, 179.
9. Nagarajan, R.; Ruckenstein, E. In *Equations of State for Fluids and Fluid Mixtures*; Sengers, J. V., Kayser, R. F., Peters, C. J., White, H. J., Jr., Eds.; Elsevier Science: Amsterdam, 2000; Chapter 15, pp 589–749.
10. Nagarajan, R. *Chem. Eng. Commun.* **1987**, *55*, 251.
11. de Gennes, P. G. In *Solid State Physics*; Liebert, J., Ed.; Academic Press: New York, NY, 1978; Suppl. 15, p 1.
12. Leibler, L.; Orland, H.; Wheeler, J. C. *J. Chem. Phys.* **1983**, *79*, 3550.
13. Noolandi, J.; Hong, M. H. *Macromolecules* **1983**, *16*, 1443.
14. Whitmore, D.; Noolandi, J. *Macromolecules* **1985**, *18*, 657.
15. Nagarajan, R.; Ganesh, K. *J. Chem. Phys.* **1989**, *90*, 5843.
16. Flory, P. J. *Principles of Polymer Chemistry*; Cornell University Press: Ithaca, NY, 1962.
17. Semenov, A. N. *Soviet Phys. JETP* **1985**, *61*, 733.
18. de Gennes, P. G. *Scaling Concepts in Polymer Physics*; Cornell University Press: Ithaca, New York, 1979.
19. Defay, R.; Prigogine, I.; Bellemans, A.; Everett, D. H. *Surface Tension and Adsorption*; Wiley: New York, 1966.
20. Stockmayer, W. H. *J. Polym. Sci.* **1955**, *15*, 595.
21. Jacobsen, H.; Stockmayer, W. H. *J. Chem. Phys.* **1950**, *18*, 1600.
22. Nagarajan, R. *Colloids Surf., B* **1999**, *16*, 55.
23. Malcolm, G. N.; Rowlinson, J. S. *Trans. Faraday Soc.* **1957**, *53*, 921.
24. Bahadur, P.; Sastry, N. V.; Marty, S.; Riess, G. *Colloids Surf.* **1985**, *16*, 337.
25. Nagarajan, R. *Polym. Adv. Technol.* **2001**, *12*, 23.

Chapter 2

Pyridinium Amphiphiles in Gene Delivery – Present and Perspectives

Marc A. Ilies,[*] Tiffany V. Sommers, Li Ching He, Adrian Kizewski, and Vishnu Dutt Sharma

Department of Pharmaceutical Sciences, Temple University School of Pharmacy, 3307 N Broad Street, Philadelphia, PA 19140
[*]mailies@temple.edu

The success of gene therapy as a revolutionary method to treat diseases relies on finding delivery systems that can efficiently transfer and express (transfect) DNA into target cells while producing reduced cytotoxic effects. Recently, pyridinium amphiphiles were shown to generate promising transfection systems that can fulfill these requirements. The impact of various structural characteristics (hydrophobic chain type, linkage type, counterion) on the transfection activity and cytotoxicity among various classes of pyridinium amphiphiles (detergents, lipids, gemini surfactants, lipophilic polycations) is discussed, emphasizing the most important structural parameters, efficient designs and formulations.

Introduction: Gene Therapy - Premises and Promises

Gene therapy constitutes the revolutionary form of therapy of the new millennium, promising to cure hereditary and/or acquired diseases at their core (cellular) level by using DNA as a drug. It consists of the transfer and expression (transfection) of genetic material into specific cells of a patient in order either to replace a defective gene or to introduce a new function to the cells. The completion of the Human Genome Project and the constant decrease of sequencing costs, together with major advancements in genomics, proteomics, and stem cell research had established a solid base for this new form of therapy (*1–5*).

In contrast with traditional therapies, the success of gene therapy relies not only on the drug itself (DNA) (*6*), but especially on finding safe and efficient

© 2011 American Chemical Society

delivery vectors for the genetic material (*7*). Other key factors include the specificity and persistence of DNA expression, and the side effects associated with the process (*1–5*, *8*, *9*).

Reprogramming cells and tissues is not an easy task, since higher organisms evolved intricate protection mechanisms of their genome against foreign DNA insertion. These delivery barriers must be overcome for successful nucleic acid transfection (*10*, *11*). One general strategy relies on the use of brute force, effectively pushing the DNA into the target cells through temporary membrane poration. The force can either be electric (electroporation (*12–14*)), magnetic (magnetofection (*15*)), an ultrasound wave (*16*), or simply mechanical (the gene gun technology (*17*, *18*), hydrodynamic injection (*19–21*), or manual massage of the tissue (*22*)). A second major strategy relies on the use of delivery systems, either biological or chemical (*23*, *24*).

Gene delivery systems modify the pharmacokinetics of nucleic acids, reducing their size and masking their negative charge. They enhance the mechanical properties of the genetic cargo, being able to maintain the structural integrity of the nucleic acids in a diverse, harsh environment (e.g. high shear stress, presence of various DNAses/RNAses, amphiphilic proteins and sequestering pathways in liver and spleen). Delivery systems can also provide targeting via ligands with affinity to specific target determinants, can optimize delivery into desired cellular compartments or across cellular barriers by exploiting existing cellular pathways or artificial ones, and can prolong and control the duration of the therapeutic effect (*10*, *11*, *25*, *26*). Importantly, the design of each particular delivery system must fit the nature of its target, the size and nature of the genetic material, and the pathological processes aimed to be treated.

Viral vectors are still the most frequently used and efficient transfection systems available. Millions of years of evolution have created perfect transfection systems, able to overcome all delivery barriers. The large proportion of non-coding regions in the human DNA is believed to originate in part from viral transfections encountered during our own evolution. Similarly to other hosts, humans developed a very efficient immune system and this is why use of viral vectors for therapeutic purposes is associated with strong immune responses that usually prevent repeated administration, and can sometimes be fatal (*27*). Other major disadvantages are their mutagenicity, their tropism to other organs than the targeted ones, the limited size of the plasmid that can be incorporated (especially for DNA delivery), and the practical difficulties associated with their GMP production and storage (*28*, *29*). The *top-down* approach to gene delivery systems involves deletion of immunogenic features from the structure of known viruses in order to reduce immunogenicity while retaining key structural elements responsible for transfection efficiency.

An alternative *bottom-up* approach to gene delivery systems involves the use of synthetic amphiphiles such as cationic lipids (*10*, *30–38*), dendrimers (*39*, *40*), and polymers (*23*, *26*, *40–45*). These cationic amphiphiles can interact electrostatically with the negatively charged nucleic acid strands forming DNA complexes (lipoplexes, polyplexes) (*46*) in which the genetic material is highly compacted and protected from the action of endogenous nucleases. Both lipoplexes and polyplexes have much lower immunogenicity and cytotoxicity than viral vectors. They allow the use of plasmids of practically unlimited size, and can be manufactured and stored in bulk quantities under GMP-compliant norms using platforms already existent in the pharmaceutical industry (*29*). Their transfection efficiency must be improved by adapting the structure of these DNA complexes to the delivery barriers encountered in vivo (*10*, *11*). This scientific goal requires a better understanding and better control of the supramolecular interactions between the genetic material and the synthetic delivery system and their dynamics in vitro and in vivo, similar to viral capsid packing and unpacking.

Lipoplexes: Formation, Stability, and Delivery Barriers

Cationic lipids are synthetic amphiphiles comprising of a cationic (or pro-cationic) polar head attached via a linker to a hydrophobic tail (usually comprising two alkyl chains or a cholesteryl moiety). Due to their similarity with naturally-occurring lipids they generally possess a good biocompatibility and low cytotoxicity. These positively-charged lipids can be formulated into cationic liposomes either alone or in mixtures with other lipids (co-lipids) such as cholesterol (Chol) or dioleoylphasphatidylethanolamine (DOPE). In this form, cationic lipids can spontaneously associate with the DNA through the electrostatic attraction between the cationic polar head of the lipid and the negative phosphate backbone of the DNA. The process of lipoplex formation is triggered by the DNA-mediated fusion of liposomes and involves a large scale lipid rearrangement, driven by co-operative hydrophobic association between the hydrophobic tails of the lipids and also by counterion release from both lipids and DNA (entropically favored) (*47–51*).

The structure of lipoplexes remains a subject of intense investigations (*35*, *52–55*) since it determines the new pharmacokinetics of the nucleic acid and its transfectability. It was shown (*53*) that the DNA can be either sandwiched in between cationic lipid bilayers in a multilamellar structure (L_{α}^{C}), or it can be encapsulated within inverse cylindrical micelles (H_{II}^{C}) (*56*). The transfection properties of these two structures are different, being influenced by many factors (vide infra).

The pharmacokinetics of genetic material encapsulated in lipoplexes depends heavily on the route of administration, which can be systemic (iv injection), local (intramuscular, intratumor, subcutaneous, percutaneous, intracerebral, intraarticular, and intraperitoneal injection), or topical (delivery through skin, cornea, intranasal/intratracheal instillation, inhalation of an aerosol) (*32*). For example, major hurdles must be overcome when lipoplexes are administered systemically, *via* the bloodstream (*10*, *11*, *57–59*). Their biodistribution is complex, depending on the colloidal properties of the lipoplex (size, shape, charge, etc) and on the interaction with serum and other blood components such as erythrocytes (*60*), negatively charged plasma proteins (*61–65*), lipoproteins (*66*), endogenous DNA-ses (*67*), antigens (*11*, *67*), and the complement system (*68–72*). Particularities of the target tissue, especially tumor tissue (poorly formed blood vessels with fenestrations, impaired lymphatic drainage, etc), can also dramatically change the biodistribution of lipoplexes through the enhanced permeability and retention (EPR) effect (*10*, *73*).

After reaching the desired tissue, lipoplexes have to pass through the target cell membranes. This process primary occurs *via* endocytosis (*74–77*), although receptor-mediated endocytosis, pinocytosis and other mechanisms can contribute in specific cases (*32*). Important factors influencing this step are the lipid composition of the target cell membrane (*76*, *78*, *79*) and the fusogenicity of the lipoplex, both accounting for the different transfection profiles observed *in vitro* for the same lipoplex preparation on various cell lines. The association and membrane fusion involves both electrostatic interactions (e.g. in between sialic acid groups from the glycoproteins in the host's membrane and excess cationic lipids in the lipoplex) and hydrophobic interaction between the lipoplex and the membrane lipid bilayer. Fusogenicity is influenced by the structure of the cationic lipid and also by the nature and relative amount of the neutral helper lipid used in lipoplex formulations (*32*, *33*, *37*, *80–82*).

Following endocytosis, the lipoplexes are retained in endosomes. The normal process of acidification of these compartments can degrade the complexes, including the encapsulated DNA, so their ability to fuse and/or disrupt the vesicle's membrane and release the plasmid in the cytoplasm is essential for achieving high transfection efficiency (*83*). Fusogenicity of the lipoplex and therefore the structure and properties of its lipid components are essential for overcoming this important intracellular delivery barrier. It was shown that the use of DOPE as a co-lipid in the lipoplex formulation generates endosomal membrane disruption by inducing a structural change to the inverted hexagonal II phase in the acidic medium of the endosome (*55*, *74*, *75*, *84*). Moreover, all cationic lipids must completely dissociate from the plasmid at this point since the genetic material cannot be expressed in complexed form (*75*). Lipids containing reduction-sensitive moieties (*85–88*) can dissociate faster, under the influence of the high glutathione concentration in the cytoplasm.

The DNA released into the cytoplasm must reach the nucleus in order to be expressed (*75*, *89*), otherwise it is normally digested by cytoplasmic nucleases within 90 minutes (*90*). Active transport is required for large plasmids; nuclear localization signal peptides (*91–93*), histones (*94*, *95*) or histone fragments (*96*) were proved to mediate transport and nuclear entry, greatly enhancing the transfection efficiency. Final expression of the foreign nucleic material was found to depend on the properties of the plasmid (type of promoters, enhancers, etc. (*97*, *98*)) and also on the cell cycle.

Particularly appealing was the use of cationic lipids in nonviral siRNA delivery systems for silencing mRNAs expressing various proteins involved in different human diseases using RNA interference (*99–101*). Since mRNA silencing is done in the cytoplasm, the efficiency of the process is greatly enhanced in comparison with DNA delivery because the nuclear entry is not necessary.

Intrinsic toxicity of the cationic lipids used for transfection and their metabolites is an important parameter that has to be assessed and it was an issue of constant concern. Natural or biodegradable building blocks were preferentially used in the structure of cationic lipids from the very beginning of the field. Other efficient ways to decrease toxicity of these synthetic vectors imply the correct selection of the dosage, delivery method, and lipoplex generation, the use of targeting moieties to obtain the desired biodistribution, etc (*32*, *33*).

Cationic Lipids: Structures and Therapeutic Applications

Two decades after the introduction of tetraalkylammonium cationic lipid DOTMA (**1**) by Felgner (*30*), a plethora of lipids bearing various cationic and pro-cationic polar heads such as trialkyl- and tetralkylammonium, polyamines, amidine, guanidine, imidazolium, pyridinium, etc (structures **1-14**, Chart 1) attached to different hydrophobic anchors (bis-alkyl and acyl, cholesteryl) were synthesized and biologically tested for transfection, in vitro and in vivo. Many reviews covering their structure-activity relationship studies are available (*10*, *31–38*). Several representatives (e.g. structures **1-6**, **8**, **10**, **13**, Chart 1) were formulated into commercially-available transfection systems by various manufacturers.

Pyridinium Amphiphiles – Efficient Gene Delivery Vectors

Heterocyclic cationic lipids in general and pyridinium cationic lipids in particular proved to be very efficient, reaching or surpassing the transfection efficiency of commercial cationic lipid formulations, both in vitro and in vivo, while maintaining a low cytotoxicity (*23*, *32–34*, *59*, *102–114*). These attributes are believed to be due to the delocalization of the positive charge on several atoms of the heterocyclic polar head that effectively dilutes the electronic density deficit. This "soft" charge is particularly beneficial to the delivery process since transfection involves *binding* (and compaction of) the nucleic acid outside the cell and the unwrapping of the complex with *release* of the DNA/siRNA, etc in the cytoplasm. A very strong binding interaction may impede the release process and, vice-versa, very facile unwrapping due to weak binding may compromise the complex integrity and trigger premature inactivation of the genetic material. Heterocyclic cationic lipids seem to provide a good balance between these two antagonistic processes.

Pyridinium cationic lipids were introduced by Engberts, in collaboration with Hoekstra and Ruiters (*104–106*). They synthesized, characterized, and assessed for transfection efficiency the cationic amphiphiles SAINT (Synthetic Amphiphiles INTerdisciplinary), comprising cationic lipids **13**, **15-20** and their congeners (Chart 2). The monocationic species **13**, **15**, **16** outperformed the dicationic lipids **17** and **18** that possessed tetraalkylammonium and guanidinium positively charged moieties in addition to delocalized pyridinium ones, probably due to difficulties related with genetic material unwrapping from the lipoplex (*104–106*). Introduction of biodegradable ester linkages improved the transfection profile, as shown for lipids **19** and **20**. Interestingly, the orientation of the pyridinium ring was also important, lipid **20** being more efficient than amphiphile **19** in transfecting COS-7 cells (*108*). Mention must be made that all these pyridinium lipids were made by alkylation of the corresponding pyridines, a relatively low-yield process due to the relatively low basicity of this heterocycle. The availability of starting pyridines is another limiting factor of this technology.

Balaban, Ilies, and colaborators proposed a new strategy to access pyridinium cationic lipids via reaction of primary amines with pyrylium salts (*32*, *33*, *111–114*). The main advantage of this strategy is that the pyridinium polar head and the linker can be obtained in a single, high yield, step. A large variety of SPYRIT (Synthetic PYRIdinium for Transfection) amphiphiles such as **14**, **21-26** can be generated in this way (Chart 2).

The new amphiphiles have proved very efficient in transfecting lung carcinoma NCI-H23, breast carcinomas MCF-7, MDA-MB231, prostate carcinoma DU-145, and glioma SWB-95. We also proved their efficiency in vivo (*113*). The success of this new synthetic strategy was confirmed by Mahato's group, who synthesized several congeners of lipids **24** and **25** and showed their good transfection properties (*115*).

The same synthetic strategy also allowed access to Gemini surfactants and lipophilic polycations **28-36** (Chart 3) via reaction of lipophilic pyrylium salts with di- and polyamines (*114*). Engberts's group proposed gemini surfactants **27** via quaternization of lipophilic pyridines with dibromides of various sizes (Chart 3) (*105*, *106*).

The transfection efficiency of these polycationic species proved to be rather heterogeneous, with small structural variations generating substantial differences in biological activity. Thus, gemini surfactants **27** generally showed a transfection efficiency similar or higher than their corresponding lipid analogs **16** (*105*). Elongation of linker from three to four carbon atoms substantially increased the transfection efficiency, without a significant increase in cytotoxicity. Adding one more methylene units in the linker increases the cytotoxicity of the vector without any significant increase in the gene delivery efficiency (*105*). A similar dependence of transfection efficiency on the linker length was observed by Ilies et al. (*114*) for pyridinium gemini surfactants **28** formulated with cholesterol at 1:1 molar ratio. Best transfection efficiency was displayed by the compound having an ethylene linker, structurally resembling the monomeric lipid **23**. A dramatic drop in transfection efficiency was observed for its superior homolog (n = 3), after which transfection efficiency was increasing monotonously while elongating the linker from n = 3 to n = 8 (*114*). Making the linker more hydrophilic had a beneficial effect on transfection, as proved by compound **29**, which was more efficient than its pentamethylene congener **28**. However, introduction of procationic amino moieties in the structure of the linkers (compounds **31** and **33**) diminished the transfection activity, presumably due to a very strong association of the amphiphiles with the DNA that slows down the release of the plasmid in vivo. In support of this hypothesis, the Boc-precursors **30** and **32** proved to be 5-10 times more active than their deprotected polycationic congeners **31** and **33**. This observation prompted the two groups from Texas A&M and UTMB Galveston to synthesize the lipophilic polycations **34-36**. Biological testing of these formulations has revealed that the best transfection efficiency was obtained when compound were co-formulated with DOPE as co-lipids, at 1:1 molar ratio, the tapered-shaped **35** being the most active. Importantly, the trimeric surfactant

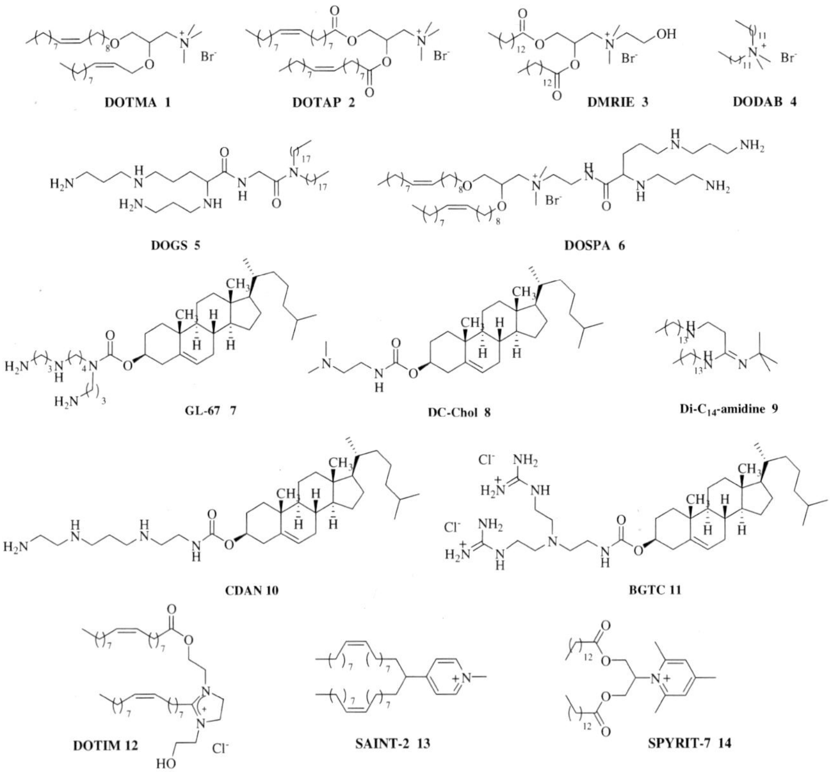

Chart 1. Representative cationic lipids used in DNA and RNA delivery.

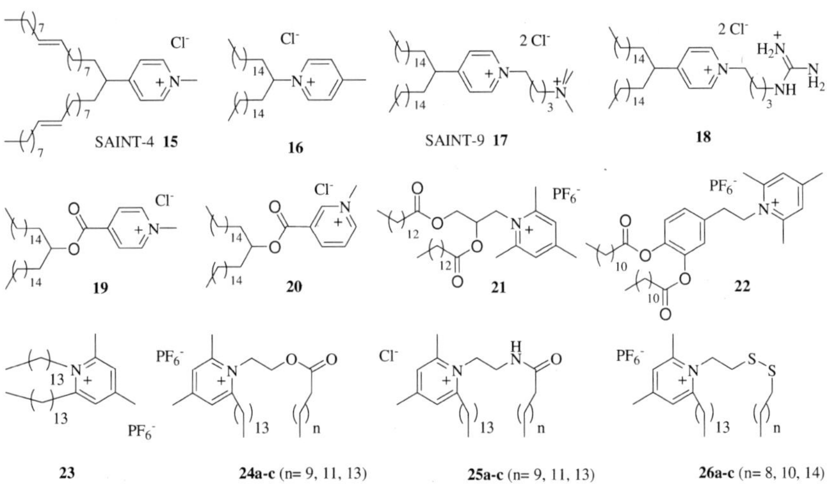

Chart 2. Pyridinium cationic lipids synthesized by different groups, with proved efficiency towards nucleic acid delivery.

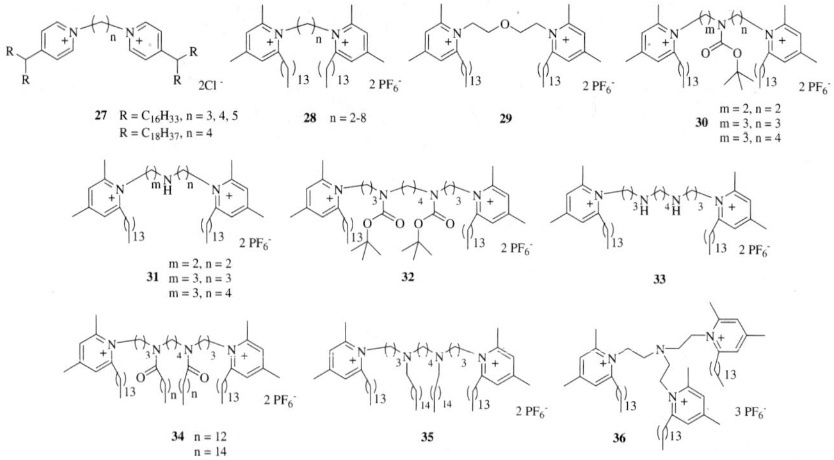

Chart 3. Pyridinium gemini surfactants and lipophilic polycations as novel gene delivery systems

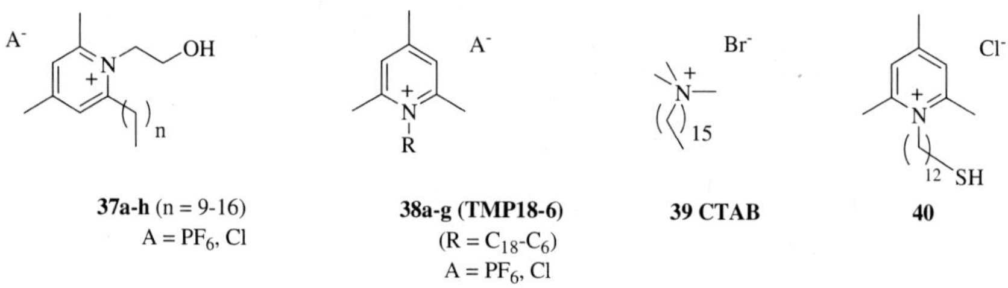

Chart 4. Novel pyridinium surfactants synthesized by our group

36 was able to transfect the NCI-H23 cell line alone, displaying about half of the Lipofectamine® transfection power (*105*, *106*).

At this point it must be stressed that Gemini surfactants and lipophilic polycations have a higher charge/mass ratio than their corresponding monocationic lipids. Consequently, their complexes with nucleic acids are usually smaller than the ones generated from lipid congeners. Since size of the lipoplexes is an important physicochemical parameter that influences the transfection efficiency in vitro and in vivo (vide supra), it is not surprising that this class of compounds is intensively investigated for generating efficient gene delivery systems (*35*, *37*, *109*).

However, the highest charge/mass ratio of cationic amphiphiles is displayed by positively charged surfactants, which were used successfully in transfection (*116–120*). Detergents were also used for preparing liposomes, using dialysis methods (*10*, *121*). Since their size is the smallest of all amphiphiles, pyridinium surfactants were expected to generate even smaller nucleic acid complexes than gemini surfactants. During our research in oligomeric surfactants (*114*) we have noticed that an intermediate in the synthesis of pyridinium lipids **24**, namely the pyridinium surfactant **37e** (n = 13), has displayed significant transfection efficiency and a low cytotoxicity. Prompted by these potential advantages, we have decided to systematically investigate the gene delivery abilities of pyridinium surfactants by synthesizing two sets of amphiphiles, the 2,4,6-trimethylpyridinium derivatives **38a-g** (**TMP18-TMP6**, with even number of carbon atoms in the hydrophobic tails) and a set of homologs of **37e** (n = 9-12, 14-16). Two counterions, PF_6^- and Cl^-, previously found to possess an optimal transfection efficiency/cytotoxicity ratio (*112*), were used in these studies (Chart 4).

Formulation of amphiphiles **38** was found to depend heavily on the counterion type, the lipophilic hexafluorophosphate anion making these compounds difficult to disperse. Changing the counterion to a more hydrophilic chloride eased considerably the formulation of the surfactants. Recent investigations on molecular structure of cationic lipids **24** revealed that the PF_6^- counterion has a tendency to bind tightly to the lipophilic pyridinium ring, efficiently "bridging" in between two surfactant molecules and stabilizing the solid phase, thus opposing hydration and dispersion of the amphiphile in water (*122*). The elevated degree of counterion binding (β parameter (*123*)) for PF_6^- counterion as compared with Cl^- is also responsible for the smaller average zeta potentials obtained for supra-molecular assemblies of **38** PF_6^- (+ 5-10 mV) as compared with their **38** Cl^- congeners (+ 50-70 mV).

The pyridinium chlorides were formulated in water either alone, or with co-lipids cholesterol (C) or DOPE (D) at 1:1 molar ratio. Only the C18-C12 representatives **38a-d** (**TMP18-TMP12**) were able to self-assemble and to form stable micelles and liposomes; inferior homologs precipitated shortly after formulation. Structurally related tetraalkylammonium surfactant cetyltrimethylammonium bromide (**CTAB**) **39**, formulated alone, or with cholesterol or DOPE at 1:1 molar ratio was used as reference. The cationic amphiphile to DNA charge ratio was optimized using the C16 representative **38b** (**TMP16**). Efficient compaction could be achieved when the positive charge of the cationic amphiphiles was "diluted" with either cholesterol or DOPE. Size of lipoplexes decreased as the amphiphile/DNA charge ratio increased. Compaction was more efficient when cholesterol was used as co-lipid for pyridinium surfactants **38**, probably due to a better size match between the packing parameters of pyridinium amphiphiles **38** and this co-lipid. Zeta potential raised with the increase of amphiphile/DNA charge ratio, leveling around + 30 mV at 5/1 +/- charge ratio (Figure 1). The DNA compaction by these amphiphile formulations was confirmed by gel electrophoresis. Mention must be made that surfactant **38b** formulated alone was not able to reliably compact the DNA, irrespective of charge ratio, probably due to strong electrostatic repulsion between the charged polar head of the amphiphiles in complex with DNA (data not shown).

Taking into consideration the results of the charge titration experiment, we have assessed the transfection efficiency of all **TMP** representatives **38a-d** and **CTAB 39** reference, in parallel, at 5/1 cationic amphiphile/DNA charge ratio. The reporter gene plasmid gWiz-Luc encoding the firefly luciferase was used to quantify the transfection efficiency in lung carcinoma NCI-H23 cell line (Figure 2). Luminescence data was normalized for protein amount.

Data from Figure 2 confirmed that both ammonium and pyridinium surfactants formulated alone are unable to transfect DNA into target cells, thus confirming the preliminary analysis done on **TMP16** representative **38b**. With

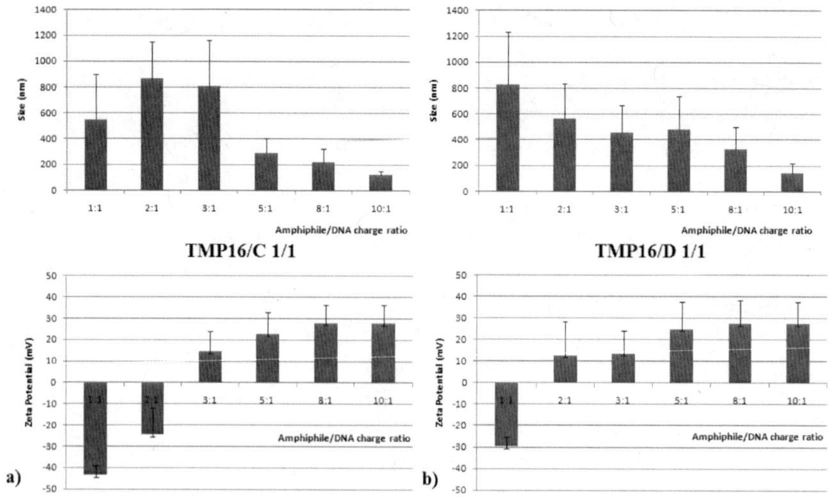

*Figure 1. Size (top graphs) and zeta potential (bottom graphs) for lipoplexes generated with pyridinium amphiphile **38b (TMP16)** co-formulated with cholesterol (a) and DOPE (b) at 1:1 molar ratio, at different amphiphile/DNA charge ratios*

the exception of **38a** (**TMP18**) all lipoplexes had a negative zeta potential at 5/1 +/- charge ratio that would prevent internalization via negatively charged sialic acid residues on the surface of the target cell membrane. Interestingly, the zeta potential of **TMP** lipoplexes increases monotonously with the increase of hydrophobic tail length from C12 to C18, while the size of lipoplexes decreases, pleading for the important role of hydrophobic effect in lipoplex formation and their physicochemical properties.

Transfection efficiency improved when surfactants were co-formulated with cholesterol or DOPE (1/1 molar ratio). For ammonium surfactant **CTAB 39** DOPE was more efficient than cholesterol though both lipoplex populations had positive zeta potential. This is probably due to a better size match and structural compatibility between **CTAB** and DOPE (*117*). The **CTAB**/DOPE had displayed a 2,712,132 RLU/mg protein average transfection efficiency and was taken as reference. In the case of pyridinium surfactants **38** both co-lipids were efficient, with cholesterol being more efficient than DOPE for **38a** (**TMP18**) and DOPE being more efficient than cholesterol for **38c** (**TMP14**). Maximum transfection efficiency was obtained with surfactant **38b** (**TMP16**), where the two formulations had equal efficiency, being about three times more efficient than **CTAB**/DOPE 1/1. For the shortest **38d** (**TMP12**) both formulations were essentially inefficient, proving that size/packing parameter (mis)match plays an obviously important role in the formation, physicochemical stability, and transfection efficiency, together with the hydrophobic effect.

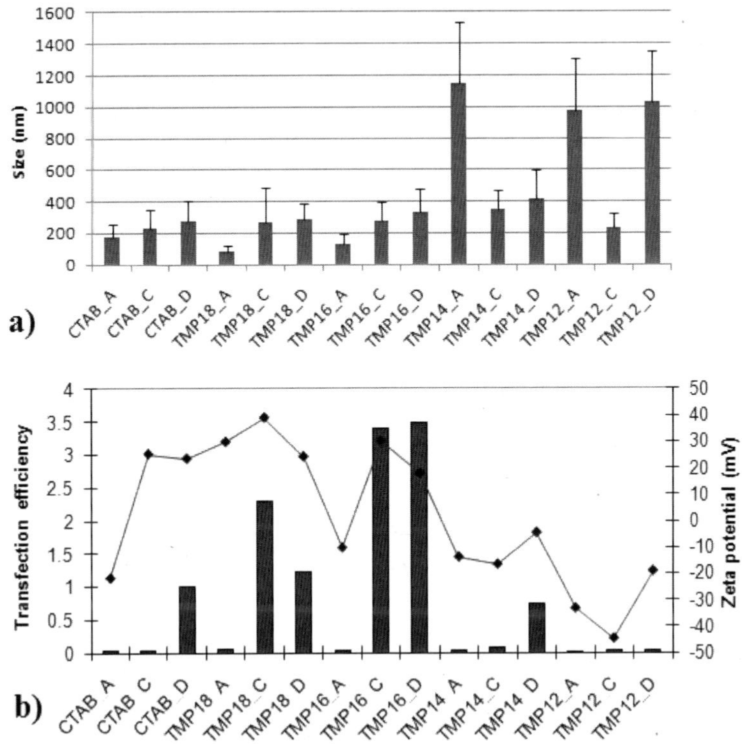

*Figure 2. Size (a), relative transfection efficiency on NCI-H23 cell line (b, left scale, efficiency of CTAB/DOPE = 1) and zeta potential (b, right scale) for lipoplexes generated with pyridinium amphiphiles **38a-d (TMP18, TMP16, TMP14** and **TMP12**), formulated alone (A) or co-formulated with cholesterol (C) and DOPE (D) at 1:1 molar ratio, at 5/1 amphiphile/DNA charge ratios.*

Conclusions and Future Studies

The use of pyridinium polar head in the design of cationic amphiphiles for nucleic acid delivery has been proved efficient by several research groups, including ours. It was shown that delocalization of the positive charge on the aromatic heterocyclic moiety confers a good balance between binding and releasing of the nucleic acid, which is essential for superior transfection efficiency. As a consequence of this soft charge the pyridinium polar head possesses special hydration properties, and enhanced lipophilicity, allowing better tissue penetration for the synthetic gene delivery system (*113*). Moreover, since the (substituted) pyridinium polar head is larger than the quaternary ammonium one, the packing parameter of these amphiphiles is very sensitive to elongation of hydrophobic tail of the amphiphile, as proved by the study presented. Related ongoing studies of self-assembled monolayers of pyridinium amphiphiles such as **40** on gold solid substrates have also confirmed these trends. Co-formulation with co-lipids is usually required, and the choice of the co-lipid depends on the particular packing parameter of the pyridinium amphiphile. Thus, preliminary results from our group have shown that pyridinium surfactants **37** follow the same trends in terms of transfection efficiency dependence on hydrophobic tail length and co-lipid formulation. A detailed transfection study, doubled by a thorough investigation of cytotoxicity of both pyridinium surfactant series is currently underway. The most efficient amphiphiles and their formulations will be tested in vivo, together with other efficient pyridinium vectors described above. These studies will gather valuable data towards systemic lipofection with pyridinium amphiphiles – a promising but rather underexploited class of synthetic gene delivery systems.

Acknowledgments

The financial support of the Temple University – Provost's Office, College of Science and Technology Undergratuate Research Program, and Temple University School of Pharmacy – Dean's Office is greatfully acknowledged.

References

1. Dotti, G.; Savoldo, B.; Okur, F.; Rousseau, R. F.; Brenner, M. K. *Gene Cell Ther.* **2009**, 1001.
2. Murphy, S. L.; High, K. A. *Br. J. Haematol.* **2008**, *140*, 479.
3. Lyon, A. R.; Sato, M.; Hajjar, R. J.; Samulski, R. J.; Harding, S. E. *Heart* **2008**, *94*, 89.
4. Cavazzana-Calvo, M.; Lagresle, C.; Hacein-Bey-Abina, S.; Fischer, A. *Annu. Rev. Med.* **2005**, *56*, 585.
5. Lin, J.; Li, M. *Expert Rev. Mol. Diagn.* **2008**, *8*, 263.
6. Templeton, N. S. *Biosci. Rep.* **2002**, *22*, 283.
7. Greco, O.; Scott, S. D.; Marples, B.; Dachs, G. U. *Front. Biosci.* **2002**, *7*, d1516.
8. Wu, C. H.; Wilson, J. M.; Wu, G. Y. *J. Biol. Chem.* **1989**, *264*, 16985.
9. Fischer, A.; Cavazzana-Calvo, M. *Lancet* **2008**, *371*, 2044.
10. Li, W.; Szoka, F. C., Jr. *Pharm. Res.* **2007**, *24*, 438.
11. Nishikawa, M.; Huang, L. *Hum. Gene Ther.* **2001**, *12*, 861.
12. Neumann, E.; Schaefer-Ridder, M.; Wang, Y.; Hofschneider, P. H. *EMBO J.* **1982**, *1*, 841.
13. Bodles-Brakhop, A. M.; Heller, R.; Draghia-Akli, R. *Mol. Ther.* **2009**, *17*, 585.
14. Wells, D. J. *Gene Ther.* **2004**, *11*, 1363.
15. Plank, C.; Schillinger, U.; Scherer, F.; Bergemann, C.; Remy, J. S.; Krotz, F.; Anton, M.; Lausier, J.; Rosenecker, J. *Biol. Chem.* **2003**, *384*, 737.
16. Liu, Y.; Miyoshi, H.; Nakamura, M. *J. Controlled Release* **2006**, *114*, 89.
17. Armaleo, D.; Ye, G. N.; Klein, T. M.; Shark, K. B.; Sanford, J. C.; Johnston, S. A. *Curr. Genet.* **1990**, *17*, 97.
18. Sato, H.; Hattori, S.; Kawamoto, S.; Kudoh, I.; Hayashi, A.; Yamamoto, I.; Yoshinari, M.; Minami, M.; Kanno, H. *Biochem. Biophys. Res. Commun.* **2000**, *270*, 163.

19. Kobayashi, N.; Kuramoto, T.; Yamaoka, K.; Hashida, M.; Takakura, Y. *J. Pharm. Exp. Ther.* **2001**, *297*, 853.
20. Zhang, G.; Budker, V.; Wolff, J. A. *Hum. Gene Ther.* **1999**, *10*, 1735.
21. Zhang, G.; Song, Y. K.; Liu, D. *Gene Ther.* **2000**, *7*, 1344.
22. Liu, F.; Huang, L. *Hepatology* **2002**, *35*, 1314.
23. Midoux, P.; Pichon, C.; Yaouanc, J. J.; Jaffres, P. A. *Br. J. Pharmacol.* **2009**, *157*, 166.
24. Seow, Y.; Wood, M. J. *Mol. Ther.* **2009**, *17*, 767.
25. Ditto, A. J.; Shah, P. N.; Yun, Y. H. *Expert Opin. Drug Delivery* **2009**, *6*, 1149.
26. De Laporte, L.; Cruz Rea, J.; Shea, L. D. *Biomaterials* **2006**, *27*, 947.
27. Thomas, C. E.; Ehrhardt, A.; Kay, M. A. *Nat. Rev. Genet.* **2003**, *4*, 346.
28. Fischer, A.; Cavazzana-Calvo, M. *PLoS Med.* **2005**, *2*, e10.
29. Li, S. D.; Huang, L. *J. Controlled Release* **2007**, *123*, 181.
30. Felgner, P. L.; Gadek, T. R.; Holm, M.; Roman, R.; Chan, H. W.; Wenz, M.; Northrop, J. P.; Ringold, G. M.; Danielsen, M. *Proc. Natl. Acad. Sci. U.S.A.* **1987**, *84*, 7413.
31. Miller, A. D. *Angew. Chem., Int. Ed.* **1998**, *37*, 1768.
32. Ilies, M. A.; Balaban, A. T. *Expert Opin. Ther. Pat.* **2001**, *11*, 1729.
33. Ilies, M. A.; Seitz, W. A.; Balaban, A. T. *Curr. Pharm. Des.* **2002**, *8*, 2441.
34. Niculescu-Duvaz, D.; Heyes, J.; Springer, C. J. *Curr. Med. Chem.* **2003**, *10*, 1233.
35. Kostarelos, K.; Miller, A. D. *Chem. Soc. Rev.* **2005**, *34*, 970.
36. Martin, B.; Sainlos, M.; Aissaoui, A.; Oudrhiri, N.; Hauchecorne, M.; Vigneron, J. P.; Lehn, J. M.; Lehn, P. *Curr. Pharm. Des.* **2005**, *11*, 375.
37. Bhattacharya, S.; Bajaj, A. *Chem. Commun.* **2009**, 4632.
38. Srinivas, R.; Samanta, S.; Chaudhuri, A. *Chem. Soc. Rev.* **2009**, *38*, 3326.
39. Esfand, R.; Tomalia, D. A. *Drug Discovery Today* **2001**, *6*, 427.
40. Mintzer, M. A.; Simanek, E. E. *Chem. Rev.* **2009**, *109*, 259.
41. Hwang, S. J.; Davis, M. E. *Curr. Opin. Mol. Ther.* **2001**, *3*, 183.
42. Yu, H.; Wagner, E. *Curr. Opin. Mol. Ther.* **2009**, *11*, 165.
43. Schaffert, D.; Wagner, E. *Gene Ther.* **2008**, *15*, 1131.
44. Davis, M. E. *Adv. Drug Delivery Rev.* **2009**, *61*, 1189.
45. Kabanov, A. V.; Lemieux, P.; Vinogradov, S.; Alakhov, V. *Adv. Drug Delivery Rev.* **2002**, *54*, 223.
46. Felgner, P. L.; Barenholz, Y.; Behr, J.-P.; Cheng, S. H.; Cullis, P.; Huang, L.; Jessee, J. A.; Seymour, L.; Szoka, F. C., Jr.; Thierry, A. R. *Hum. Gene Ther.* **1997**, *8*, 511.
47. Gershon, H.; Ghirlando, R.; Guttman, S. B.; Minsky, A. *Biochemistry* **1993**, *32*, 7143.
48. Wong, F. M.; Reimer, D. L.; Bally, M. B. *Biochemistry* **1996**, *35*, 5756.
49. Fang, Y.; Yang, J. *J. Phys. Chem. B* **1997**, *101*, 441.
50. Mok, K. W.; Cullis, P. R. *Biophys. J.* **1997**, *73*, 2534.
51. Radler, J. O.; Koltover, I.; Jamieson, A.; Salditt, T.; Safinya, C. R. *Langmuir* **1998**, *14*, 4272.
52. Lasic, D. D.; Strey, H.; Stuart, M. C. A.; Podgornik, R.; Frederic, P. M. *J. Am. Chem. Soc.* **1997**, *119*, 832.

53. Radler, J. O.; Koltover, I.; Salditt, T.; Safinya, C. R. *Science* **1997**, *275*, 810.
54. Ewert, K.; Ahmad, A.; Evans, H. M.; Safinya, C. R. *Expert Opin. Biol. Ther.* **2005**, *5*, 33.
55. Lin, A. J.; Slack, N. L.; Ahmad, A.; Koltover, I.; George, C. X.; Samuel, C. E.; Safinya, C. R. *J. Drug Targeting* **2000**, *8*, 13.
56. Koltover, I.; Salditt, T.; Radler, J. O.; Safinya, C. R. *Science* **1998**, *281*, 78.
57. Pouton, C. W.; Seymour, L. W. *Adv. Drug Delivery Rev.* **1998**, *34*, 3.
58. Cohen, R. N.; van der Aa, M. A.; Macaraeg, N.; Lee, A. P.; Szoka, F. C., Jr. *J. Controlled Release* **2009**, *135*, 166.
59. Zuhorn, I. S.; Engberts, J. B.; Hoekstra, D. *Eur. Biophys. J.* **2007**, *36*, 349.
60. Sakurai, F.; Nishioka, T.; Saito, H.; Baba, T.; Okuda, A.; Matsumoto, O.; Taga, T.; Yamashita, F.; Takakura, Y.; Hashida, M. *Gene Ther.* **2001**, *8*, 677.
61. Yang, J. P.; Huang, L. *Gene Ther.* **1997**, *4*, 950.
62. Escriou, V.; Ciolina, C.; Lacroix, F.; Byk, G.; Scherman, D.; Wils, P. *Biochim. Biophys. Acta* **1998**, *1368*, 276.
63. Vitiello, L.; Bockhold, K.; Joshi, P. B.; Worton, R. G. *Gene Ther.* **1998**, *5*, 1306.
64. Tros de Ilarduya, C.; Duzgunes, N. *Biochim. Biophys. Acta* **2000**, *1463*, 333.
65. Li, S.; Tseng, W. C.; Stolz, D. B.; Wu, S. P.; Watkins, S. C.; Huang, L. *Gene Ther.* **1999**, *6*, 585.
66. Tandia, B. M.; Lonez, C.; Vandenbranden, M.; Ruysschaert, J. M.; Elouahabi, A. *J. Biol. Chem.* **2005**, *280*, 12255.
67. Takakura, Y.; Nishikawa, M.; Yamashita, F.; Hashida, M. *Eur. J. Pharm. Sci.* **2001**, *13*, 71.
68. Bradley, A. J.; Devine, D. V. *Adv. Drug Delivery Rev.* **1998**, *32*, 19.
69. Liu, S.; Ishida, T.; Kiwada, H. *Biol. Pharm. Bull.* **1998**, *21*, 390.
70. Yamada, M.; Harashima, H.; Kiwada, H. *Biol. Pharm. Bull.* **1998**, *21*, 964.
71. Huong, T. M.; Harashima, H.; Kiwada, H. *Biol. Pharm. Bull.* **1999**, *22*, 515.
72. Huong, T. M.; Ishida, T.; Harashima, H.; Kiwada, H. *Biol. Pharm. Bull.* **2001**, *24*, 439.
73. Maeda, H. *Bioconjugate Chem.* **2010**, *21*, 797.
74. Farhood, H.; Serbina, N.; Huang, L. *Biochim. Biophys. Acta* **1995**, *1235*, 289.
75. Zabner, J.; Fasbender, A. J.; Moninger, T.; Poellinger, K. A.; Welsh, M. J. *J. Biol. Chem.* **1995**, *270*, 18997.
76. Wrobel, I.; Collins, D. *Biochim. Biophys. Acta* **1995**, *1235*, 296.
77. Zelphati, O.; Szoka, F. C., Jr. *Pharm. Res.* **1996**, *13*, 1367.
78. Bailey, A. L.; Cullis, P. R. *Biochemistry* **1997**, *36*, 1628.
79. Meers, P.; Ali, S.; Erukulla, R.; Janoff, A. S. *Biochim. Biophys. Acta* **2000**, *1467*, 227.
80. Felgner, J. H.; Kumar, R.; Sridhar, C. N.; Wheeler, C. J.; Tsai, Y. J.; Border, R.; Ramsey, P.; Martin, M.; Felgner, P. L. *J. Biol. Chem.* **1994**, *269*, 2550.
81. Balasubramaniam, R. P.; Bennett, M. J.; Aberle, A. M.; Malone, J. G.; Nantz, M. H.; Malone, R. W. *Gene Ther.* **1996**, *3*, 163.
82. Elouahabi, A.; Ruysschaert, J. M. *Mol. Ther.* **2005**, *11*, 336.
83. Xu, Y.; Szoka, F. C., Jr. *Biochemistry* **1996**, *35*, 5616.

84. Safinya, C. R. *Curr. Opin. Struct. Biol.* **2001**, *11*, 440.
85. Tang, F.; Hughes, J. A. *Bioconjugate Chem.* **1999**, *10*, 791.
86. Tang, F.; Hughes, J. A. *Biochem. Biophys. Res. Commun.* **1998**, *242*, 141.
87. Hirko, A.; Tang, F.; Hughes, J. A. *Curr. Med. Chem.* **2003**, *10*, 1185.
88. Tang, F.; Wang, W.; Hughes, J. A. *J. Lipid Res.* **1999**, *9*, 331.
89. Capecchi, M. R. *Cell* **1980**, *22*, 479.
90. Lechardeur, D.; Sohn, K. J.; Haardt, M.; Joshi, P. B.; Monck, M.; Graham, R. W.; Beatty, B.; Squire, J.; O'Brodovich, H.; Lukacs, G. L. *Gene Ther.* **1999**, *6*, 482.
91. Zanta, M. A.; Belguise-Valladier, P.; Behr, J. P. *Proc. Natl. Acad. Sci. U.S.A.* **1999**, *96*, 91.
92. Branden, L. J.; Mohamed, A. J.; Smith, C. I. *Nat. Biotechnol.* **1999**, *17*, 784.
93. Subramanian, A.; Ranganathan, P.; Diamond, S. L. *Nat. Biotechnol.* **1999**, *17*, 873.
94. Hagstrom, J. E.; Sebestyen, M. G.; Budker, V.; Ludtke, J. J.; Fritz, J. D.; Wolff, J. A. *Biochim. Biophys. Acta* **1996**, *1284*, 47.
95. Haberland, A.; Knaus, T.; Zaitsev, S. V.; Buchberger, B.; Lun, A.; Haller, H.; Bottger, M. *Pharm. Res.* **2000**, *17*, 229.
96. Schwartz, B.; Ivanov, M. A.; Pitard, B.; Escriou, V.; Rangara, R.; Byk, G.; Wils, P.; Crouzet, J.; Scherman, D. *Gene Ther.* **1999**, *6*, 282.
97. Dean, D. A. *Exp. Cell Res.* **1997**, *230*, 293.
98. Wilson, G. L.; Dean, B. S.; Wang, G.; Dean, D. A. *J. Biol. Chem.* **1999**, *274*, 22025.
99. Kim, D. H.; Rossi, J. J. *Nat. Rev. Genet.* **2007**, *8*, 173.
100. Tseng, Y. C.; Huang, L. *J. Biomed. Nanotechnol.* **2009**, *5*, 351.
101. Gao, K.; Huang, L. *Mol. Pharm.* **2009**, *6*, 651.
102. Heyes, J. A.; Niculescu-Duvaz, D.; Cooper, R. G.; Springer, C. J. *J. Med. Chem.* **2002**, *45*, 99.
103. Solodin, I.; Brown, C. S.; Bruno, M. S.; Chow, C. Y.; Jang, E. H.; Debs, R. J.; Heath, T. D. *Biochemistry* **1995**, *34*, 13537.
104. van der Woude, I.; Visser, H. W.; ter Beest, M. B.; Wagenaar, A.; Ruiters, M. H.; Engberts, J. B.; Hoekstra, D. *Biochim. Biophys. Acta* **1995**, *1240*, 34.
105. van der Woude, I.; Wagenaar, A.; Meekel, A. A.; ter Beest, M. B.; Ruiters, M. H.; Engberts, J. B.; Hoekstra, D. *Proc. Natl. Acad. Sci. U.S.A.* **1997**, *94*, 1160.
106. Meekel, A. A. P.; Wagenaar, A.; Smisterova, J.; Kroeze, J. E.; Haadsma, P.; Bosgraaf, B.; Stuart, M. C. A.; Brisson, A.; Ruiters, M. H. J.; Hoekstra, D.; Engberts, J. B. F. N. *Eur. J. Org. Chem.* **2000**, 665.
107. Smisterova, J.; Wagenaar, A.; Stuart, M. C.; Polushkin, E.; ten Brinke, G.; Hulst, R.; Engberts, J. B.; Hoekstra, D. *J. Biol. Chem.* **2001**, *276*, 47615.
108. Roosjen, A.; Smisterova, J.; Driessen, C.; Anders, J. T.; Wagenaar, A.; Hoekstra, D.; Hulst, R.; Engberts, J. B. F. N. *Eur. J. Org. Chem.* **2002**, 1271.
109. Kirby, A. J.; Camilleri, P.; Engberts, J. B.; Feiters, M. C.; Nolte, R. J.; Soderman, O.; Bergsma, M.; Bell, P. C.; Fielden, M. L.; Garcia Rodriguez, C. L.; Guedat, P.; Kremer, A.; McGregor, C.; Perrin, C.; Ronsin, G.; van Eijk, M. C. *Angew. Chem., Int. Ed.* **2003**, *42*, 1448.

110. Scarzello, M.; Chupin, V.; Wagenaar, A.; Stuart, M. C.; Engberts, J. B.; Hulst, R. *Biophys. J.* **2005**, *88*, 2104.
111. Ilies, M. A.; Seitz, W. A.; Caproiu, M. T.; Wentz, M.; Garfield, R. E.; Balaban, A. T. *Eur. J. Org. Chem.* **2003**, 2645.
112. Ilies, M. A.; Seitz, W. A.; Ghiviriga, I.; Johnson, B. H.; Miller, A.; Thompson, E. B.; Balaban, A. T. *J. Med. Chem.* **2004**, *47*, 3744.
113. Ilies, M. A.; Johnson, B. H.; Makori, F.; Miller, A.; Seitz, W. A.; Thompson, E. B.; Balaban, A. T. *Arch. Biochem. Biophys.* **2005**, *435*, 217.
114. Ilies, M. A.; Seitz, W. A.; Johnson, B. H.; Ezell, E. L.; Miller, A. L.; Thompson, E. B.; Balaban, A. T. *J. Med. Chem.* **2006**, *49*, 3872.
115. Zhu, L.; Lu, Y.; Miller, D. D.; Mahato, R. I. *Bioconjugate Chem.* **2008**, *19*, 2499.
116. Ballas, N.; Zakai, N.; Sela, I.; Loyter, A. *Biochim. Biophys. Acta* **1988**, *939*, 8.
117. Pinnaduwage, P.; Schmitt, L.; Huang, L. *Biochim. Biophys. Acta* **1989**, *985*, 33.
118. Blessing, T.; Remy, J. S.; Behr, J. P. *Proc. Natl. Acad. Sci. U.S.A.* **1998**, *95*, 1427.
119. Farhood, H.; Bottega, R.; Epand, R. M.; Huang, L. *Biochim. Biophys. Acta* **1992**, *1111*, 239.
120. Dauty, E.; Remy, J. S.; Blessing, T.; Behr, J. P. *J. Am. Chem. Soc.* **2001**, *123*, 9227.
121. Hofland, H. E.; Shephard, L.; Sullivan, S. M. *Proc. Natl. Acad. Sci. U.S.A.* **1996**, *93*, 7305.
122. Balaban, A. T.; Ilies, M. A.; Eichhofer, A.; Balaban, T. S. *J. Mol. Struct.* **2010**, *984*, 228.
123. Evans, H. C. *J. Chem. Soc.* **1956**, 579.

Chapter 3

Artificial Polymers Mimic Bacteriophage Capsid Proteins To Protect and Functionalize Nucleic Acid Structures

David B. Robinson,[1,*] George M. Buffleben,[1] Michael S. Kent,[2] and Ronald N. Zuckermann[3]

[1]Sandia National Laboratories, Livermore, CA 94551-0969
[2]Sandia National Laboratories, Albuquerque, NM 87185-1411
[3]Molecular Foundry, Lawrence Berkeley National Laboratory, Berkeley, CA 94720
[*]drobins@sandia.gov

The filamentous bacteriophage m13 and related viruses encapsulate DNA with protein, forming an organic nanowire about 1 micrometer long and less then 10 nanometers wide. The length of the wire is formed from many copies of a single protein, which is a single alpha helix formed from about 50 amino acids. It can be viewed as a very sophisticated surfactant, with hydrophilic regions that interact with the DNA and form the outer surface, and hydrophobic regions that pack against each other. We have implemented these design principles in peptoids (sequence-specific N-functional glycine oligomers) and have found that they form aggregates with DNA that have adjustable properties. This approach may complement phage display methods, providing new approaches to gene transfection and nanofabrication that do not require expression in bacteria and that provide a wider range of chemical stability and functionality.

Surfactants are central to the control of size, shape, and assembly of nanoscale materials in both artificial and living systems. State-of-art inorganic nanoparticles are synthesized using readily available surfactants such as oleic acid, oleylamine,

© 2011 American Chemical Society

hexadecyltrimethylammonium bromide, and trialkyl phosphine oxides. By adjusting the relative concentrations of these, nearly monodisperse particles of metals, metal oxides, or II-VI semiconductors can be made as spheres or similar low aspect ratio shapes in the 2-20 nm range (*1*). At higher ratios of surfactant to metal precursor, ordered assembly of surfactants and/or association of surfactants with specific crystal faces allows formation of materials with higher aspect ratio (*2*, *3*). At very high concentrations, some surfactants form lyotropic liquid crystal phases that can serve as templates for porous materials (*4*).

Much more exquisite structural control is obtained in living systems, the simplest of which are viruses, which can be precisely monodisperse. Two well known examples of high aspect ratio are tobacco mosaic virus (*5*, *6*) and the filamentous bacteriophage family, the most studied of which is m13 (*7*, *8*). This control is achieved through the use of two types of sequence-specific polymers or oligomers: nucleic acids that act as templates for assembly, and peptides or proteins with precisely located hydrophilic and hydrophobic regions defined through their sequence and conformation, as well as additional chemical functionality, that assemble onto this template. This scheme can inspire strategies toward development of more sophisticated artificial nanostructured materials. Control of material on this scale is of known value for information storage (*9*), interfacing with biological systems for diagnosis and therapy (*10*), and energy storage and harvesting (*11*, *12*), among other applications.

Substantial efforts have been made to modify viruses to make them useful for these applications. Chemical modification of intact viruses has allowed growth of inorganic materials in or around them (*13*, *14*). Other chemical modifications have been used to create specific biological interactions (*15*).

To gain a greater degree of architectural control, it would be valuable to have a detailed understanding of the principles by which the viral particles are constructed. The tobacco mosaic virus can be assembled in vitro by mixing its single-stranded RNA with the 17.5 kDa capsid protein under the proper conditions (*5*, *16*). The capsid grows from both individual proteins and preformed small aggregates or discs. There is a nucleation site within the RNA sequence where a disc of capsid proteins binds. The 5′ end is threaded through the center of the disc. Growth is rapid in that direction as entire discs add on that side, whereas growth in the 3′ direction involves addition of smaller aggregates or individual proteins and is smaller.

The bacteriophage family has a much smaller major coat protein – a single alpha helix built from about 50 amino acids. The free proteins have low solubility, and their assembly is achieved in a membrane-mediated process within their bacterial hosts (*17*, *18*). When they are assembled, cationic residues near the C terminus are in close proximity to the anionic phosphate backbone of the single-stranded DNA genome, neutral and anionic hydrophilic residues face outward toward the aqueous environment, and hydrophobic residues are buried within neighboring coat proteins. The structure of one of the coat proteins as recorded in the Protein Data Bank (1IFP) is shown in Figure 1.

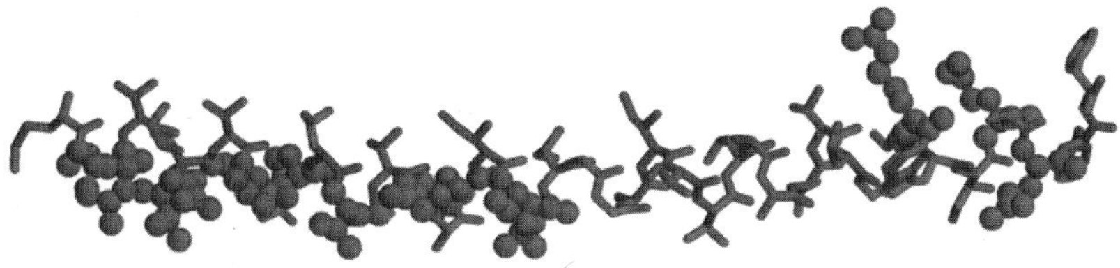

Figure 1. pf3 major capsid protein structure from the Protein Data Bank, with hydrophobic residues shown as sticks and hydrophilic residues as balls.

The sequence of the protein is easily modified by biological techniques, and substantial virion yields are still possible when major changes are made to some or all of the proteins in a particle (*19*). The structure and assembly principles of this protein are simple enough that they present a useful starting point for design of artificial polymers that test the generality of this approach to encapsulation of nucleic acids, and for development of molecules that combine the well defined structure and robust aqueous stability of the viruses with the convenience, scalability, and functionality afforded by artificial chemical synthesis.

Our proposed strategy is to synthesize sequence-specific oligomers that are known to adopt well defined helical conformations. If the conformation is known, then functionality can be introduced at well defined and known relative positions along the backbone. Peptide oligomers are known to adopt such conformations, relying on hydrogen bonding within the backbone to achieve helicity (*20*). Another approach is known that does not rely on hydrogen bonding, and is thus insensitive to agents that disrupt it, as well as to enzymes that can degrade proteins. This is a class of peptoids (peptide mimics) in which functionality appears on the amide nitrogen instead of the alpha carbon (*21*, *22*). Such N-functional glycine oligomers rely on the use of bulky chiral side groups to establish helical conformations that are stable even in 8M aqueous urea at 75 C (*23*).

Synthesis of Peptoids

The peptoids are synthesized on a solid support such as Rink amide resin. A terminal bromoacetamide group is formed by coupling bromoacetic acid to a terminal primary or secondary amine using one equivalent of diisopropylcarbodiimide (DIC) in dimethyl formamide (DMF) at room temperature for a few minutes. After rinsing with DMF, a primary amine dissolved in DMF is added, which displaces the bromide to form a terminal secondary amine in about one hour at room temperature. After rinsing, these steps repeat for each monomer in the desired sequence, as shown in Figure 2. The process can be automated by using a low-cost programmable syringe pump with an 8-port valve to feed reagents to a septum-capped fritted reaction tube containing the resin. Three two-way solenoid valves provide agitation by bubbling the solution with nitrogen, and drain spent reactants and rinsate to a waste container. Figures 3 and 4 are photographs of one version of such an apparatus. A 5 psi nitrogen relief valve is included. Reagents are stored in 50 mL centrifuge tubes with tubing

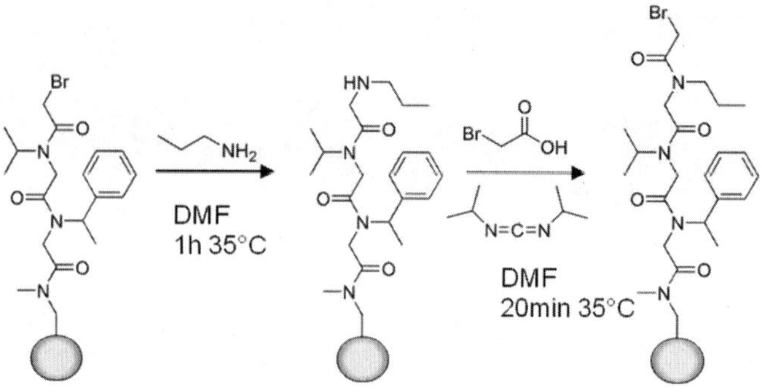

Figure 2. Reaction scheme for synthesis of N-functional glycine oligomers on a solid support.

passing through the caps (not an airtight seal). Reagents enter the reaction tube through a crimped needle that sprays reagents downward into the reaction tube.

A range of bulky chiral primary amines is available commercially, most notably from BASF, and from Chem-Impex for a range of amines with protected functional groups. The sidegroups that have been most carefully studied for establishment of helical conformations are the hydrophobic α-methyl benzylamines and the t-butyl ester of alanine (*23*, *24*). Protected amines are usually provided as hydrochloride salts, and these must be converted to the free base before use. This is achieved by dissolving or suspending the hydrochloride salt in dichloromethane, adding a nearly stoichiometric amount of 4M potassium hydroxide, and extracting the HCl into the aqueous phase. After separating, the organic phase is washed with brine, dried with magnesium sulfate, and reduced to an oily or waxy product by rotary evaporation. This product is dissolved in DMF or N-methyl pyrrolidone. Amine and bromoacetic acid solutions are typically 0.5M to 1M and the DIC is either neat or 3M. It is advisable to use a slight excess of bromoacetic acid versus DIC, so care should be taken in determining the molarity of these reagents, but the amine concentration is less critical.

The Rink amide resin initially contains a fluorenylmethyloxycarbonyl (Fmoc)-protected amine bonded to a bis (alkoxyphenyl) methyl group that is in turn tethered to a polystyrene support. Before synthesis, the Fmoc group is removed with 20% 4-methylpiperidine in DMF. At the end, the oligomer is cleaved from the resin using 95:5 trifluoroacetic acid (TFA):water, leaving a primary amide at the C terminus. Any t-butyl ester, t-butoxycarbonyl, or triphenylmethyl protecting groups on sidechains are also removed, but not necessarily with complete generality. The N terminus is protonated by the acid, unless it had been acetylated using either carbodiimide chemistry or a mixture of 0.4M acetic anhydride and 0.4M pyridine in DMF. The crude product is typically redissolved in an acetonitrile-water mixture containing about 0.1% TFA.

The crude product usually contains small amounts of byproducts that are missing a monomer, were incompletely deprotected, or are degraded. For some applications, it may be possible to use the crude product or use a simple purification procedure, but in general, it is necessary to purify the product by preparatory HPLC on a C18 or C4 column using an acetonitrile-water gradient that

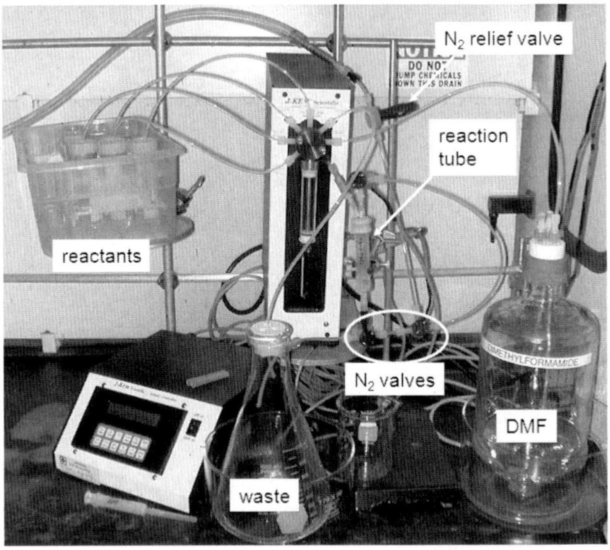

Figure 3. Peptoid synthesizer that uses a programmable syringe pump (J-Kem) and valve assembly to dispense reagents and solvent to a fritted reaction tube.

Figure 4. Closer view of syringe pump, reaction tube, and valves.

contains 0.1% TFA. Product fractions are checked using LC-MS or MALDI-MS, concentrated using vacuum centrifugation, and freeze dried.

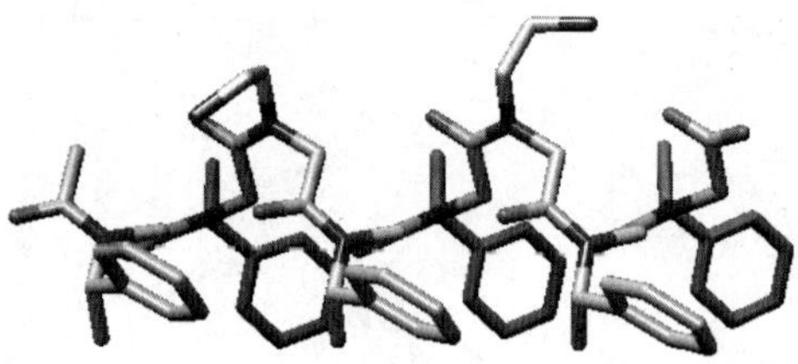

Figure 5. Structure of Ac(Nspe₂Nae)₂Nspe₂.

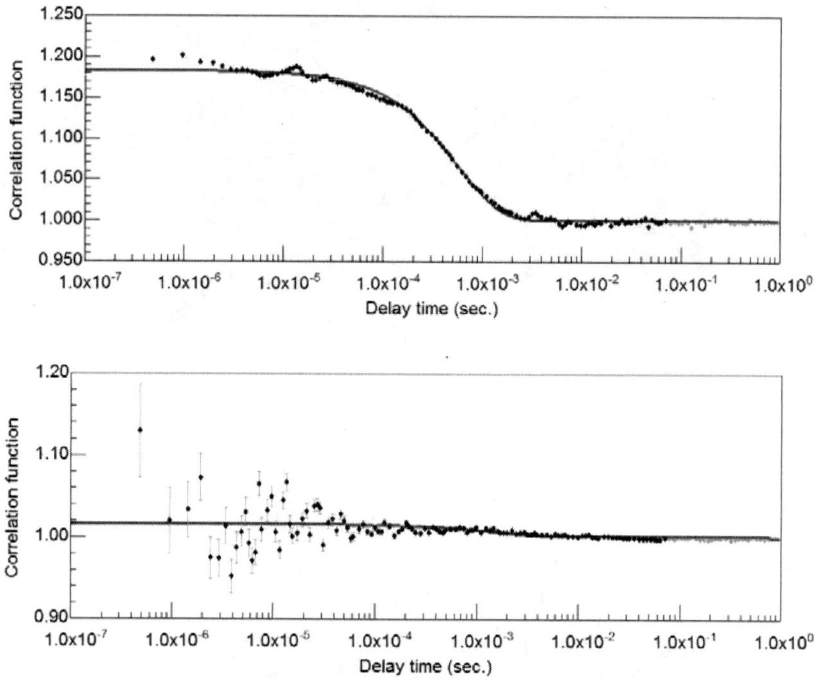

Figure 6. Dynamic light scattering correlation functions for $(Nspe_2Nae)_3Nspe_2$ in water (top) and 2-ethylhexanol (bottom).

Helical Peptoid Surfactants

If in a thought experiment the m13 coat peptide is split in the middle, one obtains two fragments that each have surfactant character. The right half in Figure 1 has a longitudinal hydrophobic stripe and a wider longitudinal hydrophilic stripe. This arrangement is similar to the water-soluble helical peptoid studied by Sanborn et al (*23*), who denote their sequence as $(NsceNsceNspe)_{12}$. The Nsce monomer has an *S*-1-carboxyethyl sidegroup, and Nspe has an *S*-1-phenylethyl sidegroup. It is believed that helical peptoids adopt a pitch of three monomers per turn (*21*), which would align the Nspe monomers longitudinally.

The left half of m13 has a narrow domain of hydrophilic groups approximately on the opposite side, and with opposite charge. Peptoids with similar motifs have been studied as mimics of lung surfactants that are used to treat premature infants

(*25*). We have synthesized several of this family, of the form $(Nspe_2Nae)_nNspe_2$, where Nae is an aminoethyl sidegroup. Variants have an acetylated N terminus, or diisopropyl aminoethyl sidegroups (Ndipae), or Nrpe, the *R* enantiomer of Nspe. One of these is drawn in Figure 5.

The peptoid $(Nspe_2Nae)_3Nspe_2$ forms aggregates at 2 mM concentrations in water, as determined from the timescale of the autocorrelation function of the dynamic light scattering signal, as shown in Figure 6. The hydrodynamic radius derived from this is 98 nm, but this number is derived assuming a spherical aggregate; a rodlike aggregate can be expected to diffuse slowly and give a radius that is similarly much larger than the length of a single molecule. Figure 6 also shows that the peptoid does not aggregate in 2-ethylhexanol. The circular dichroism spectra of $Ac(Nrpe_2Ndipae)_2Nrpe_2$ was measured in that solvent, showing evidence of a helical conformation: the characteristic pair of peaks above and opposite peak below 200 nm in Figure 7 (*26, 27*). Aggregation of this peptoid, which may aid adoption of helical conformations, was observed by light scattering in 2-ethylhexanol.

Bacteriophage Mimic Peptoids

To construct an initial peptoid that bears the motifs of the bacteriophage coat protein, we have joined three segments into a sequence: on the N-terminal end, the $(Nspe_2Nae)_x$ segment, an intermediate $Nspe_{3y}$ segment, and a $(NsceNsceNspe)_z$ segment. We designate this series as "Phage xyz", where each of x, y, and z are 1 or 2. These peptoids are much smaller than the phage coat protein, but are water-soluble and easy to synthesize, allowing for initial characterization to ensure that properties are following trends toward those of useful materials. Phage 112 is illustrated in Figure 8.

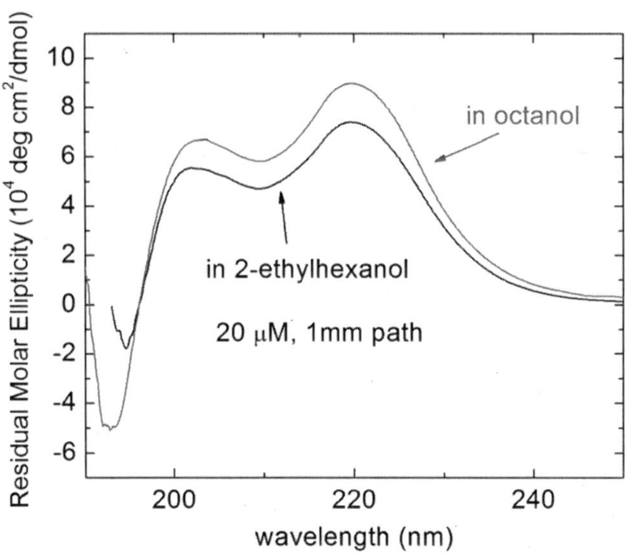

Figure 7. Circular dichroism of $Ac(Nrpe_2Ndipae)_2Nrpe_2$ in long-chain alcohols.

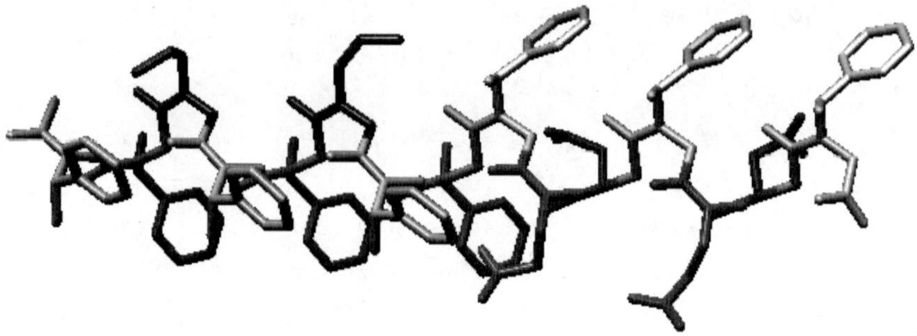

Figure 8. Model of phage 112 peptoid, with sequence (Nspe$_2$Nae)$_1$Nspe$_3$(NsceNsceNspe)$_2$

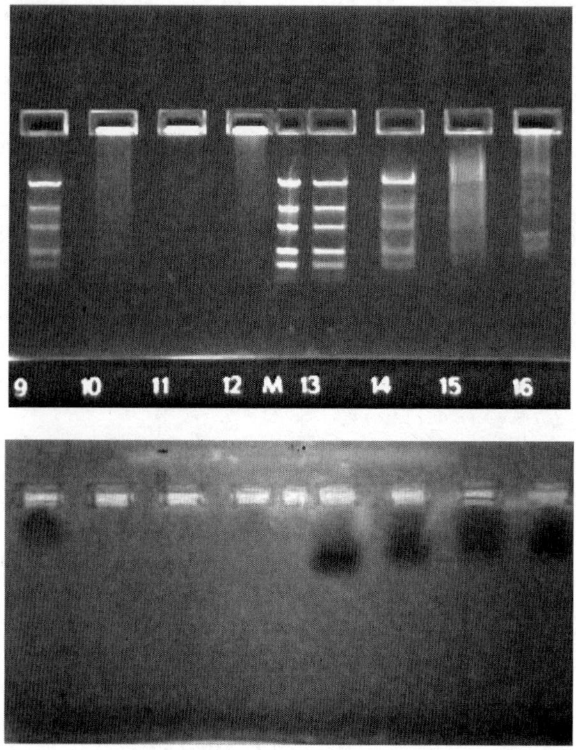

Figure 9. 0.8% Agarose gel electrophoresis of peptoid-DNA ladder mixtures in TAE buffer. The top image shows fluorescence of ethidium bromide bound to double-stranded DNA. The bottom image shows Coomassie-stained peptoids. The left 4 lanes are with Phage 212, and the right 4 are with Phage 112. For these, pH decreases from left to right. The center lane contains only the DNA ladder.

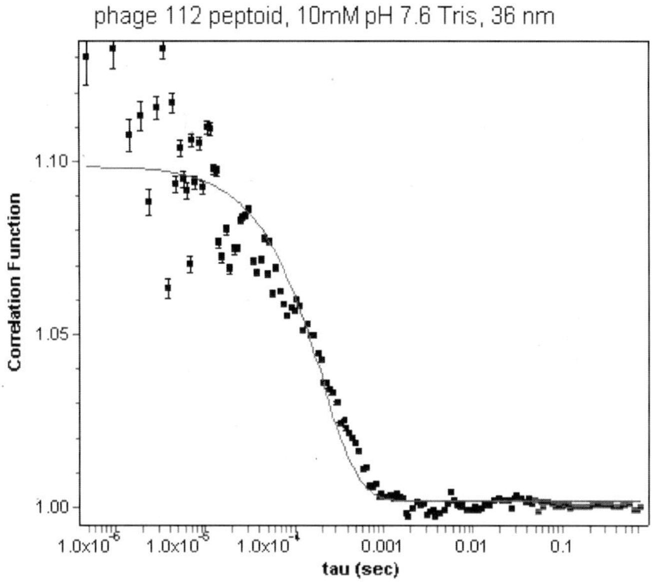

Figure 10. Autocorrelation function of dynamic light scattering signal from solution of Phage 112.

Our goal with this peptoid family is to create an aggregate with nucleic acids that are rigid rodlike and water-soluble. Compact DNA-peptoid aggregates have been previously studied for gene transfection applications (*28*, *29*). Rodlike structures may prove useful for this application, and also may help stabilize DNA origami and related structures under a broader range of solution conditions without inducing a disordered structure (*30–32*).

To evaluate the performance of phage 212 and phage 112, the peptoids were combined with double-stranded DNA ladders (about 4 peptoid molecules per DNA base) and run through a 0.8% agarose electrophoresis gel in TAE buffer (Figure 9). The peptoids showed no interaction with DNA at pH 7, as shown in the images of fluorescence of ethidium bromide intercalated with DNA. However, after reducing the pH with acetic acid or hydrochloric acid so that the carboxyl groups were partially protonated, phage 212 formed an aggregate with DNA, whereas phage 112 did not. Coomassie staining of the peptoids (Invitrogen SimplyBlue SafeStain) after capturing the fluorescence image shows that phage 112 still has a net negative charge under the more acidic conditions, traveling in the same direction as the DNA, but decreasing due to the pH reduction. The phage 212-DNA aggregate did not stain or did not stay in the gel during the staining procedure.

As with the simpler cationic surfactants, the phage mimics form aggregates in aqueous solution, as shown in Figure 10 for phage 112. The derived hydrodynamic radius of 36 nm is greater than the length scale of an individual peptoid molecule. This is corroborated by transmission electron microscopy of grids that were dipped into this solution. A drop of 1% sodium tungstosilicate (pH between 7 and 8) was placed on the grid and then wicked off, acting as a negative stain. Figure 11

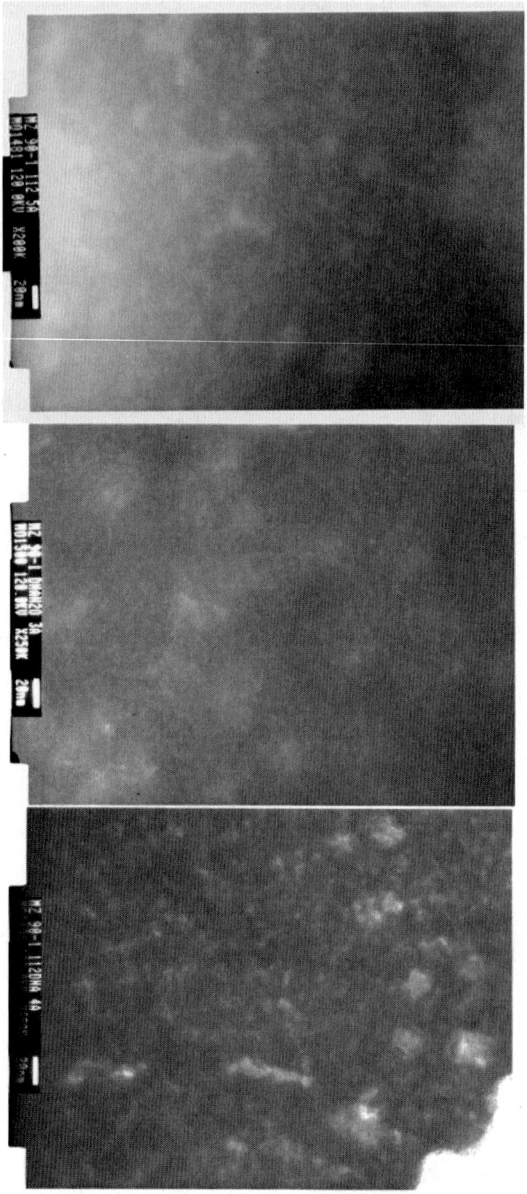

Figure 11. Transmission electron micrographs of phage 112 peptoid (top), a single-stranded DNA 30mer (middle), and a mixture of the two (bottom), stained with sodium phosphotungstate. Scale bars are 20 nm.

shows partially filamentous structures that are larger when the peptoid and DNA are combined.

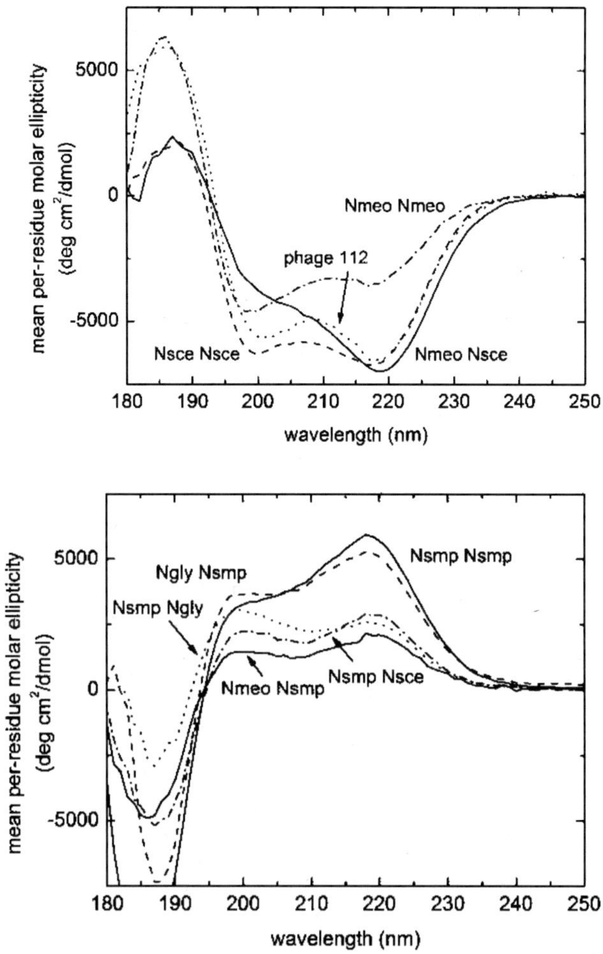

Figure 12. Circular dichroism spectra of (XYNspe)$_4$ (top) and (XYNrpe)$_4$ (bottom) where X and Y are the monomers noted in the labels.

As a strategy to obtain a phage mimic that will form an aggregate at neutral pH, we sought a neutral hydrophilic sidechain that can substitute for some of the carboxyl groups and increase the net charge on the peptoids at neutral pH. We explored the use of amides, but found that unprotected primary amide sidegroups undergo dehydration reactions during the cleaving step, and that adding protecting groups to α-amino amides is challenging. Instead, we turned to the commercially available chiral *S*-1-methoxy-2-aminopropane, which can form the monomer Nsmp. Methoxy groups are relatively hydrophilic, and peptoids containing them are known to be water-soluble. The *R* enantiomer is not commercially available, and the *S* enantiomer rotates polarized light in the opposite direction from the Nspe sidechain. To determine whether helical conformations result from the use of Nsmp, we prepared peptoids of the form (XYNspe)$_4$ and (XYNrpe)$_4$ where X and Y are combinations of Nsmp, Nsce, and the achiral methoxyethyl (Nmeo) or carboxymethyl (Ngly). Figure 12 shows that some of these show more prominent peaks in their circular dichroism spectra. Note that this was a screening study using crude products, so the abscissa units are approximate and impurities may have influenced these results.

Apparently, helicity is preserved with considerable flexibility to add neutral hydrophilic substituents when Nsce is also used, and Nsmp can be used alone or with achiral carboxyl groups with varying results. The peptoids $(Nsmp_2Nspe)_4$ and a version of phage 112 where Nsce is replaced with Nsmp were studied by gel electrophoresis as above.

As shown in Figure 13, an Nsmp-containing peptoid with no cationic sidegroups (and is cationic only at the N terminus) does not interact with DNA, and is either not hydrophobic enough to be stained by Coomassie, or is extracted from the gel during staining. Phage 112-smp interacts with DNA at neutral pH, decreasing the extent to which it binds to ethidium bromide. Excess peptoid appears in the Coomassie stain in the opposite direction from the travel of DNA, indicating that the peptoid is cationic. Approximately three peptoid molecules were present per DNA base.

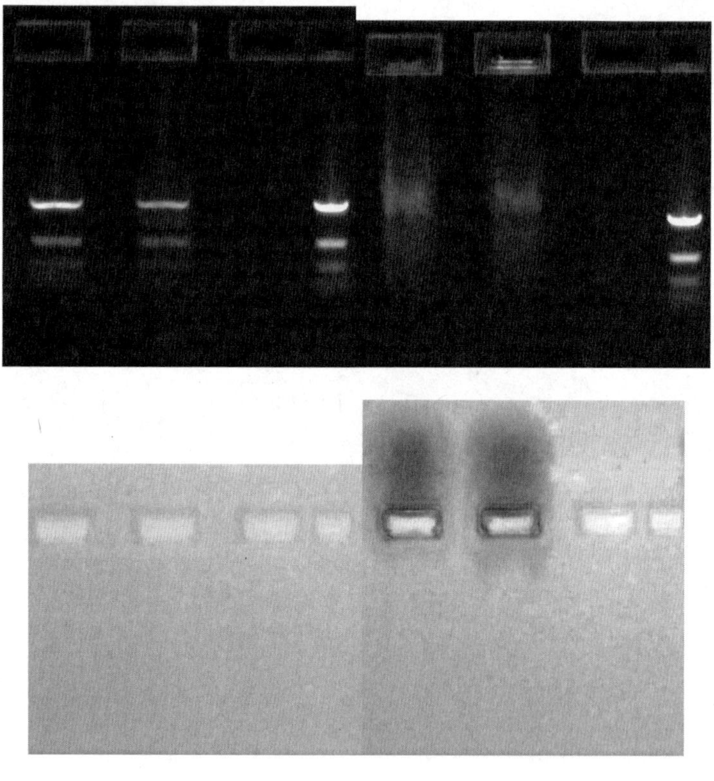

Figure 13. Gel electrophoresis of $(Nsmp_2Nspe)_4$ (left) and phage 112-smp (right). Ethidium bromide fluorescence is shown in the top images, and Coomassie-stained gels are on the bottom. In each quadrant, from left to right, are peptoid + DNA ladder, peptoid + DNA ladder at reduced pH, an empty lane, and DNA ladder with no peptoid.

Summary

We have designed a family of helical peptoids with structural motifs similar to the m13 bacteriophage, and have found that we can tune the interaction of peptoids with double-stranded DNA through variation of the sidegroup sequence. Increasing the number of cationic amine sidegroups from 0 to 2 increases the interaction with DNA, and decreasing the anionic charge by either protonation of carboxylate sidegroups or substitution with neutral hydrophilic sidegroups increases interaction. We have not yet demonstrated formation of rigid filaments, but have observed high aspect ratio aggregates by electron microscopy. Synthesis of longer peptoids – between the 12- and 15mers prepared by us and the approximately 50mers of the bacteriophage coat proteins – may result in more ordered structures that can still be prepared in vitro, and then tested for gene delivery and structural DNA applications.

Acknowledgments

This work was supported by the Laboratory-Directed Research and Development Program at Sandia National Laboratories, a multiprogram laboratory operated by Sandia Corporation, a Lockheed Martin Company, for the United States Department of Energy's National Nuclear Security Administration under contract DE-AC04-94AL85000. Work at the Molecular Foundry was supported by the Office of Science, Office of Basic Energy Sciences, of the U.S. Department of Energy under Contract No. DE-AC02-05CH11231.

References

1. Zeng, H.; Li, J.; Liu, J. P.; Wang, Z. L.; Sun, S. H. *Nature* **2002**, *420* (6914), 395–398.
2. Sau, T. K.; Murphy, C. J. *Langmuir* **2004**, *20* (15), 6414–6420.
3. Simmons, B. A.; Li, S. C.; John, V. T.; McPherson, G. L.; Bose, A.; Zhou, W. L.; He, J. B. *Nano Lett.* **2002**, *2* (4), 263–268.
4. Attard, G. S.; Bartlett, P. N.; Coleman, N. R. B.; Elliott, J. M.; Owen, J. R.; Wang, J. H. *Science* **1997**, *278* (5339), 838–840.
5. Butler, P. J. G. *Proc. R. Soc. London, Ser. B* **1999**, *354* (1383), 537–550.
6. Stubbs, G. *Rep. Prog. Phys.* **2001**, *64* (11), 1389–1425.
7. Marvin, D. A.; Hale, R. D.; Nave, C.; Citterich, M. H. *J. Mol. Biol.* **1994**, *235* (1), 260–286.
8. Smith, G. P.; Petrenko, V. A. *Chem. Rev.* **1997**, *97* (2), 391–410.
9. Park, S.; Lee, D. H.; Xu, J.; Kim, B.; Hong, S. W.; Jeong, U.; Xu, T.; Russell, T. P. *Science* **2009**, *323* (5917), 1030–1033.
10. Robinson, D. B.; Persson, H. H. J.; Zeng, H.; Li, G. X.; Pourmand, N.; Sun, S. H.; Wang, S. X. *Langmuir* **2005**, *21* (7), 3096–3103.
11. Tao, Y. F.; McCulloch, B.; Kim, S.; Segalman, R. A. *Soft Matter* **2009**, *5* (21), 4219–4230.
12. Long, J. W.; Dunn, B.; Rolison, D. R.; White, H. S. *Chem. Rev.* **2004**, *104* (10), 4463–4492.

13. Huang, Y.; Chiang, C. Y.; Lee, S. K.; Gao, Y.; Hu, E. L.; De Yoreo, J.; Belcher, A. M. *Nano Lett.* **2005**, *5* (7), 1429–1434.
14. Knez, M.; Sumser, M.; Bittner, A. M.; Wege, C.; Jeske, H.; Martin, T. P.; Kern, K. *Adv. Funct. Mater.* **2004**, *14* (2), 116–124.
15. Kehoe, J. W.; Kay, B. K. *Chem. Rev.* **2005**, *105* (11), 4056 4072.
16. Klug, A. *Proc. R. Soc. London, Ser. B* **1999**, *354* (1383), 531–535.
17. Wickner, W.; Killick, T. *Proc. Natl. Acad. Sci. U.S.A.* **1977**, *74* (2), 505–509.
18. Marvin, D. A.; Welsh, L. C.; Symmons, M. F.; Scott, W. R. P.; Straus, S. K. *J. Mol. Biol.* **2006**, *355* (2), 294–309.
19. Roth, T. A.; Weiss, G. A.; Eigenbrot, C.; Sidhu, S. S. *J. Mol. Biol.* **2002**, *322* (2), 357–367.
20. Oneil, K. T.; Degrado, W. F. *Science* **1990**, *250* (4981), 646–651.
21. Armand, P.; Kirshenbaum, K.; Falicov, A.; Dunbrack, R. L.; Dill, K. A.; Zuckermann, R. N.; Cohen, F. E. *Folding Des.* **1997**, *2* (6), 369–375.
22. Figliozzi, G. M.; Goldsmith, R.; Ng, S. C.; Banville, S. C.; Zuckermann, R. N. Synthesis of N-substituted glycine peptoid libraries. *Comb. Chem.* **1996**, *267*, 437–447.
23. Sanborn, T. J.; Wu, C. W.; Zuckerman, R. N.; Barron, A. E. *Biopolymers* **2002**, *63* (1), 12–20.
24. Wu, C. W.; Sanborn, T. J.; Huang, K.; Zuckermann, R. N.; Barron, A. E. *J. Am. Chem. Soc.* **2001**, *123* (28), 6778–6784.
25. Seurynck, S. L.; Patch, J. A.; Barron, A. E. *Chem. Biol.* **2005**, *12* (1), 77–88.
26. Wu, C. W.; Sanborn, T. J.; Zuckermann, R. N.; Barron, A. E. *J. Am. Chem. Soc.* **2001**, *123* (13), 2958–2963.
27. Wu, C. W.; Kirshenbaum, K.; Sanborn, T. J.; Patch, J. A.; Huang, K.; Dill, K. A.; Zuckermann, R. N.; Barron, A. E. *J. Am. Chem. Soc.* **2003**, *125* (44), 13525–13530.
28. Murphy, J. E.; Uno, T.; Hamer, J. D.; Cohen, F. E.; Dwarki, V.; Zuckermann, R. N. *Proc. Natl. Acad. Sci. U.S.A.* **1998**, *95* (4), 1517–1522.
29. Huang, C. Y.; Uno, T.; Murphy, J. E.; Lee, S.; Hamer, J. D.; Escobedo, J. A.; Cohen, F. E.; Radhakrishnan, R.; Dwarki, V.; Zuckermann, R. N. *Chem. Biol.* **1998**, *5* (6), 345–354.
30. Jungmann, R.; Liedl, T.; Sobey, T. L.; Shih, W.; Simmel, F. C. *J. Am. Chem. Soc.* **2008**, *130* (31), 10062.
31. Rothemund, P. W. K. *Nature* **2006**, *440* (7082), 297–302.
32. Yang, H.; McLaughlin, C. K.; Aldaye, F. A.; Hamblin, G. D.; Rys, A. Z.; Rouiller, I.; Sleiman, H. F. *Nat. Chem.* **2009**, *1* (5), 390–396.

Chapter 4

Screening of Basic Properties of Amphiphilic Molecular Structures for Colloidal System Formation and Stability

The Case of Carbohydrate-Based Surfactants

Hary Razafindralambo,* Christophe Blecker, and Michel Paquot

University of Liege, Gembloux Agro-Bio Tech, Passage des déportés, 2, B-5030 Gembloux, Belgium
***h.razafindralambo@ulg.ac.be**

Colloidal systems are bubbles, droplets, or/and particles dispersed in a continuous media, which are encountered in nature, biological system, and manufactured products. As site of interfaces between at least two bulk phases, their formation and stability need the presence of amphiphilic molecules. Carbohydrate-based surfactants (CBS) take part of these groups of compound having numerous attractive features: an abundance of precursor raw material sources, an excellent environmental compatibility, and a possible quasi-unlimited design of their molecular structure and geometry by different preparation routes. Numerous properties and activities including the aptitude to form and stabilize colloidal systems are therefore expected from a wide range of CBS compounds. Screening of their interfacial behaviors, which govern their functionalities in colloidal system properties, are then relevant in fundamental and practical point of view. In this chapter, interfacial properties of uronic acid-based surfactants are presented as examples of screening approaches. Experimental measurements of both dynamic and equilibrium properties at air-liquid and liquid-liquid interfaces, as well as a computational approach are reported.

© 2011 American Chemical Society

Introduction

Colloidal systems are dispersions of at least one bulk phase (gas, liquid, solid) reduced in small size structure (bubbles, droplets, fine particles) into another, the continuous medium. Consequently, one or more kinds of interfaces, i.e. the boundary limits between different immiscible phases, take place inside. By combining gas, liquid, and solid bulk phases two by two, nine kinds of interface grouped in fluids (only fluid bulk phases) and solids (containing at least one solid bulk phase), can be encountered in nature, biological, and manufactured product colloidal systems (Table I).

At any interfaces, whatever their nature, molecules of each bulk phase are in perpetual motion that are originally the interfacial free energy. For dispersing molecules of one phase into the other one, that is, for forming any colloidal systems, an amount of energy should be supplied. Whereas, in order to maintain dispersed molecules within the continuous phase, their condensation again in bulk phase should be restricted. Therefore, any mixed immiscible systems are thermodynamically unstable over the time. For forming and stabilizing them naturally, the interfacial energy is to be reduced, and maintained as low as possible. Amphiphilic molecules acting as "catalysts" are then required to achieve that, because of their structure containing two opposite entities having each one a greater affinity for each bulk phase. According to their structure, size, and geometry, amphiphilic molecules can be classified in different groups ranging from small surfactants (i.e.: polar lipids), to polymers (i.e.: proteins, heteropolysaccharides) through oligomeric-based structures (i.e.: peptide-, oligosaccharo-lipids). Carbohydrate-based surfactants (CBS) take part of these classes of attractive molecules with particular interests. First, these surface-active compounds can be produced from the most abundant renewable materials allowing large product concept possibilities (*2*). Second, CBS may occur in a large structural range thanks to numerous chemical functional groups. These features make possible the design of a quasi-unlimited compounds by (bio)-synthesis of new structures in different ways, or by modifying existing natural molecules. CBS structures can be varied in the hydrophilic head group (mono-, oligo-, or polysaccharides), in the hydrophobic tail (mono-, di-, tricatenar), but also in the linker/spacer between these two entities (Table II).

A few CBS geometries are shown in figure 1. Owing to the structural diversity, one can expect a wide range of CBS properties, which could be developed in food and non-food applications.

In this context, it is valuable to find out the most suitable molecular structures for post-development of CBS according to the required specific properties in various industrial application areas.

Table I. Examples of colloidal system involving different interfaces. SOURCE: Adapted with permission from reference (*1*). Copyright 1999 Wiley – VCH

			Examples		
Dispersed phase	*Dispersing phase (medium)*	*Name*	*In Nature*	*In Biological system*	*In manufactured products*
Solid	Liquid	Colloidal sol	River water	?	Paint, ink
Liquid	Liquid	Emulsion	Latex	Biological membranes	Drug delivery
Gas	Liquid	Foam	Polluted rivers	Insect excretions	Fire extinguishers
Solid	Gas	Aerosol	Volcanic smoke	Pollen	Inhalation of solid pharmaceuticals
Liquid	Gas	Aerosol	Clouds	Result of coughing	Smog
Gas	Gas	Aerosol ?	?	?	?
Solid	Solid	Solid suspension	Wood	Bone	Composites
Liquid	Solid	Porous material	Oil reservoir rock	Pearl	High impact plastics
Gas	Solid	Solid foam	Pumice	Loofah	Styrofoam

Table II. Main classes of CBS compounds

CBS	*Polar head group*	*Hydrophobic tail*	*Linker*
Mono-, bi-, tricatenars	One polar head	One, two, three chains	
Glycoglycerolipids		Acylglycerol	
Glycosphingolipids		Sphingoïdes	Ester, amide, ether,…
Bolaforms	Two polar heads	One chain	
Geminis		Two chains	
Alkylpolyglucosides	Several polar heads	One or more chains	

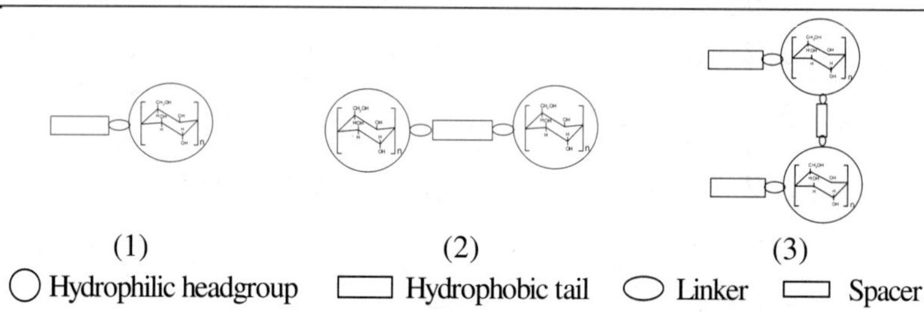

Figure 1. Main structures and geometries of CBS: (1)monocatenar, (2)bolaform, and (3)gemini

By screening interfacial properties of CBS, it is then possible to pre-select a group of molecular structures for suitable functional properties to be investigated. Indeed, through several research works in last decades, it has been demonstrated that those fundamental properties including dynamic and equilibrium interfacial tensions, interfacial rheology, and Langmuir monolayer characteristics, govern the colloidal system formation and stability (*3–5*) like foams, emulsions, and suspensions. Moreover, the direct assessment of amphiphilic molecules for their aptitudes in forming and stabilizing colloidal systems require relatively large amounts of sample, which are expensive to prepare. Therefore, a pre-selection of potential compounds from a set of molecular structures according to their fundamental or "microscopic" properties requiring only small amounts, or even no sample by theoretical approaches, could be beneficial prior to characterize their functional properties qualified as "macroscopic" aptitudes. On the other hand, this approach allows a better understanding of relationships between fundamental and more practical properties.

In this chapter, screening strategies concern, especially, monocatenar and bicatenar uronic acid derivatives, a well representative class of carbohydrate monomers widely distributed in natural plant polysaccharides (*6*). Uronic acids are monosaccharides bringing a carboxylic group instead of the primary alcohol (*7*). Analogs and homologous series of derivative compounds have been prepared via chemical synthesis routes by connecting one or two hydrophobic alkyl chains to glucuronic (4-OH in equatorial position) and galacturonic (4-OH in axial position) acids through various linkers including ester, amide, or /and ether bonds.

Screening general strategies for CBS interfacial properties, followed by illustrative examples with uronic acid derivative surfactants, are for the first time reported. A brief description of basic interfacial parameters and their roles in colloidal system properties is first presented.

Basic Interfacial Properties

Surface and Interfacial Tensions

Surface and interfacial tensions (γ) are the main basic parameters characterizing any interfaces. These do not exist when measured surface or interfacial tension values are close to zero (only one phase exists), that is, dispersed and continuous bulk phases are quasi-totally miscible. According to measurement modes, dynamic or equilibrium surface/interfacial tensions can be distinguished. By static mode measurements, the interfacial area is in a steady-state, and the γ reaches its minimum value, i.e. the equilibrium surface tension (γ_e), after any time depending molecule intrinsic properties (interface nature, amphiphilic structure, size,...), and environmental chemical and physical parameters concentration, temperature, ionic strength,...). By determining γ_e as a function of amphiphilic molecule concentration under well-defined conditions, one can assess other parameters like the critical micellar concentration (CMC), surface excess or maximum interfacial concentration (Γ_{max}), and minimum molecular area (A_{min}) using the Gibbs adsorption isotherm:

$$\Gamma_{max} = -\frac{1}{RT}\left(\frac{d\gamma}{d\ln A}\right)_{cmc} \quad \text{and} \quad A_{min} = \frac{1}{\Gamma_{max}}$$

where $R = 8.314$ J.K^{-1}.m^{-2} is the gas constant; and T, the absolute temperature.

On the other hand, dynamic surface/interfacial tensions correspond to values measured when the interfacial area is being grown during the measurement itself. This allows the determination of actual adsorption kinetics parameters such as diffusion rate (D), adsorption/desorption time or rate (t), interfacial tension of mi-equilibrium (γ_m), for instance, using the basic equation of Ward-Tordai (*8*):

$$\Gamma(t) = 2c_o\sqrt{\frac{Dt}{\pi}} - 2\sqrt{\frac{D}{\pi}}\int_0^{\sqrt{t}} c_s d(\sqrt{t-\tau})$$

where c_o is the bulk concentration, D the monomer diffusion coefficient, c_s the concentration in the subsurface, $\pi = 3.142$ and τ is a dummy variable of integration, and that of Hua-Rosen (*9*):

$$\gamma(t) = \gamma_m + \frac{\gamma_o - \gamma_m}{1+\left(\frac{t}{t^*}\right)^n}$$

where t^* and n are constants having units of time and dimensionless, respectively.

Surface Pressure

Surface pressure (Π) is defined as the difference between interfacial tension of pure liquid and that of liquid with adsorbed or spread molecules at the interface:

$$\Pi = \gamma_o - \gamma$$

Π is commonly used for characterizing monolayer formed by amphiphilic molecules at the surface, known as Langmuir monolayers. When amphiphilic compounds are slightly or totally insoluble in the bulk phase (aqueous or not), their interfacial properties can be characterized by spreading them directly onto the surface, using a solvent or a mixture of solvent in which they are totally soluble. Π is measured as a function of delimited area where amphiphilic molecules are spread. Unlike Gibbs adsorption approach, which requires numerous sample concentration measurements, a surface pressure-area isotherm can be established with one sample concentration, and under well defined conditions, by reducing in continuous the available surface. From this plot, it is possible to deduce numerous physical characteristic parameters such as the limiting area (A_o) from which a Π value becomes detectable, collapse pressure (Π_c) corresponding to the maximum surface pressure developed by the monolayer, and different intermediate characteristic points that characterize various monolayer states (transition molecular area and surface pressure). By analogy, Π_c and corresponding molecular area (A_c) values obtained by spreading technique could give the same indication than γ_{cmc} and A_{min} from Gibbs monolayer formed by adsorption. As Π values of pure liquid-liquid interface are always equal to zero, they can then be used to compare the effect of the same amphiphilic molecule at different interfaces, and its affinity for each interface.

Elasticity/Compressibility Parameters

The film elasticity (*E*) or compressibility (*C*) at fluid interfaces is defined by the following equation:

$$E = -\frac{d\gamma}{d \ln A} \quad \text{and} \quad C = \frac{1}{E}$$

These parameters can be determined from surface pressure *vs.* area plots of Langmuir monolayers. They measure the extent of interfacial tension variation after small area changes due to mechanical dilation or compression. The film elasticity gives an indication on the ability of amphiphilic molecules to maintain the changes of interfacial tension as low as possible, whereas the film compressibility is related to the resistance of interfacial film under compression.

Viscoelasticity Parameters

Film viscoelasticity parameters are determined by applying compression/ dilation cycles to the interface. These parameters allow the measurement of both elastic (real part E′) and viscosity modulus (imaginary part E″) of amphiphilic molecule film adsorbed at the interface.

$$E = E' + iE''$$

Viscoelastic properties indicate the resistance of amphiphilic molecule monolayer to deformations (elastic modulus) and the associated relaxations in the surface region (surface viscosity). These parameters are often measured at different frequencies (sinusoidal fluctuations of bubble or droplet areas) and surfactant concentrations.

Colloidal System Properties

All basic parameters described above can be used to predict amphiphilic compounds able for forming and stabilizing colloidal system properties according to dispersed and dispersing bulk phases (gas, liquid, solid), and the type of created interfaces (gas-liquid, liquid-liquid, gas or liquid-solid). The equilibrium parameter like γ_e is a global indicator of the energy required for dispersing one bulk phase to another. For γ_e close to zero, the dispersion is nearly spontaneous, and lower energy is necessary for forming the colloidal system. It is for instance the case of microemulsions. As for CMC and Γ_{max}, they indicate the minimum and maximum concentrations of surfactant for forming micelles (monomer aggregation), and saturating the interface, respectively. These parameters are predictive of surfactant solubilizing power of bulk phases (liquid or solid) inside the aggregated structure. By contrast, dynamic interfacial tensions, as well as related parameters like D, t^*, and γ_m are rather associated to the time-dependent of colloidal system formation, and measure the effectiveness of amphiphilic molecules to reduce interfacial tensions while the growing of interfacial areas. On the other hand, all characteristic parameters of amphiphilic molecule monolayers, such as elasticity, compressibility, and viscoelasticity allow the prediction of colloidal system stability. Amphiphilic molecules forming interfacial film with high elasticity, low compressibility, and high viscoelasticity are generally required as stabilizers.

Screening Strategies

Screening strategies involve the choice of appropriate materials, techniques, and methodologies to be used for reaching the objectives of prediction. The preparation of amphiphilic molecules in laboratory scale using chemical, enzymatic, or chemo-enzymatic synthesis routes, or by modifying potential existing structures is suitable for generating a set of similar compounds for rational design. For CBS, analogous series of mono-, oligo-, and polysaccharides varying in polar head group, hydrophobic tail, and linker/spacer may be prepared, and characterized. As highly purified compounds are needed to do that, instruments using small amounts of sample like drop and bubble tensiometers are required. Moreover, computational approaches for modeling a single and an assembly of molecules at the interface are useful for a better understanding of mechanism at

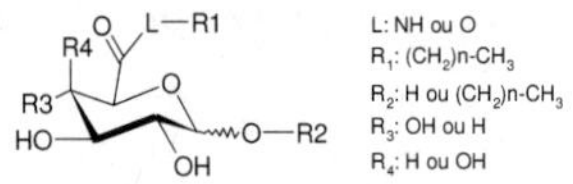

Figure 2. General structure of uronic acid-based surfactants

Table III. Equilibrium surface parameters of various CBS at 25°C

CBS	γ_{cmc} *[mN/m]*	*CMC* *[mM]*	A_{min} *[Å²/molecule]*
C8-O-GlcA (Ester)	28.0	10.7	43.1
C8-N-GlcA (Amide)	24.4	3.3	25.0
C8-N-GalA (Amide)	20.5	4.0	33.8

a molecular level, but also as a good control tool of experimental results without using any samples and characterization instruments.

Interfacial Properties of Uronic Acid Derivative-Based Surfactants

Monosaccharide-based surfactants belonging to uronic acid derivatives well illustrate the screening of interfacial properties of CBS for forming and stabilizing colloidal systems. Uronic acids are monosaccharides in which the primary alcohol is substituted by a carboxylic acid. From this starting compound which can be prepared from renewable vegetable materials (*6*), numerous derivative structures varying in polar head group, hydrophobic tail, and linker are easily available after chemical synthesis in laboratory scale. Their chemical structures fully characterized by a series of spectroscopic techniques are summarized in Figure 2. The polar head is constituted by either a glucuronic acid (Glc-A) or a galacturonic acid (Gal-A), which are stereochemically different regarding the position of the OH group at the fourth carbon. The former has OH-4 in equatorial position, whereas the latter contains OH-4 in axial position. The hydrophobic tail is monocatenar (with single chain) or bicatenar (with double chains). Both antagonist entities are connected by ester, amide, or/and ether bonds.

Equilibrium Surface Properties

Table III compares the equilibrium surface parameters of three monocatenar uronic acid derivatives having different polar head groups and linkers, and constituted by the same hydrophobic tail. These results allow the evaluation of OH-4 stereochemistry and linker effects on equilibrium surface properties of short monocatenar uronic acid derivatives. Here, interfacial properties result from the adsorption of surfactant molecules, from water bulk phase to the interface under static conditions.

Table IV. Dynamic interfacial parameters of uronamide-based surfactants (1.3 ± 0.3 mM) at fluid interfaces

CBS	*Interfacial pressure [mN/m] 20°C*			
	Time [s]	*Air/Water*	*Dodecane/Water*	*Rapeseed oil/Water*
C8-Gal-N	0	0.0	0.0	0.0
	13	27.4	21.8	7.8
	25	28.3	22.8	8.5
C8-Glc-N	0	0	0.0	0.0
	13	11.2	14.8	4.2
	25	11.6	14.9	4.5
C12-Gal-N	0	0.0	0.0	0.0
	13	13.3	7.9	4.5
	25	19.7	10.2	5.3
C12-Glc-N	0	0.0	0.0	0.0
	13	24.2	9.7	5.9
	25	30.1	14.1	8.4

An amide bond as linker allows surfactants more efficient than the ester bond regarding both γ_{cmc} and CMC. However, the ester bond induces a larger minimum molecular area than the amide one. In the same way, galacturonamide appears slightly more surface-active than glucuronamide. However, their CMC values are in the same range. Thus, the axial position of the OH-4 at the fourth carbon is more favorable than the equatorial one for uronamide-based surfactants under study. Moreover, the OH in axial position also gives a larger minimum area occupied per molecule, which could be beneficial in terms of concentration required for reaching the same effect, that is, for the same interfacial area. The effect of OH stereochemistry on interfacial properties is easily understandable, and could be connected to the difference in H-bonding within molecules as suggested in previous works (*10*, *11*).

Dynamic Surface Properties

Table IV compares dynamic interfacial pressures of uronamide-based surfactants with different polar head group and hydrophobic chain length at air-water (A/W), dodecane-water (D/W), and rapeseed oil-water interfaces (O/W). As pure fluid interfaces are different regarding the interfacial tension value (W>D>O), the use of the interfacial surface pressure as dynamic adsorption parameter is more convenient for comparative studies. The interfacial pressure values were obtained by measuring interfacial tensions within 30 s by a drop volume tensiometer, and subtracting these to those of corresponding pure liquids recorded at 20°C. It appears that: (1) Galacturonamide surfactants with a shorter alkyl chain length are more efficient than Glucuronamide ones, whereas the opposite effect occurs for higher alkyl chain derivatives, whatever the tested interfaces. Thus, OH-4 in axial position with a shorter alkyl chain length seems to have more favorable effects on dynamic properties, whereas the OH-4 in equatorial position appears interesting, when it is linked to a longer alkyl chain length. (2) In all cases, uronamide-based surfactants have much more affinity for A/W and D/W than O/W. Dynamic interfacial pressure can be used as a physical parameter for the evaluation of the surfactant affinity to any interface including an aqueous phase, in which surfactants were previously dissolved. The higher the dynamic interfacial pressure value, the more important the surfactant affinity for the kind of interface. As all surfactants adsorb from the aqueous phase, they normally diffuse to the interface with the same rate, whatever the nature of the other bulk liquid phase. The molecular area occupied by surfactants, and their configuration for each interface probably play the key role for dynamic interfacial properties as reported in our previous works (*12*). These results also suggest the importance of the optimum combination of both entities for achieving the best performance on dynamic interfacial pressures. These parameters are very relevant especially for the colloidal system formation, which is a dynamic process (*13*).

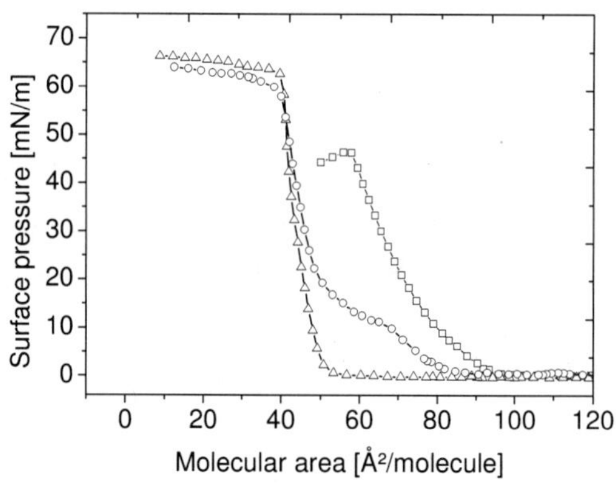

Figure 3. (π-A) isotherms of bicatenar uronic acid derivative-based surfactants: (○) $C_{14/14}$-Glc-A; (Δ) $C_{14/14}$-Gal-A; (□) $C_{10/10}$-Gal-A established at 20°C

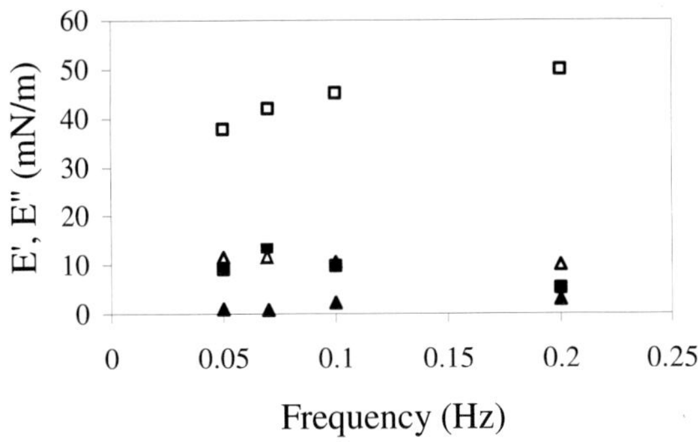

Figure 4. The real E'(square) and imaginary E"(triangle) components of glucose octanoate (open symbols) and octyl glucuronate (filled symbols) films at 0.33 mM with various oscillation frequencies. (Reproduced from reference (12). Copyright 2009 American Chemical Society.)

Spread Film Properties

Figure 3 shows surface pressure *vs.*molecular area isotherms of bicatenar uronic acid derivatives spread on pure water at 20°C. As these CBS are insoluble in aqueous solution because of the two hydrophobic chains within their structure, the Langmuir monolayer approach was used in order to evaluate their interfacial properties. Samples were solubilized into a mixture of dichloromethane/methanol (9/1), and directly spread onto the surface. By comparing π-A isotherms of bicatenar glucuronic acid and galacturonic acid derivatives with the same and different alkyl chains, one can deduce the effect of the OH-4 stereochemistry, and that of the hydrophobic alkyl chain length. Bicatenar derivatives with longer alkyl chains give higher monolayer stability according to the collapse pressure values exceeding 60 mN/m, whereas shorter ones form a more expanded film regarding the limiting area, i.e. when the surface pressure begins to rise. Moreover, a bicatenar surfactant monolayer is more stable than a monocatenar one (data not shown), and the OH-4 group in equatorial position (Glc-A) exhibits a transition state, which does not occur for an OH-4 in axial position. The existence (or not) of transition phases for closely related molecules forming monolayers at the air-water interface under well-defined conditions has been reported previously with hexadecanoic acid and its derivatives, and was interpreted in terms of differences in inter- and/or intra-molecular interactions by H-bonding at the head group entity (*14*).

Film Viscoelastic Properties

Figure 4 illustrates the changes of surface compex viscoelasticities constituted by the real (E') and imaginary (E'') parts of octyl glucuronate and glucose octanoate as a function of the surface oscillation frequency. A small change of the ester

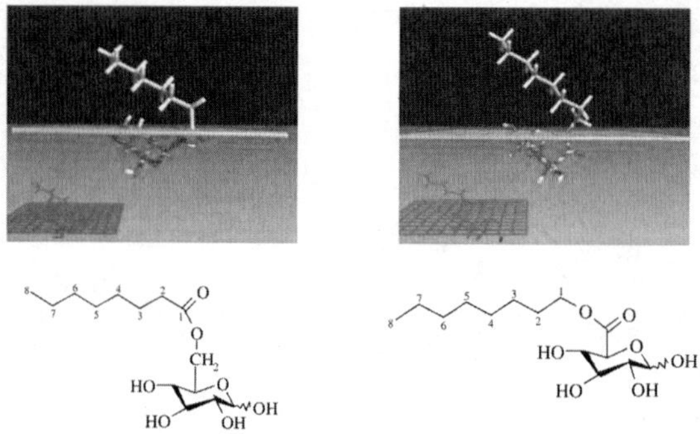

Figure 5. Molecular conformation models at a polar/apolar interface: glucose octanoate (left) and octyl glucuronate (right). (Reproduced from reference (12). Copyright 2009 American Chemical Society.)

bond direction induces significant differences in the complex modulus of surface viscoelasticity with a predominance of the elasticity contribution over that of the viscosity in both cases, for a wide range of oscillation frequency. The relationship between the surfactant molecule structure and surface reheological properties has been attributed to the difference in the adsorption rate and the surface excess concentration of the sugar ester surfactants (*12*). Glucose octanoate adsorbing more slowly at the air-water interface can generate a higher interfacial gradient, which strengthens the monolayer response against the surface perturbation.

Molecule Computational Modeling

Computational modeling of amphiphilic molecules at the interface is a complementary approach to experimental measurements. It allows a theoretical prediction of the conformation adopted by single and assembly surfactant molecules at the interface. Figure 5 illustrates the results obtained with monocatenar short chain sugar surfactants, which only differ in the orientation of the ester bond between the polar head group and hydrophobic tail between glucose and octanoic acid, and glucuronic acid and octanol. The effect of the linker "direction" on the molecular area occupied by short chain derivative sugar surfactants at the A/W interface is observable by a single molecule model at the interface. For glucose octanoate, the carbonyl group (CO) is closer to the alkyl chain length so as the polar head group is assumed to be smaller than that of octyl glucuronate for which the same CO is near the polar head group. As listed in Table V, experimental and theoretical approaches lead to the same values of the interfacial molecular area. Differences observed between the two analogous surfactants on the occupied molecular area have an important impact on the dynamic surface tension, film viscoelasticity, and CMC (*12*).

Concerning CMCs, the difference could arise from the interruption or not of the alkyl chain length by the CO group. Consequently, glucose octanoate is less hydrophobic than octyl glucuronate since its alkyl chain is interrupted by CO, so its CMC value is higher than that of its analogous compound.

Table V. Interfacial molecular areas determined by computational and experimental approaches. SOURCE: Reproduced from Reference (*12*). Copyright 2009 American Chemical Society

CBS	*Interfacial molecular area calculation (Å²)*	
	From computing data	*From experimental data*
Glucose octanoate	38	37
Octyl glucuronate	35	44

Table VI. Main trends of amphiphilic molecular structure from screening investigations of uronic acid-based surfactant interfacial properties[a]

Structural entities	*Interfacial properties*				*Techniques*
	Dynamic	Equilibrium			Drop, bubble, and Wilhelmy plate tensiometers
	t/E*	*CMC*	γ_{cmc}	A_{min}	
Polar head	+	+	+	+	
Apolar chain (*C=8 to 18*)	+++	+++	++	++	
Linker	+/+	++	++	++	
	Molecular area		Monolayer stability		Langmuir film balance
Apolar chain (*n=1 to 2*)	++		+++		

[a] *t**: *adsorption time parameter; n: number of alkyl chain; E: complex viscoelastic modulus*

Conclusion

Table VI summarizes the main trends of the screening investigations using small amounts of uronic acid-based surfactant samples, as CBS illustrative examples, and various appropriate techniques used. Each structural entity of this class of amphiphilic molecule such as the polar head group stereochemistry (OH-4), hydrophobic chain length and number, and linker, affect their dynamic and equilibrium interfacial properties. The strongest effect is that of the alkyl chain number and length on dynamic adsorption time, CMC, and monolayer stability. The kind of linker has also significant effects, mainly, on the CMC. Such screening results allow us to predict the most effective structures to be applied especially for forming and stabilizing foams and emulsions, for detergent performance, and for other applications such as biological activities, as well as to get insight into the relationships between structure and activities. As main conclusions, uronic acid-based surfactants with shorter monocatenar alkyl chains could be tested as foaming and emulsifying agents thanks to their best dynamic interfacial properties, whereas longer monocatenar alkyl chains are more appropriate for detergents and biocide properties due to their relevant equilibrium surface properties, in particularly, their lowest CMC values. By contrast, bicatenar surfactants having sufficient alkyl chain length could be evaluate as stabilizer in emulsions because of their very stable and less compressible monolayer at the interface. The knowledge on the linker effect allows the choice of the appropriate one for improving overall properties.

Acknowledgments

The authors thank Dr Aurore Richel and Dr Pascal Laurent for the synthesis and spectral characterization of uronic acid-based surfactants. Mr. Alexandre Schandeler is acknowledged for technical assistance. This work receives a financial support from the Belgian Walloon Region within DG06 research project of excellence "TECHNOSE".

References

1. Fennell Evans, D.; Wennerström, H. *The colloidal domain*; Wiley-VCH: New York, 1999; pp 1−515.
2. Ruiz, C. C. *Sugar-based surfactants fundamentals and applications*; CRC Press: New York, 2009; Vol. 143, pp 1−639.
3. Gaines, G. L. *Insoluble monolayers at liquid-gas interface*; Interscience Publishers: New York, 1966; pp 1−386.
4. Adamson, A. W.; Gast, A. *Physical chemistry of surfaces*; Wiley: New York, 1990.
5. Langevin, D. *Structure and dynamic properties of surfactant systems*; 1998; Vol. 117, pp 129−134.
6. Langguth, H.; Benet, L. Acyl Glucuronides Revisited: Is the Glucuronidation Proces a Toxification as Well as a Detoxification Mechanism? *Drug Metab. Rev.* **1992**, *24* (1), 5–47.
7. Laurent, P.; Razafindralambo, H.; Wathelet, B.; Blecker, C.; Wathelet, J.-P.; Paquot, M. *J. Surfactants Deterg.* **2011**, *14* (1), 51–63.
8. Ward, A.; Tordai, L. *J. Chem. Phys.* **1946**, *14*, 453–462.
9. Hua, X.; Rosen, M. *J. Colloid Interface Sci.* **1988**, *124*, 652–659.
10. Hato, M.; Minamikawa, H.; Tamada, K.; Baba, T.; Tanabe, Y. *Adv. Colloid Interface Sci.* **1999**, *80*, 233–270.
11. Hinz, H. J.; Kuttenreich, H.; Meyer, R.; Renner, M.; Fründ, R.; Koynova, R.; Boyanov, A. I.; Tenchov, B. G. *Biochemistry* **1991**, *30*, 5125–5138.
12. Razafindralambo, H.; Blecker, C.; Mezdour, S.; Deroanne, C.; Crowet, J.; Brasseur, R.; Lins, L.; Paquot, M. *J. Phys. Chem. B* **2009**, *113*, 8872–8877.
13. Miller, R.; Joos, P.; Fainerman, V. B. *Adv. Colloid Interface Sci.* **1994**, *49*, 249–302.
14. Yue, X.; Steffen, P.; Dobner, B.; Brezesinski, G.; Möhwald, H. *Colloids Surf., A* **2004**, *250*, 57–65.

Chapter 5

Levo vs. Dextro

Effect of the Headgroup Chirality on Nanoassemblies

P. Lo Nostro,* N. Peruzzi, L. Giustini, and P. Baglioni

Department of Chemistry and CSGI, University of Florence, 50019 Sesto Fiorentino (Firenze), Italy
***Email: pln@csgi.unifi.it**

Surfactants that bear chiral headgroups form a variety of supra self-assembled nanostructures and can be used for different applications. In particular, ascorbic acid possesses two chiral centers, a rigid planar ring, a powerful redox active moiety, two acidic –OH residues, and a primary –OH group, and therefore is one of the most versatile polar headgroups for its peculiar properties.

Here we report our studies on the nanoassemblies produced by L-(+)-ascorbyl-alkanoates (L-ASCn) and D-(-)-isoascorbyl-alkanoates (D-ASCn), carried out through DSC, SAXS, XRD, and surface tension experiments. The results suggest that the different configuration of the headgroups induces relevant changes in the structural and physico-chemical properties of the aggregates, due to the different hydration of the two epimers L-ascorbic and D-isoascorbic acid. The mixtures of D-ASC12 and L-ASC12 surfactants produce interesting phase diagrams, revealing the existence of a 1:1 compound, and of two eutectic points.

Introduction

The production of nanostructures with specific properties is crucial for many different applications, such as electronics, material science, food products, pharmaceutics, environmental protection, and so forth. Their size, shape and physico-chemical properties are involved in the design, synthesis and use of new "smart" chemical species. For example, innovative starting materials with redox,

© 2011 American Chemical Society

thermal, pH, photochemical and chemical features are used as building blocks for higher hierarchy architectures. Therefore, the introduction of specific moieties and functional groups in a molecule can modify its nature and improve or simply change its physico-chemical and chemical behavior and its biological effects, as desired.

Extended conjugated unsaturated bonds, large linear or branched hydrophobic regions, polar ionizable groups, and acceptor/donor residues for hydrogen or coordination bonding determine the behavior of the building blocks, for example their capacity to self-assemble and produce supramolecular structures in water. This is directly due to a variety of non-covalent interactions that warrant the necessary reversibility and capacity to react to environmental changes.

Chiral surfactants usually contain a sugar or an amino acid in the hydrophilic region, and are commonly studied for enantiomeric separations (*1*). Chirality affects significant processes such as molecular recognition and self-organization, which are common in biomembranes and other biological supramolecular structures (*2*). In fact, chirality is one of the main factors that contribute to the formation of supramolecular structures from building blocks (*3*, *4*). For example, a bolaform surfactant that carries two L-ascorbic acid rings as polar headgroups forms monodisperse nanorods in water due to the presence of (*5*):

- chiral headgroups
- bulky headgroups
- several residues that are involved in intermolecular hydrogen bonding
- relatively flexible hydrophobic tail

These are the main structural features that lead to the formation of an ordered supramolecular structure with a well defined topology.

In the following sections we will explore and compare in detail the effect of the headgroup chirality on the physico-chemical properties of L-ascorbic acid and D-isoascorbic acid amphiphilic derivatives and of the self-assembly properties of their aqueous dispersions.

Ascorbic Acid and Its Derivatives

L-(+)-ascorbic acid or vitamin C possesses several appealing structural features (see Figure 1). The presence of the HO-C=C-OH group in the lactone ring makes it an efficient reducing agent (E° = +0.23 V). On the other hand, the presence of several hydrophilic residues (hydroxyl groups and the lactone ring), and of the stereogenic centers (C_4 and C_5) results in a rigid conformation. Intermolecular and intramolecular hydrogen bonds are established with the solvent and other molecules. Due to its reducing properties, vitamin C can protect delicate species such as other vitamins (tocopherols), polyunsaturated chemicals (carotenes, omega-3 and omega-6 acids) and other materials that are particularly sensitive to oxidation and more generally to radical degradation.

Although vitamin C acts as an efficient, powerful, biocompatible and cheap antioxidant, however its use is drastically limited to aqueous media, because of its very poor solubility in almost all organic polar or apolar solvents. The solubility at

25° C in water is about 334 g/L, in ethanol 10 g/L, and in methanol 1.8 g/L (*6*). The addition of of a hydrophobic segment to the non-redox active side of vitamin C can improve its solubility in non-aqueous solvents, without jeopardizing the precious antioxidant behavior.

Moreover, the formation of amphiphilic molecules that carry ascorbic acid as the polar headgroup can lead to the production of nanostructures that can solubilize valuable lipophilic molecules in the inner core and protect them against oxidative degradation. With these premises, vitamin C-based single or double chained surfactants and bolaamphiphiles have been synthesized, and their phase behavior, self-assembly properties and antioxidant capacity have been investigated.

The most promising chemical site for a covalent permanent bond of the hydrophobic segment to the ascorbic acid moiety appears to be the primary –OH group in position 6, in fact it can easily form esters with high yields, while the reducing properties of the ring remain unaltered. Other derivatives can be obtained, by using the –OH groups in different positions (mainly the acidic hydroxyls in 2 and 3), but the synthetic procedure is much more complicated (*7*), and the redox activity can be altered.

Figure 2 shows the pathway used for the synthesis of 6-O-L-ascorbyl derivatives. The single chained surfactants are abbreviated as L-ASCn, where "n" is the number of carbon atoms in the side chain (see Figure 3). L-ascorbyl palmitate (L-ASC16), the surfactant that was first synthesized, is used either alone or in combination with α-tocopherol to stabilize oral pharmaceutical formulations and food products (*8*). The synthetic procedure depicted in Figure 2 leads only to the formation of the 6-O-ascorbyl derivative (*9*).

Ascorbic acid and its derivatives are stable in the solid state, while they are readily oxidized in solution, because of the dissolved oxygen and traces of metal ions. For this reason the solid surfactants were stored in the dark at 4° C and their aqueous dispersions were investigated immediately after the sample preparation. The addition of a few drops of Br_2 dissolved in dichloromethane to a solution of ASC-n in acetone brings about instantaneous discoloration of bromine, showing that the vitamin C double bond is intact. The DPPH method (*10*) confirms the presence of the ascorbic acid ring in the derivative. The purity of the ASC-n derivatives was assessed through TLC, NMR, IR, mass spectrometry, and elemental analysis (*5, 10*). After purification, no trace of the original fatty acid was found in the final product. DSC measurements confirmed the absence of subproducts or unreacted reagents in the synthesized materials. Within the temperature and pH intervals used in these studies, and for the time required to perform the experiments, the ester bonds in position 6 are not cleaved in the aqueous dispersions, and no degradation of the samples was detected.

The alkyl chain length determines the physico-chemical properties of the derivatives, and in particular the phase behavior in water dispersions. The addition of a second chain in the surfactant increases by far its hydrophobicity (see Figure 3), so that the final product can act as a low-molecular-mass organic gelator (LMOG) (*4*). These molecules self-assemble in selective solvents where they produce thermoreversible organogels (see Figure 4). Again, the ascorbic

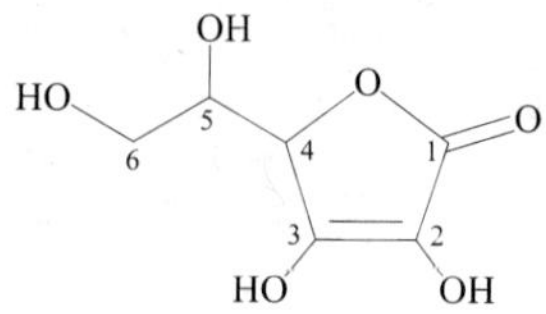

Figure 1. Structure of ascorbic acid.

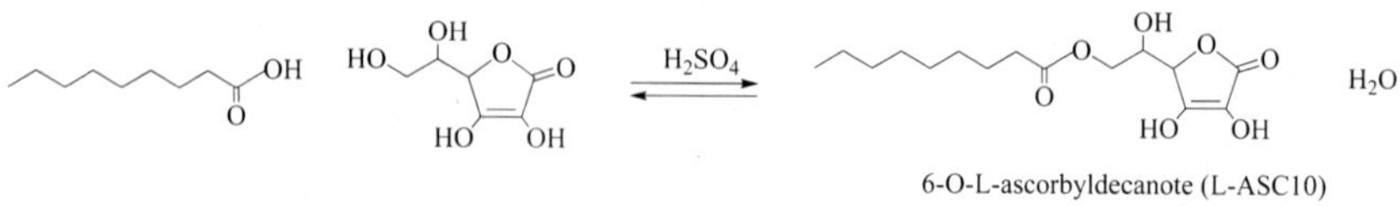

Figure 2. Synthesis of L-ASC10 from decanoic acid and L-ascorbic acid.

acid moiety provides here some of the structural requirements that an LMOG must possess in order to form an organogel (*10*). These are:

- high melting point
- low solubility in the solvent
- structural anisotropy
- presence of donor/acceptor groups for hydrogen bonding

On the other hand, the synthesis of a bolaform surfactant, bearing two vitamin C moieties at the ends of a single alkyl chain (see Figure 3), improves the hydrophilicity of the molecule and different structures are formed, such as nanotubes, nanofibers and helical ribbons (*11–13*).

In aqueous dispersions above 0.5% w/w, BOLA12 forms opaque semitransparent condensed phases at room temperature. Upon heating at about 34 °C, these systems turn into a transparent homogeneous micellar phase. The critical aggregation concentration is about 10 mM (*5*).

In the condensed phase, cryo-TEM experiments showed the presence of quite monodisperse hollow nanotubes (Figure 5). Their width is about 28 nm, the mean internal diameter is 15 nm, and their length spans approximately between 200 and 1000 nm (*5*).

XRD experiments carried out on the dry sample showed a spacing of 25 Å, that corresponds to the Corey-Pauling-Koltun (CPK) length of the molecule, 26 Å. This finding suggests the presence of compact monolayers of BOLA12 with the hydrophobic blocks fully extended and perpendicular to the planes formed by the headgroups. In the hydrated sample the spacing increases of about 15 Å, which corresponds to the thickness of the sandwiched aqueous interlayer, similar to that found in single chained L-ASCn and with phospholipids (*14*).

The microstructure of the nanorods detected by cryo-TEM micrographs was investigated through SAXS, which indicated the presence of a lamellar structure with a repetition spacing of 33 Å (see Figure 6). This result suggests the presence of hydrated molecules in a lamellar array, with a thin layer (8 Å) of hydrating water molecules bound to the headgroup of BOLA12 as depicted in Figure 7.

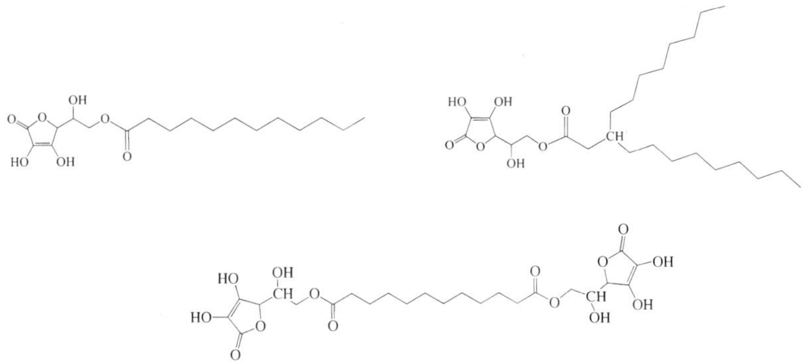

Figure 3. From top left: chemical structure of the single chained surfactant L-ASC12, of the double chained surfactant 8ASC10, and of the bolaamphiphile BOLA12.

Figure 4. Organogels produced from 8ASC10 and cyclohexane. Reproduced from ref. (10). Copyright 2009 American Chemical Society

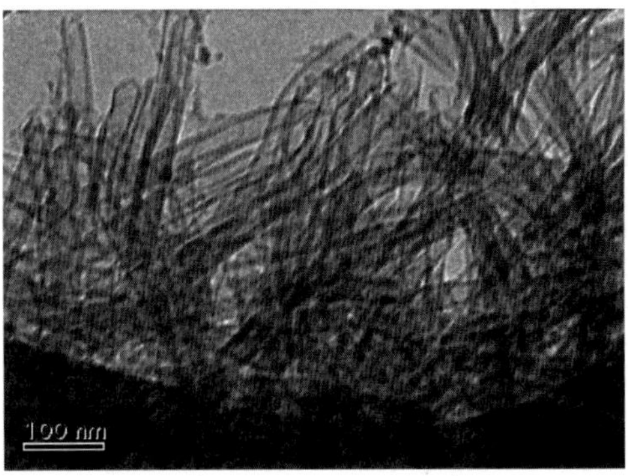

Figure 5. Cryo-TEM micrograph of a 5% w/w aqueous dispersion of BOLA12 in the condensed phase.

SAXS experiments on a 5% w/w sample of BOLA12 in water reveals the presence of micellar aggregates above 34 °C. The assemblies are comprised of small spherical particles with an effective radius of 26 Å, an aggregation number of 240 monomers, and an area per polar group of 53 Å^2 (see Figure 8). These data indicate that the BOLA12 molecules in the micelles are closely packed in a stretched or in a bent conformation (*5*).

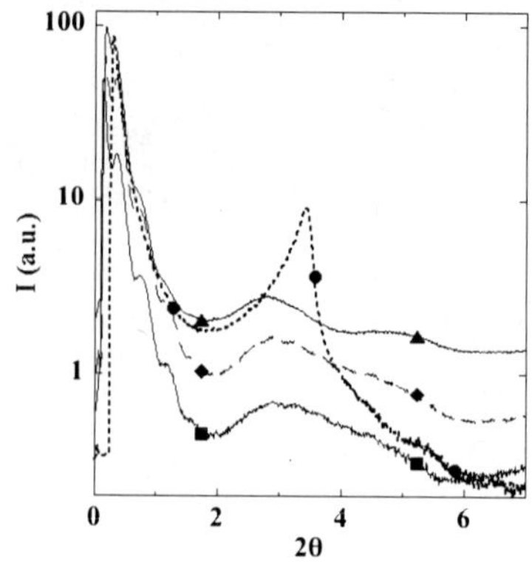

Figure 6. SAXS intensity distribution for BOLA12 and its coagels in water at 25° C: anhydrous sample (●); 10% (■); 20% (◆); 40% (▲). Reproduced from ref. (5). Copyright 2006 American Chemical Society

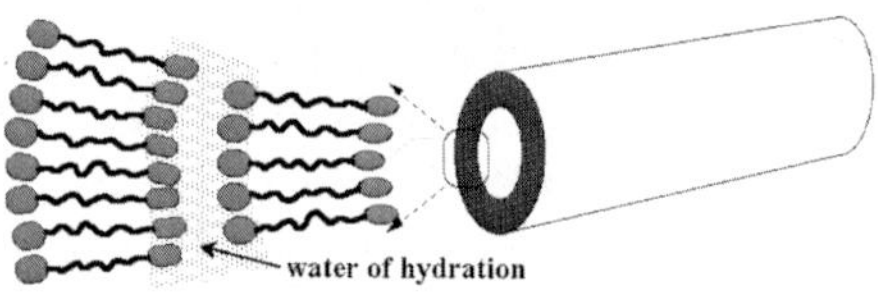

Figure 7. Schematic structure of a nanotube produced by BOLA12: a hollow cylinder comprises 2-3 layers of hydrated surfactant molecules.

Of particular interest is the monodispersity of the nanorods found in the condensed phase (*15*, *16*). The shape of the aggregates is controlled by the non-covalent interactions that are related to the chemical structure of the building blocks. The interplay of the different structural features, i.e. stereochemistry, bulkiness, flexibility and rigidity of different parts of the molecule, presence of hydrogen bonded groups, and degrees of lateral freedom of the aliphatic chains, will impose strict limits to the molecular orientation and arrangement on the nanoscale, thus controlling the size and monodispersity of the sample.

Properties of L-Ascorbic Acid and D-Isoascorbic Acid

Another structural feature of vitamin C that can be exploited is stereochemistry. Since C_4 and C_5 are both stereogenic centers, four epimers can be drawn. Figure 9 shows the structure of L-ascorbic and D-isoascorbic acids.

In D-(-)-isoascorbic acid the configuration of C_5 is inverted with respect to vitamin C. Although the antioxidant activity is retained (*17*), this epimer has only 1/20 of the biological actiity of L-ascorbic acid (*18*).

The apparently small structural change at C_5 will actually result in large effects of the physico-chemical properties of the parent acids and of their derivatives.

Table I shows some physico-chemical properties of the two parent acids. The D-epimer is less soluble and a little more acidic than L-ascorbic acid, and it melts at a lower temperature.

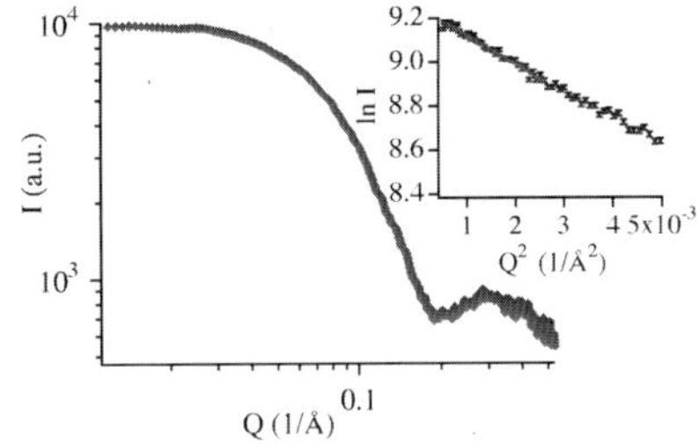

Figure 8. SAXS intensity distribution for a 5% w/w sample of BOLA12 in water at 50° C. Inset: Guinier plot (lnI vs Q^2) and linear fit. Reproduced from ref. (5). Copyright 2006 American Chemical Society

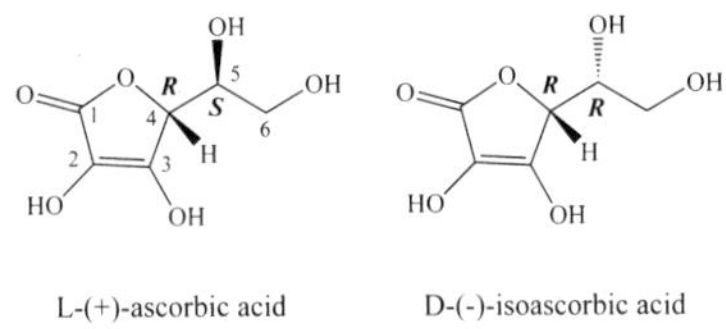

Figure 9. Structure of L-ascorbic acid and D-isoascorbic acid. Reproduced from ref. (19). Copyright 2009, American Chemical Society

All differences are ascribed to the different set of inter- and intramolecular hydrogen bonding network in their structure, as depicted in Figure 10. These interactions involve different residues: the –OH in position 1 or 3 (for the two tautomeric forms), and the C=O in position 1 and the –OH in 6. These bonds are present in both species. Instead, an intramolecular hydrogen bond (HB) is established between the –OH group in position 5 with the acidic –OH residue in 3 only in D-isoascorbic acid. Such interaction is not allowed in L-ascorbic acid, since the hydrogen atom in position 4 points in the same direction as C_5–OH.

The different set of HB results in a different capacity to interact with adjacent molecules in the crystal, and with water molecules in solution. As a matter of fact, L-ascorbic acid is three times more soluble in water and less acidic than the D-epimer. The latter effect is presumably due to the fact that the C_3-O^- donor residue strengthens the intramolecular hydrogen bond with the C_5-OH. Moreover, the intermolecular interactions in D-isoascorbic are less strong than those in L-ascorbic acid, as the lower mp and ΔH_{melt} show.

The formation of inter- and intramolecular hydrogen bonds has some consequences on the structure and interactions established in the solid and in the solution states:

1. In the crystal lattice the L-ascorbic acid molecules are involved in eight intermolecular HB with neighbor molecules and no intramolecular interaction is present. All oxygen atoms, except for the ether atom in the ring, participate in the formation of the HB: those in position 2, 5, and 6 act both as donor and acceptor, while that in position 3 can only be a donor and that in 1 only an acceptor (*20–23*).

Table I. Physico-chemical properties of L-ascorbic and D-isoascorbic acid. Reproduced from ref. (*19*). Copyright 2009 American Chemical Society

	L-(+)-asc	*D-(-)-isoasc*
Melting point (° C)	192.1°	169.5°
Enthalpy change of melting (kJ/mol)	43.54	40.86
Solubility in water (g/L) at 25° C	334	100
pK_{a1}	3.36	3.29

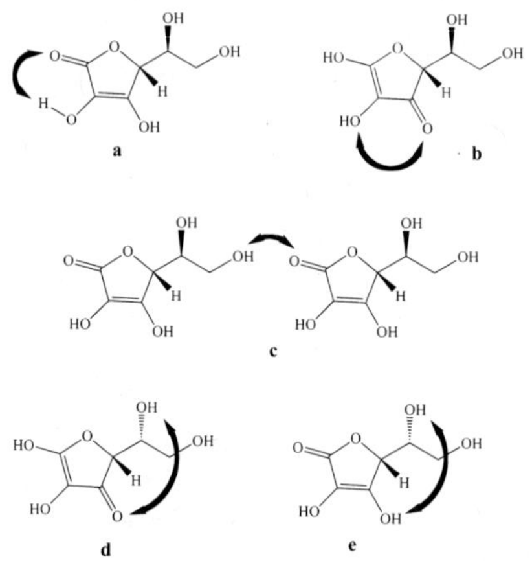

Figure 10. Examples of the inter- and intramolecular hydrogen bonds established in L-ascorbic acid (a, b, c) and in D-isoascorbic acid (d, e). Reproduced from ref. (19). Copyright 2009, American Chemical Society

2. In solution, L-ascorbic acid lacks of intramolecular interaction as in the solid state, while a strong hydration shell is created by the solvent (*24*).
3. The structural features of the two epimers, i.e. bond distances and angles in the ring and of C_2-OH and C_3-OH groups, do not change markedly. Instead the angles of the –CHOH-CH_2OH side chain do differ significantly. This effect modifies the ionization capacity of the molecules (*25*).
4. The interactions with water molecules in solution are quite different for the two species, in fact D-isoascorbic acid is supposed to have a significant impact of the dynamical structure of the solvent although its hydrogen bonding with water is reduced by the presence of the intramolecular interactions (*24, 26–29*).
5. Near-infrared (NIR) spectra confirm the presence of an intermolecular hydrogen bond for L-ascorbic acid. This peak (at 7837 cm^{-1}) was not found in D-isoascorbic acid (*19, 30*). Medium-infrared (MIR) spectroscopy provides more interesting information in the region characteristic of O-H stretching modes. L-ascorbic acid shows four intense peaks that correspond to the intermolecularly bound –OH groups

in position 5 and 6. The same peaks in D-isoascorbic acid are shifted to lower wavenumbers (see Figure 11).

Experimental Details

Surface Tension

Surface tension (γ) measurements on the ASCn aqueous solutions were carried out using a Du Noüy ring at 30° C. The critical micellar concentration (cmc, in mol/L) was determined from plots of γ vs Log c as the intersection point of the two measured straight lines for low and high concentrations.

The polar headgroup area of the monomer (a_p, in Å^2/molecule) was calculated from the Gibbs equation for adsorbed monolayers at the air/water interface for surfactant concentrations lower than the cmc, according to the formula $a_p = -2.303 \cdot RT/[6.023 \cdot (\partial\gamma/\partial Logc)]$ where R and T are the gas constant (8.31 J/mol·K) and the absolute temperature, respectively.

Differential Scanning Calorimetry

The coagel-to-gel phase transition temperature and the corresponding enthalpy change were obtained through DSC runs with a Q1000 differential scanning calorimeter (TA Instruments), using hermetic aluminum pans, sealed under nitrogen atmosphere. The transition temperatures were taken as the temperature of the endothermic peaks. The runs were made at the rate of 5° C/min. For measurement of the frozen, strongly bound water, the samples were first cooled to -60° and then heated up to 20° C at 0.5° C/min.

Small Angle X-ray Scattering

SAXS measurements were carried out with a HECUS SWAX-camera (Kratky) equipped with a position-sensitive detector (OED 50M) containing 1024 channels of width 54 μm. Cu K_α radiation of wavelength $\lambda = 1.542$ Å was provided by a Seifert ID-3003 X-ray generator (sealed-tube type), operating at a maximum power of 2 kW. A 10 μm thick nickel filter was used to remove the Cu K_β radiation. The sample-to-detector distance was 273 mm. The volume between the sample and the detector was kept under vacuum during the measurements to minimize scattering from the air. The Kratky camera was calibrated using silver behenate (d = 58.48 Å). Scattering curves were obtained in the Q-range between 0.01 and 0.54 Å^{-1}, Q being the scattering vector $Q = (4\pi \sin\theta)/\lambda$, and θ the scattering angle. Solid samples were placed into 1 mm demountable cells having Kapton films as windows. Liquid samples were filled into a 1 mm quartz capillary. The temperature was controlled by a Peltier element, with an accuracy of ±0.1° C. All scattering curves were corrected for the solvent and the empty cell contribution considering the relative transmission factor. SAXS curves from liquid samples were iteratively desmeared using the procedure reported by Lake (*51*).

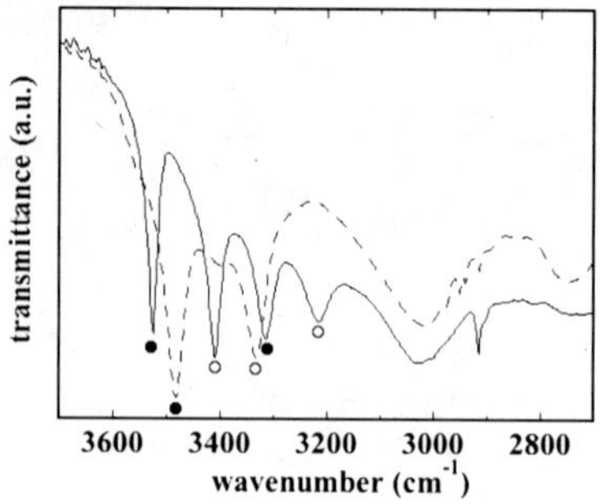

Figure 11. MIR spectra for L-ascorbic acid (full line) and D-isoascorbic acid (dotted line) between 3700 and 2700 cm^{-1}. The full circles correspond to the stretching of the –OH group in position 6, and the open circles to the –OH group in position 5. Reproduced from ref. (19). Copyright 2009, American Chemical Society

X-ray Diffraction

XRD diffractograms were recorded by a powder Bruker D8 Advance diffractometer (BRUKER axs) using Bragg-Brentano geometry, $\lambda = 1.54$ Å (Cu K_α). Experiments were carried out in the range $1.5° \leq 2\theta \leq 40°$, with a step size of 0.04° and a time/step of 1 s, setting voltage and current at 40 kV and 30 mA, respectively.

Cryo-Transmission Electron Microscopy

The cryo-TEM samples were prepared in a controlled environment vitrification system (CEVS) under controlled temperature and humidity. 5 μL of the sample were placed on a holey carbon Cu-grid which had previously been glow discharge treated. After blotting excess liquid with a filterpaper, the sample was left on the grid for a few minutes before plunging into the cryogene. The samples were imaged under cryogenic conditions with a Philips CM120 Biotwin Cryo electron microscope and images recorded with a CCD camera (Gatan 791).

Medium- and Near-Infrared Spectra

MIR spectra were acquired in the wavenumber interval 4000-650 cm^{-1}, with a Nexus 970-FTIR (Thermo-Nicolet), on KBr pellets, using KBr beam splitter and MCT/A detector, with a resolution of 8 cm^{-1} and coadding 64 scans. NIR spectra were acquired in the wavenumber range 8000-4000 cm^{-1}, with a Nexus 870-FTIR (Thermo-Nicolet) and a FT-IR Continuum microscope in diffuse reflectance mode (beam splitter: CaF_2; detector: MCT/A) with a resolution of 8 cm^{-1} and coadding 64 scans.

Sample Preparation

The coagel samples were prepared by weighing appropriate amounts of surfactant and water in a vial. In order to make sure that the delicate redox active ascorbic moiety remains intact, the samples were quickly annealed (four heating-cooling cycles around the phase transition temperature and storage in a refrigerator at 4° C) and then immediately tested. D-ASC12/L-ASC12 mixed samples were prepared by dissolving appropriate amounts of the two surfactants in acetone. The solvent was slowly evaporated at 20° C and then stripped off by freeze-drying at 25 mTorr for 48 h.

Aqueous Dispersions

All ASCn derivatives dispersed in water show a typical phase behavior: at low temperature they form *coagels*, i.e. hydrated semicrystalline nanoassemblies in equilibrium with an aqueous dilute solution of the monomers. Coagels are also described as "opaque suspensions of crystals", "poorly hydrated multilamellar polycrystalline suspensions", "hydrated solids" or "biphasic mixtures containing crystals" (*31*). Upon heating, these supramolecular structures produce either a *micellar* dispersion if the surfactant chain length is short (n = 8, 10), or a viscous "*gel*" phase if the alkyl chain is longer ($n \geq 12$) (see Figure 12). The surfactants that bear a very long chain ($n \geq 18$) are very poorly soluble in water and form Langmuir monolayers at the air/water interface (*32–34*).

Figure 13 schematically depicts the phase transitions, in terms of hydration and partial melting of the alkyl chains. In the coagel state the lamellae are separated by a small layer of strongly bound water molecules, about 10 Å thick. The interdigitated aliphatic chains are packed in a compact way, perpendicularly to the interface. In the gel phase the hydrophobic tails rearrange, and acquire some freedom of lateral motion. The aqueous pool becomes thicker and includes the so-called "intermediate" water, whose properties are different from those of the bulk free water and of the strongly bound solvent.

Figure 14 shows the surface tension (γ) of aqueous dispersions of L-ASCn and D-ASCn (with n = 8, 10, 12) at 30° C as a function of the surfactant concentration. From the plots the values of the critical micellar concentration (cmc) and of the area per headgroup (a_p) were determined.

The cmc was obtained from the intersection point of the two fitting lines for high and low concentrations. For both series, the cmc decreases with the length of the aliphatic chain (see Table II). The cmc of the D-ASCn derivatives is almost five times that of the corresponding L-ASCn with the same chain length.

The area per polar group (in Å^2/molecule) was calculated from the slope of the γ/Logc plot (see Experimental Details). The a_p values (see Table II) for the D-ASCn surfactants are always lower than those of the corresponding L-ASCn, presumably because of the different hydration and ionization capacity of the polar heads (*19*).

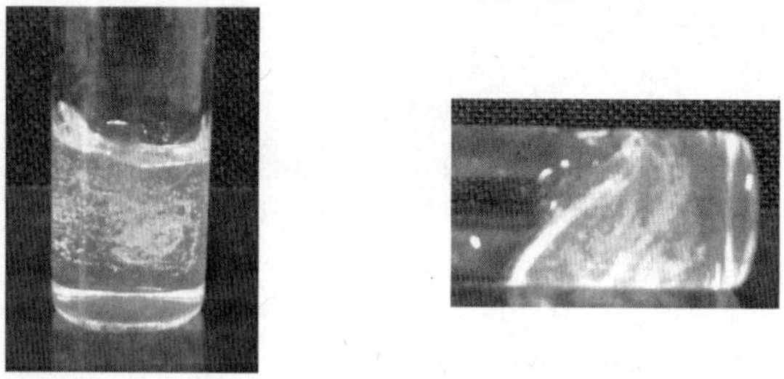

Figure 12. A gel sample obtained from L-ASC12 in water (10% w/w).

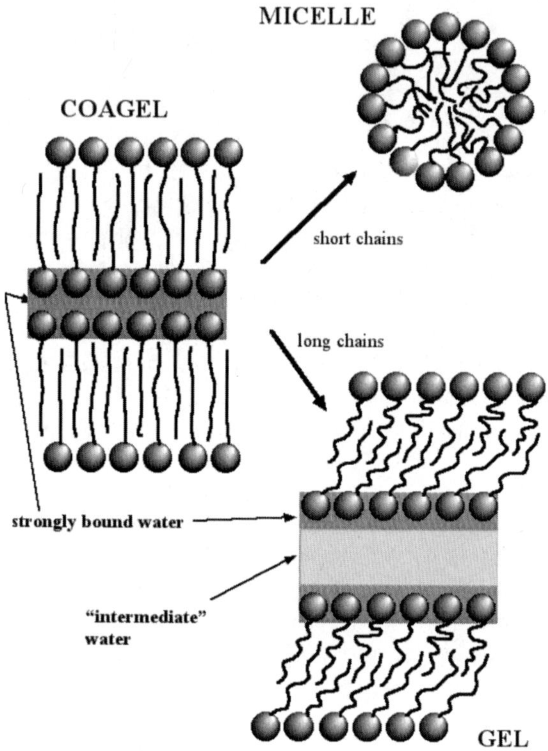

Figure 13. Transition from the coagel structure to the micellar or the gel state, depending on the surfactant chain length. The dark gray area indicates the strongly bound water layer, and the light gray region represents the "intermediate" water compartment.

The thermodynamic properties of the coagel-to-micelle and of the coagel-to-gel phase transitions are determined through differential scanning calorimetry by measuring the transition temperature and the transition enthalpy change.

The results indicate that the transition temperature is nearly constant up to a surfactant concentration of 50% w/w, and then increases almost linearly. Instead, the enthalpy change is proportional to the surfactant concentration. All D-ASCn

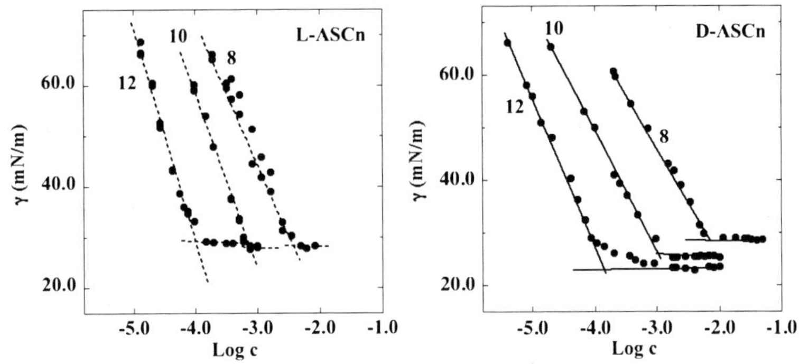

Figure 14. Surface tension (γ) of L-ASCn (left) and D-ASCn (right) as a function of the surfactant concentration, for n = 8, 10, and 12, at 30° C.

Table II. Critical micellar concentration and area per polar headgroup of D-ASCn and L-ASCn. Reproduced from ref. (*19*). Copyright 2009 American Chemical Society

n	*cmc (mol/L)*		*a_p (Å²/molecule)*	
	D-ASCn	*L-ASCn*	*D-ASCn*	*L-ASCn*
8	5.38	1.06	46	49
10	1.41	0.34	45	47
12	0.12	0.03	39	47

derivatives show a lower transition temperature, and a higher ΔH respect to the L-ASCn analogues (*19*).

According to Kodama and Seki, the phase transition occurs through three different steps (*35*):

1. first the headgroups separate from their counterions (in our case H^+) and the corresponding electrostatic contribution to the enthalpy change (ΔH_{el}, endothermic step) is taken up
2. the headgroups are hydrated by the surrounding water molecules, and the hydration enthalpy is released (ΔH_{hydr}, exothermic step)
3. the conformational/packing change of the hydrophobic chains take place with the absorption of the corresponding heat (ΔH_{chain}, endothermic step).

In this simplified scheme, ΔH_{el} and ΔH_{hydr} are related to the interactions between the surfactant heads and the solvent, while the ΔH_{chain} is ascribed only to the lipophilic tails. As the number of carbon atoms in the chain increases, the contribution of ΔH_{chain} becomes more significant and determines the formation of a micellar solution or of a gel phase.

The thermal behavior of L-ASCn dispersions depends on the presence of electrolytes (*36*) as in other Hofmeister phenomena (*37*), and neutral cosolutes (*38*).

The data indicate that the hydration of the shorter surfactants, that undergo a coagel-to-micelle phase transition, is similar regardless of the stereochemistry of the polar headgroup. For the longer D-ASC12 and L-ASC12 (coagel-to-gel) the results are different, and this may be ascribed to the different ionization degree of the ascorbic acid rings.

Furthermore, DSC experiments can be used to measure, indirectly, the amount of strongly bound water molecules by subtracting from the known stoichiometric amount of water in the sample the amount of water which freezes around 0° C (free bulk water) calculated from the area of the DSC peak. It was found that the number of strongly bound water molecules per surfactant headgroup was always about 8 for L-ASCn with n spanning between 8 and 16 in a 40% w/w aqueous dispersion of the surfactant (*39*, *40*). In the case of D-ASCn the amount of strongly bound water per surfactant moiety is considerably larger (about 30). However this large value may be affected by the strong perturbation of the dynamic structure of water induced by the D-ascorbic rings (*39*). A similar investigation conducted on BOLA12 dispersions showed that on the average 10 water molecules are strongly bound to each headgroup of the bolaform amphiphile.

From the number of strongly bound water molecules per surfactant (N_b), and from the spacing distance (d) detected through XRD experiments in the coagel state (see Figure 15), we estimated the surfactant area per polar group as $a_p = 2(v_w N_b + v_p)/d_w$, where v_w and v_p are the volume of a water molecule and that of a single headgroup (*19*, *39*). Here $d_w = d - l_H$, where l_H is the length of the aliphatic chain according to Tanford's rule (*19*) (see Figure 16).

The values of a_p are about 40 Å^2 for L-ASCn, and 50-65 Å^2 for D-ASC, suggesting that L-ASCn are bound more tightly in the coagel state, through intermolecular hydrogen bonding, while the D-isoascorbic headgroup adopt a more open structure. Similar data for a_p were obtained for L-ASC8 and L-ASC10 that form micelles (*19*), indicating that the packing properties of the surfactants remain almost unaltered in the two states, and confirms that the hydrophobic chains determine the stability of the surfactant assembly at higher temperature, i.e. either in a micellar dispersion or in the gel phase.

The phase behavior of mixtures of diastereoisomers that possess amphiphilic structure and properties has not been investigated extensively. In our case the presence of a different set of intra- and intermolecular hydrogen bonds in L-ascorbic and D-isoascorbic acid produces an interesting behavior of the derived surfactants both in the solid state and in aqueous dispersions.

We found that D-ASC12 and L-ASC12 are always miscible in the liquid phase and partially miscible in the solid state, depending on the mixture composition. They form two eutectics and a 1:1 molecular compound. Figure 17 shows the phase diagram of the two diastereoisomers in the solid and in the coagel state, with the solid-liquid transition temperature in the case of the solid mixtures, or the coagel-to-gel transition temperature in the case of the coagel state, as a function of the mole fraction of L-ASC12 (x_L).

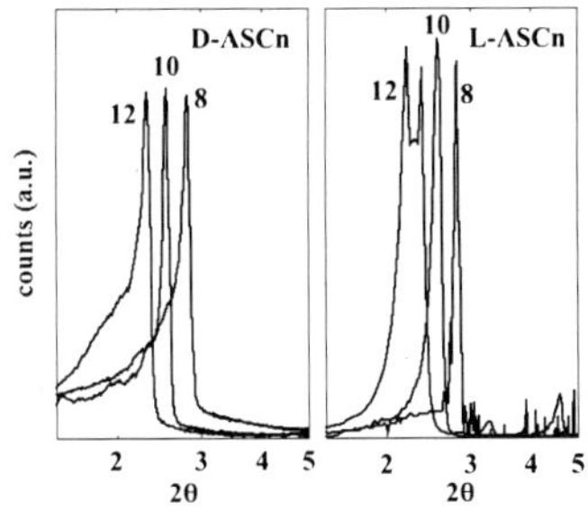

Figure 15. XRD profiles for D-ASCn (left) and L-ASCn (right) coagels (40% w/w) with n = 8, 10, and 12. Reproduced from ref. (19). Copyright 2009, American Chemical Society

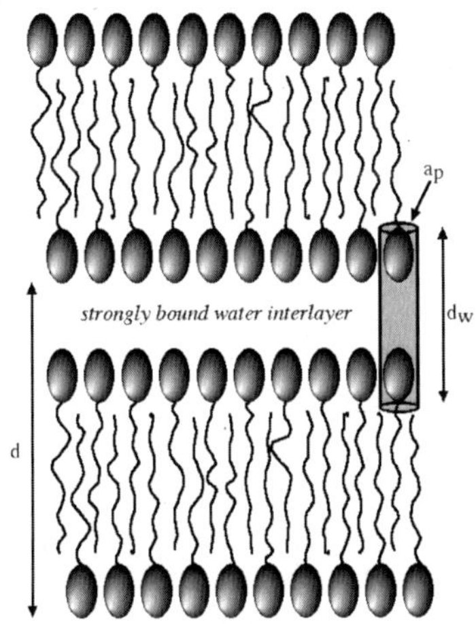

Figure 16. Schematic structure of a coagel with the ASCn molecules packed in a compact interdigitated lamellar structure. The gray cylinder contains two headgroups and the water molecules strongly bound to the polar groups. The height of the cylinder is d_w, and the basis is a_p.

Solid State

The disappearance of the eutectic effect suggests partial miscibility in the solid state for x_L lower than 0.02 and above 0.80. The phase diagram reveals the presence of two eutectics, one richer in D-ASC12, e_D, and the other richer in L-ASC12, e_L (see Table III), and of a 1:1 molecular compound that melts at 91.3 °C. The exact location of the eutectics was obtained from the Tamman plots by plotting the ΔH of melting of the liquidus curve and the enthalpy change for the eutectics (*39*).

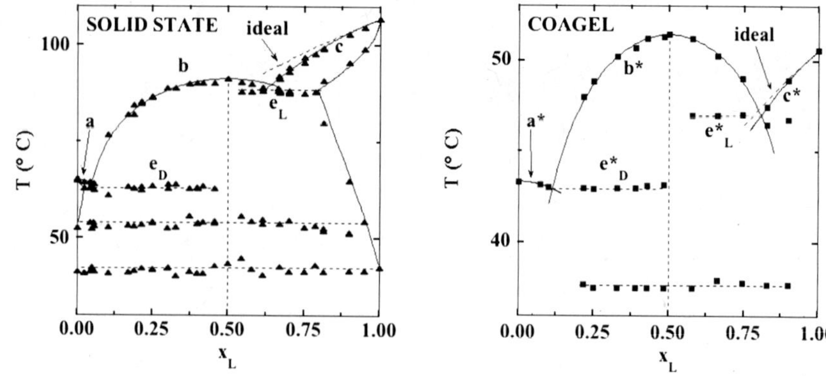

Figure 17. Phase diagram of a L-ASC12/D-ASC12 mixture in the solid state (left) and in the 10% w/w coagel state (right). The lines represent the fitting curves obtained from the ideal miscibility (dotted lines), the regular solution (a, c, a and c*), and the Prigogine-Defay (b and b*) models. e_D and e_L indicate the D-ASC12 rich and the L-ASC12 rich eutectics formed in the pure solid mixture. e_D* and e_L* mark the D-ASC12 rich and the L-ASC12 rich eutectics obtained in the coagel state. Reproduced from ref. (39). Copyright 2009, American Chemical Society*

In order to calculate the excess thermodynamic functions ΔG^E, ΔH^E and ΔS^E, the activity coefficients (γ) need to be estimated. For an ideal mixture, γ is unity, and neglecting the value of ΔC_p between the melt and the solid, the simplified Schröder-van Laar equation describes the change of the melting temperature (T) with the composition (x_1) in terms of the enthalpy change and temperature of melting of the pure component "1" in the mixture (ΔH_1 and T_1, respectively):

$$ln\, x_1 = \frac{\Delta H_1}{R}\left(\frac{1}{T_1} - \frac{1}{T}\right) \tag{1}$$

This relationship is valid when the mole fraction x_1 spans between the pure component and the eutectic.

When the two components of a binary mixture possess similar molecular arrangements and molecular volumes – such as in the present case – the deviation from ideality is well described by the regular solution theory, that accounts for the intermolecular interactions between like and unlike molecules. In particular, the deviation from ideality is expressed by the enthalpic term and the transition temperature and is given by the following formula:

$$T = T_1 \frac{\Delta H_1 + \omega(1 - x_1)^2}{\Delta H_1 - RT_1\, ln\, x_1} \quad ; \quad T = T_2 \frac{\Delta H_2 + \omega x_1^2}{\Delta H_2 - RT_2\, ln(1 - x_1)} \tag{2}$$

where T_2 and ΔH_2 are the temperature and enthalpy change of melting of component "2". ω represents the interaction parameter of the mixture, defined as:

$$\omega = z\left(\varepsilon_{12} - \frac{\varepsilon_{11} + \varepsilon_{22}}{2}\right) \tag{3}$$

here, z is the coordination number, while ε_{12}, ε_{11}, and ε_{22} are the interaction energies for the 1-2, 1-1, and 2-2 pairs in the mixture, respectively. When ω is negative the mixed system has a lower energy than the pure components, indicating an attractive interaction between the two species in the mixture. Eq 2 describes then the liquidus line comprised between each pure component and the closer eutectic.

The entropy change of mixing instead is given by the value for ideal mixing:

$$\Delta S_{mix,id} = R\sum_i x_i \; ln\, x_1 \tag{4}$$

When the two components produce a molecular compound, such as in the case of a racemate from two pure enantiomers – the coexistence curve is described by the Prigogine-Defay formula:

$$ln\left[4x_1x_2\right] = \frac{2\Delta H_c}{R}\left(\frac{1}{T_c} - \frac{1}{T}\right) \tag{5}$$

here, T_c and ΔH_c are the temperature and enthalpy change of melting for the molecular compound, respectively. This equation describes the experimental data between the two eutectics.

The lines drawn in Figure 17 show the fitting analysis of the data according to the Schröder-van Laar (eq 1), the regular solution (eq 2), and the Prigogine-Defay (eq 5) models. The results suggest that D-ASC12 and L-ASC12 do not mix ideally. The deviation from ideality is expressed by a positive value of ω (+0.4 kJ/mol) when the mixture is richer in D-ASC12, and a negative value of ω (-3.0 kJ/mol) when L-ASC12 is main component in the mixture.

The activity coefficient of each component in the liquid phase, γ_i^L, is calculated by equating the chemical potential in the liquid and in the solid phase:

$$\mu_i^{0,L} + RT\; ln\left(x_i^L\gamma_i^L\right) = \mu_i^{0,S} + RT\; ln\left(x_i^S\gamma_i^S\right) \tag{6}$$

and if $x_i^S\gamma_i^S \approx 1$ we obtain:

$$ln\left(x_i^L\gamma_i^L\right) = \frac{\Delta H_i}{R}\left(\frac{1}{T_i} - \frac{1}{T}\right) \tag{7}$$

for $\gamma_i^L = 1$, i.e. for an ideal system, eq 7 reduces to eq 1.

The small difference between the experimental and calculated melting points for the two eutectics e_D and e_L confirms the validity of the assumption made for γ in the solid phase (*41*).

At this point we can finally calculate the excess thermodynamic functions of mixing for the real system, that describe the deviation from the ideal behavior of the mixture, in terms of the intermolecular interactions that involve components

"1" and "2". At constant temperature and pressure the excess thermodynamic functions are given by:

$$\Delta G^E = RT\sum_i x_i \ln\gamma_i \tag{8}$$

$$\Delta H^E = -T^2\left[\frac{\partial\left(\Delta G^E/T\right)}{\partial T}\right] = -RT^2\sum_i x_i \frac{\partial \ln\gamma_i}{\partial T} \tag{9}$$

$$\Delta S^E = -\left(\frac{\partial\Delta G^E}{\partial T}\right) = -R\sum_i\left(x_i \ln\gamma_i + x_i T\frac{\partial \ln\gamma_i}{\partial T}\right) \tag{10}$$

Table III shows the value of the excess functions for the two eutetics. The excess Gibbs free energy is always positive, indicating a weak interaction between the two components in the eutectic melt, and a stronger association between like molecules (D-D and L-L). Instead, the excess entropy, that reflects the configurational energy change due to a change in the potential energy, is negative because of the more ordered structure in the mixture.

Coagel State

The thermal behavior of mixtures of D-ASC12 and L-ASC12 in the coagel state was investigated keeping the same total surfactant concentration (10% w/w, 0.0055 in mole fraction units). The phase diagram (see Figure 17) shows the existence of a 1:1 molecular compound that melts at 51.4 °C, and of two eutetics, e_D* and e_L* (see Table III).

The analysis of the data according to the procedure already adopted for the solid mixture indicates that in the coagel state the system behavior deviates from ideality, but at a lesser extent. The regular solution equation provides a value for ω of about +26 kJ/mol for the left side (i.e. when D-ASC12 is the main component) and of -3.3 kJ/mol for the right side (when L-ASC12 dominates in the mixture). These values suggest that when the mixture is richer in D-ASC12 strong repulsive interactions exist between the components in the coagel state, while when L-ASC12 is the main component attractive interactions dominate.

The excess thermodynamic functions (see Table III) were evaluated through eqs 8–10. The value of ΔG^E is positive for both eutectics, indicating a weaker interaction between the components D and L in the eutectic melts, and a stronger association between like molecules (D-D and L-L). The excess entropy change is negative, suggesting an increment in the structure order. In both cases, for the solid and the coagel mixture, the results indicate that the addition of small amounts of D-ASC12 to L-ASC12 brings about a small disturbance of the structure, that is L-ASC12 can accommodate quite easily the presence of a few D-ASC12 molecules. On the other hand, D-ASC12 reacts more significantly to the presence of L-ASC12, with stronger repulsive interactions. Moreover, the excess thermodynamic functions indicate that the two diastereoisomers form clusters in the eutectic mixture. These effects have been related to the presence

Table III. Melting temperature (in K), composition, activity coefficients, and excess functions (ΔG^E and ΔH^E in kJ/mol, ΔS^E in J/mol·K) of the eutectics. Reproduced from ref. (*39*). Copyright 2009 American Chemical Society

eutectic	*T*	x_L	$ln\gamma_L$	$ln\gamma_D$	ΔG^E	ΔS^E	ΔH^E
e_D	336	0.058	1.225	-0.042	89	-122	-41
e_L	361	0.630	-0.188	1.913	1768	-116	-40
e_D*	316	0.114	1.638	0.097	715	-221	-69
e_L*	320	0.805	-0.032	1.919	926	-205	-64

of different hydrogen bonding networks in the two compounds. In fact the propensity of L-ASC12 to form intermolecular hydrogen bonding with similar molecules, and the tendency of D-ASC12 to establish stronger intramolecular interactions through its polar headgroups, provide a reasonable explanation for the observed effects.

NIR (see Figure 18) and MIR (see Figure 19) spectra and X-ray diffraction studies (see Figure 20) performed on different D-ASC12/L-ASC12 mixtures in the solid and in the coagel state, as a function of the mixture composition, confirm the experimental findings obtained through DSC runs, and the presence of different hydrogen bonds in the pure samples and in their mixtures (*19*, *39*).

In conclusion, the chirality of the hydrophilic headgroups in ascorbic acid amphiphilic derivatives affects not only the physico-chemical properties of the surfactants in the pure solid state, but also their behavior in aqueous dispersions, due to the different interplay of intra- and intermolecular interactions.

Formation of Dimers

An interesting concluding remark concerns the formation of pairs in the solid and liquid phase. It is known that long chain carboxylic acids form hydrogen bonded dimers that involve their COOH residues (*42–44*). These dimers can then aggregate and form clusters with an aggregation number ranging between 10 and 1000 molecules. In these assemblies, held together by the hydrogen bonded carboxyl groups, the hydrocarbon chain are fully interdigitated (*43*). This implies that the carboxylic and the ending methyl groups are aligned alternately in a same lateral plane (*43*, *45*). It has been shown that different fatty acids form heterodimers in the liquid state when they are completely immiscible in the solid state (*46–48*).

In our case the components, D-ASC12 and L-ASC12, possess several hydrophilic residues that can establish intra- and intermolecular hydrogen bonds, and that therefore can form homodimers (D-D and L-L) in the solid state. Once the mixture is melted, heterodimers (D-L) may be produced. The results obtained in this study confirm that clusters are formed in the binary eutectic melts, both in the solid and in the coagel states.

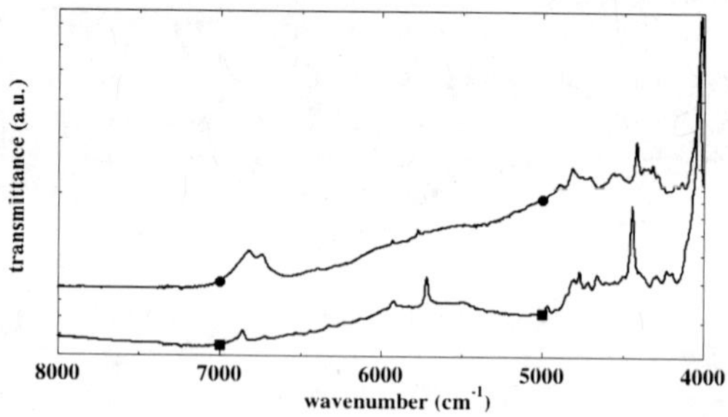

Figure 18. NIR spectra of D-isoascorbic acid (●) and L-ascorbic acid (■). Reproduced from ref. (19). Copyright 2009, American Chemical Society

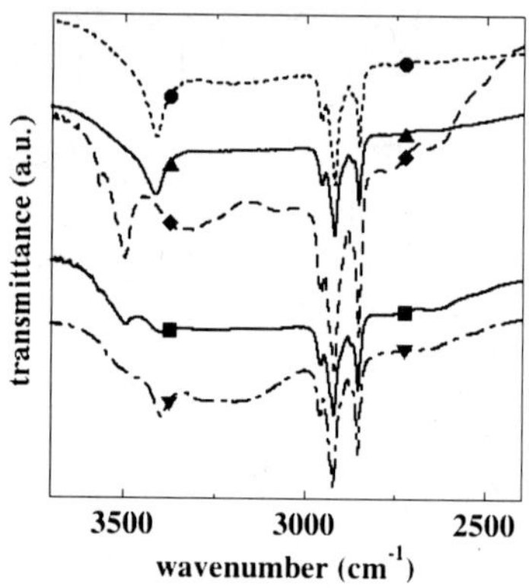

Figure 19. MIR spectra of D-ASC12/L-ASC12 mixtures at different mass ratios: 100:0 (●), 37:63 (■), 50:50 (◆), 94:6 (▲), 0:100 (▼). Reproduced from ref. (19). Copyright 2009, American Chemical Society

Furthermore, the formation of self-associated, hydrogen bonded structures from ascorbic acid dissolved in hydrophobic media such as *n*-heptane was suggested by Gupta (*49*). A similar situation is expected to occur in the interdigitated lamellar nanostructures of the coagels where intermolecular hydrogen bonds are established between the polar headgroups of ASC12 (see Figure 16) (*43*, *45*). This picture would explain reasonably the formation of homo- and heterodimers in the solid and in the coagel mixtures, held by hydrogen bonds. The results reported here indicate that the interaction parameter (ω) is small and negative when L-ASC12 is the major component, suggesting that the contact between D-D and L-L dimers is slightly favored in the liquid state, respect to the contact between homodimers, i.e. between D-D and D-D or between L-L and L-L. The opposite behavior, with a large and positive value of ω is found when D-ASC12 is the main component in the mixture, indicating that the contact between L-L and D-D dimers is strongly inhibited.

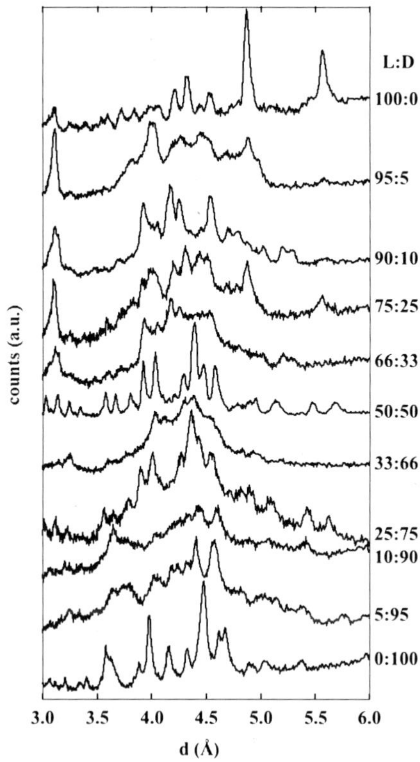

Figure 20. XRD profiles for L-ASC12 (top), D-ASC12 (bottom), and their mixtures at different L:D ratios. Reproduced from ref. (19). Copyright 2009, American Chemical Society

Conclusions

Supramolecular structures are produced through self-association of chemical building blocks. These can be surfactants, lipids, amino acids, sugars, peptides, or other species, and their arrangement is the key function of some of the most important phenomena that occur in biological systems, such as molecular recognition in enzyme-substrate or antigen-antibody complexes.

Non-covalent interactions – e.g. hydrogen bonding, dispersion forces, hydrophobic and π-π interactions – play the dominant role in setting the stability, size, shape and other properties of such associated systems (*50*). In turn, the interactions are determined by the structural features of the building blocks and of the supramolecular entity at a different hierarchical level.

Surfactants that bear chiral headgroups possess peculiar physico-chemical properties, depending on the configuration of the stereogenic center. Ascorbic acid, that carries two chiral centers, is an ideal candidate for the study of the effect of chirality of the headgroup on the self-assembly properties of its amphiphilic derivatives. In fact ascorbic acid possesses a quite rigid structure due to the lactone ring, and bears hydrophilic groups that can build up inter-

and intramolecular hydrogen bonds. These structural features are transferred to its amphiphilic ascorbyl-alkanoate derivatives and are the key factors which determine the formation of different phases and promote the formation of nano self-assembled structures with strict structural constraints.

In this contribution we report a review of the main physico-chemical properties of L-ascorbyl-alkanoates (L-ASCn, where n indicates the number of carbons in the side chain) and of D-isoascorbyl-alkanoates (D-ASCn), and the results are discussed in terms of the different stereochemistry of the surfactants' headgroups, that determine the inter- or intramolecular interactions that are established in the solid state or in aqueous dispersions.

Moreover, we investigated the thermal behavior of mixtures of L-ASC12 and D-ASC12 through DSC measurements, infrared spectra, and X-ray diffraction experiments in the solid and in the coagel states. The two components are fully miscible in the melt state and only partially miscible in the solid phase. In both cases, the phase diagrams indicate the presence of a 1:1 molecular compound and of two eutectics. The values of the excess thermodynamic functions suggest that the two diastereoisomers form clusters in the eutectic mixtures.

In conclusion, the different physico-chemical properties in the solid and in the dispersed state of the pure individual diastereoisomers are related to the chirality of the headgroups that controls the inter- and intramolecular interactions, and the different solvation properties in aqueous dispersions. Chirality also determines the different thermal behavior of mixtures of D-ASC12 and L-ASC12.

References

1. Boyd, B. J.; Krodkiewska, I.; Drummond, C. J.; Grieser, F. *Langmuir* **2002**, *18*, 597–601.
2. Bombelli, C.; Bernardini, C.; Elemento, G.; Mancini, G.; Sorrenti, A.; Villani, C. *J. Am. Chem. Soc.* **2008**, *130*, 2732–2733.
3. Cornelissen, J. J. L. M.; Rowan, A. E.; Nolte, R. J. M.; Sommerdijk, N. A. J. M. *Chem. Rev.* **2001**, *101*, 4039–4070.
4. Terech, P.; Weiss, R. G. *Chem. Rev.* **1997**, *97*, 3133–3160.
5. Ambrosi, M.; Fratini, E.; Alfredsson, V.; Ninham, B. W.; Giorgi, R.; Lo Nostro, P.; Baglioni, P. *J. Am. Chem. Soc.* **2006**, *128*, 7209–7214.
6. Shalmashi, A.; Eliassi, A. *J. Chem. Eng. Data* **2008**, *53*, 1332–1334.
7. Shibayama, H.; Ueda, K.; Yoshio, K.; Matsuda, S.; Hisama, M.; Miyasawa, M. *J. Oleo Sci.* **2005**, *54*, 601–608.
8. Rowe, R. C.; Sheskey, P. J.; Owen, S. C. *Handbook of Pharmaceutical Excipients*, 5th ed.; Pharmaceutical Press: London, 2006.
9. Swern, D.; Stirton, A. J.; Turer, J.; Wells, P. A. *Oil Soaps* **1943**, *20*, 224–226.
10. Lo Nostro, P.; Ramsch, R.; Fratini, E.; Lagi, M.; Ridi, F.; Carretti, E.; Ambrosi, M.; Ninham, B. W.; Baglioni, P. *J. Phys. Chem. B* **2007**, *111*, 11714–11721.
11. Matsui, H.; Douberly, G. E. J. *Langmuir* **2001**, *17*, 7918–7922.
12. Claussen, R. C.; Rabatic, B. M.; Stupp, S. I. *J. Am. Chem. Soc.* **2003**, *125*, 12680–12681.

13. Song, J.; Cheng, Q.; Stevens, R. C. *Chem. Phys. Lipids* **2002**, *114*, 203–214.
14. Caffrey, M.; Hogan, J.; Rudolph, A. S. *Biochemistry* **1991**, *30*, 2134–2146.
15. Selinger, J. V.; Spector, M. S.; Schnur, J. M. *J. Phys. Chem. B* **2001**, *105*, 7157–7169.
16. Nyrkova, I. A.; Semenov, A. N.; Aggeli, A.; Boden, N. *Eur. Phys. J. B* **2000**, *17*, 481–497.
17. Esselen, W. B.; Powers, J. J.; Woodward, R. *Ind. Eng. Chem.* **1945**, *37*, 295–299.
18. Leffinwell, J. C. *Vitamin C (Ascorbic acid) In Chirality & Bioactivity 1, Pharmacology*; Leffingwell Reports: Canton, GA, 2003; pp 8−10
19. Ambrosi, M.; Lo Nostro, P.; Fratini, E.; Giustini, L.; Ninham, B. W.; Baglioni, P. *J. Phys. Chem. B* **2009**, *113*, 1404–1412.
20. Hvoslef, J. *Acta Crystallogr., Sect. B* **1968**, *24*, 1431–1440.
21. Hvoslef, J. *Acta Crystallogr., Sect. B* **1968**, *24*, 23–35.
22. Kim, M. S.; Lee, S. H.; Chung, U. T.; Kang, Y. K. *Bull. Kor. Chem. Soc.* **1991**, *12*, 143–148.
23. Shin, Y. A.; Kang, Y. K. *Bull. Kor. Chem. Soc.* **1991**, *12*, 61–67.
24. Migliardo, F.; Branca, C.; Faraone, A.; Magazù, S.; Migliardo, P. *Phys. B* **2001**, *301*, 138–140.
25. Azarnia, N.; Berman, H. M.; Rosenstein, R. D. *Acta Crystallogr., Sect. B* **1971**, *27*, 2157–2161.
26. Wang, Y.; Tominaga, Y. *J. Chem. Phys.* **1996**, *104*, 1–6.
27. Umehara, T.; Tominaga, Y.; Hikida, A.; Mashimo, S. *J. Chem. Phys.* **1995**, *102*–9474–9479.
28. Mashimo, S.; Miura, N.; Umehara, T. *J. Chem. Phys.* **1992**, *97*, 6759–6765.
29. Wang, Y.; Tominaga, Y. *J. Phys. Soc. Jpn.* **1993**, *62*, 4198–4201.
30. Liu, H.; Xiang, B.; Qu, L. *J. Mol. Struct.* **2006**, *794*, 12–17.
31. Sperline, R. P. *Langmuir* **1997**, *13*, 3715–3726.
32. Capuzzi, G.; Lo Nostro, P.; Kulkarni, K.; Fernandez, J. E. *Langmuir* **1996**, *12*, 3957–3963.
33. Capuzzi, G.; Lo Nostro, P.; Kulkarni, K.; Fernandez, J. E.; Vincieri, F. F. *Langmuir* **1996**, *12*, 5413–5418.
34. Capuzzi, G.; Kulkarni, K.; Fernandez, J. E.; Vincieri, F. F.; Lo Nostro, P. *J. Colloid Interface Sci.* **1997**, *186*, 271–279.
35. Kodama, M.; Seki, S. *Adv. Colloid Interface Sci.* **1991**, *35*, 1–30.
36. Lo Nostro, P.; Ninham, B. W.; Ambrosi, M.; Fratoni, L.; Palma, S.; Allemandi, D.; Baglioni, P. *Langmuir* **2003**, *19*, 9583–9591.
37. Lo Nostro, P.; Peruzzi, N.; Severi, M.; Ninham, B. W.; Baglioni, P. *J. Am. Chem. Soc.* **2010**, *132*, 6571–6577.
38. Lo Nostro, P.; Ninham, B. W.; Fratoni, L.; Palma, S.; Manzo, R. H.; Allemandi, D.; Baglioni, P. *Langmuir* **2003**, *19*, 3222–3228.
39. Lo Nostro, P.; Ambrosi, M.; Ninham, B. W.; Baglioni, P. *J. Phys. Chem. B* **2009**, *113*, 8324–8331.
40. Ambrosi, M.; Lo Nostro, P.; Fratoni, L.; Dei, L.; Ninham, B. W.; Palma, S.; Manzo, R. H.; Allemandi, D.; Baglioni, P. *Phys. Chem. Chem. Phys.* **2004**, *6*, 1401–1407.
41. Beckmann, W.; Lorenz, H. *Chem. Eng. Technol.* **2006**, *29*, 226–232.

42. Lomer, T. R. *Acta Crystallogr.* **1952**, *5*, 11–14.
43. Iwahashi, M.; Yamaguchi, Y.; Kato, T.; Horiuchi, T.; Sakurai, I.; Suzuki, M. *J. Phys. Chem.* **1991**, *95*, 445–451.
44. Iwahashi, M.; Takebayashi, S.; Taguchi, M.; Kasahara, Y.; Minami, H.; Matsuzawa, H. *Chem. Phys. Lipids* **2005**, *133*, 113–124.
45. Lomer, T. R. *Acta Crystallogr.* **1963**, *16*, 984–988.
46. Inoue, T.; Hisatsugu, Y.; Ishikawa, R.; Suzuki, M. *Chem. Phys. Lipids* **2004**, *127*, 161–173.
47. Inoue, T.; Hisatsugu, Y.; Suzuki, M.; Wang, Z. N.; Zheng, L. Q. *Chem. Phys. Lipids* **2004**, *132*, 225–234.
48. Hernqvist, L. In *Crystallization and Polymorphism of Fats and Fatty Acids*; Garti, N.; Sato, K., Eds.; Marcel Dekker Inc.: New York, NY; 1988; pp 97–137
49. Gupta, S.; Sharma, R. K.; Chandra, H. *J. Appl. Spectrosc.* **2006**, *73*, 297–300.
50. Ninham, B. W.; Lo Nostro, P. *Molecular Forces and Self Assembly: In Colloid, Nano Sciences and Biology*; Cambridge University Press: Cambridge, 2010.
51. Lake, J. A. *Acta Crystallogr.* **1967**, *23*, 191–194.

Chapter 6

Oil-in-Oil-Emulsions: Tailor-Made Amphipolar Emulsifiers

M. S. Hoffmann, R. Haschick, M. Klapper,* and K. Müllen

Max-Planck-Institute for Polymer Research, Ackermannweg 10, 55128 Mainz, Germany.
***E-mail: klapper@mpip-mainz.mpg.de**

In order to obtain stable biphasic solvent mixtures such as emulsions and miniemulsions, distinct emulsifiers are required for the liquid/liquid interface. Until recently, nanoscale polymer particles have been mainly gained from waterborne heterophase techniques in radical processes, herein nonaqueous emulsions are presented. For these biphasic systems consisting of two aprotic organic nonmiscible solvents, high molecular weight amphipolar block copolymers were developed which show selective solubility in both phases and give additional stability by sterical shielding. Different emulsifiers, ranging from low molecular weight surfactats up to long chain block copolymers, were designed for the stabilization of hydrocarbon/perfluorocarbon mixtures and the metallocene-catalyzed synthesis of polyolefins therein. Furthermore, PI-*b*-PMMA emulsifiers were applied for mixtures of DMF dispersed in *n*-hexane. By these emulsions high molecular weight polyester and polyurethane particles were recieved, while further developments led to more sophisticated morphologies like core-shell structures.

Introduction

Emulsions and Emulsion Polymerization

Emulsion polymerization was first introduced in the beginning of the 20th century, and it became a well-established and widely utilized method in polymer synthesis within the last decades (*1*). This and other heterophase techniques

© 2011 American Chemical Society

such as micro- and miniemulsion polymerization require appropriate emulsifiers which yield stable biphasic solvent mixtures and getting them compatibilized (*2*). While viscosity and thermal problems due to exothermic polymerization reactions can be significantly reduced in comparison to bulk polymerization, this technique simultaneously allows access to both high molecular weights and high reaction rates (*3*). Furthermore, because of the distinct control over the product morphology, *i.e.* size and shape of the obtained material, emulsion polymerization as well as miniemulsion and microemulsion has become the most common technique for industrial synthesis of organic nanoparticles. These methods allow the control of the resulting particle diameters from a few nanometers up to the micrometer range in addition to giving good processability over the obtained latex dispersions.

Organic nanoparticles have gained tremendous interest in recent years and have found manifold applications, especially as nanofunctional materials due to their well-defined morphology, surface and size, resulting in tailor-made unique chemical and physical properties. The fields of applications are numerous, such as coatings (*4*), paints and pigments (*5*), but also in drug-delivery (*6*), diagnostics (*7*) and catalysis (*8*, *9*). As the polymerization media are usually based on an aqueous continuous phase, the emulsions are very attractive both from the economical point of view due to low costs and availability of the solvent and also from the ecological point of view due to its environmental friendliness.

Unfortunately, one can only benefit from this advantage when applying radical chain growth mechanism and using compounds which are not sensitive to an aqueous environment (*3*). Thus, classical waterborne emulsion and miniemulsion polymerizations do not allow the use of acid chlorides or isocyanates in step growth polymerizations or the application of moisture-sensitive catalysts, *e.g.* metallocenes for polyolefin synthesis. Side reactions might occur, resulting in hydrolysis of acid chlorides, formation of urea or decomposition of an active metal center. Therefore, due to the shift of the stoichiometric ratio or overall decomposition, only low molecular weights can be achieved often along with broad molecular weight distributions. Hence, nonaqueous systems needed to be developed which are suitable to carry out such sensitive polymerization methods. The most critical issue was the stabilization of two nonmiscible aprotic organic solvents which required new, optimized polymeric surfactants.

Nonaqueous Emulsions

Several examples of two nonmiscible organic solvent combinations – so-called oil-in-oil emulsions – were described in the literature based on alcohols *etc.* (*10*), but they do not solve the problem of side reactions with acidic protons which were mentioned above. Therefore, emulsions consisting of two nonmiscible *and* aprotic organic solvents needed to be designed. First examples of solvent mixtures consisting of DMF / *n*-hexane or cyclohexane / acetonitrile, *resp.*, were reported by Riess *et al.* already in 1970 (*11*). These mixtures were stabilized by PS-*b*-PI and PS-*b*-PMMA block copolymers, *resp.*, yielding droplet diameters not smaller than 1 μm. Stabilization of this kind of solvent mixtures – possessing a lower interfacial tension than aqueous systems – can hardly be achieved with low molecular weight surfactants, but it is reported in the literature that this can be accomplished by high molecular weight copolymers (20,000 up to 100,000 g mol^{-1}) (*12, 13*). These emulsifiers can be adapted precisely to the employed solvents according to the Hansen parameters; specific affinity and solubility in one of the phases can be achieved by tuning the monomer polarity, the segment lenth and the block length ratio (*14, 15*). Especially block copolymers turned out to be capable stabilizers for these heterogeneous systems (*2*). In contrast to low molecular weight tensides, the polymeric surfactants provide kinetic stability as the long chains are much more restricted in their mobility. Furthermore, the entanglement leads to "frozen" assemblies, whereby formation of the micellar core is the main enthalpic driving force (*16*). Above the critical micelle concentration (cmc), the molecules start to associate with the insoluble block in the core and the soluble block dissolved in the surrounding solvent. Usually the value of cmc is reported to be significantly lower than those needed for low molecular weight amphiphiles (*17, 18*). Additionally, block copolymers provide a much better sterical shielding against coalescence than low molecular weight emulsifiers because of their increased spatial demanding properties (*19*).

The biphasic mixtures of aprotic solvents reported so far provide good emulsion stability whilst we continued this concept in order to perform highly sensitive polymerizations in the confined geometry of the dispersed droplets. A decisive issue within our development of new types of nonaqueous emulsions was the synthesis of suitable emulsifiers. Additionally, in order to gain control over the product morphology, these biphasic systems need to result in heterogeneous solvent mixtures with narrow size-distributed droplets in a tunable size range to obtain well defined and uniform nanoparticles.

Results and Discussion

Perfluorinated Carbon / Hydrocarbon Emulsions – Polyolefin Synthesis

For the synthesis of polyolefins, by far the most produced commodity plastics, metallocene catalysts have recently attracted great interest. They allow for the synthesis of previously inaccessible polymers and feature excellent activities, high stereoselectivities as well as high molecular weights together with low polydispersities (*20*). A major drawback of these catalysts is their high moisture sensitivity. Furthermore, the product morphology (size and shape of the obtained polymer particles) cannot be controlled properly when applied under homogeneous conditions. Additionally, reactor fouling is commonly observed due to local overheating of the resulting product. These problems can be overcome by supporting the catalyst on inorganic or organic supports, however, this demands additional preparation of the carrier material, *e.g.* thermal treatment to remove hydroxyl groups *etc.* (*21*) Alternatively, aqueous emulsions were proven to work for polyolefin synthesis (*22*). Though, due to the water sensitivity of the well-estbalished catalysts, only specially designed metal complexes can be used, often yielding only low activities.

In order to circumvent the aforementioned drawbacks, we developed nonaqueous emulsions and successfully applied them for this demanding purpose

(*23*, *24*). An apolar aprotic solvent was needed as the dispersed phase in order to provide sufficient solubility for both olefin monomers and metallocene/cocatalyst systems. A severe decrease in polymerization activity would be caused by the complexation of a polar solvent to the active metal species. In contrast to the first demand, a solvent which was immiscible with the first one and also chemically as far as possible inert was prerequisite.

Therefore, oil-in-oil emulsions based on a perfluorinated solvent as the continuous phase and a non-coordinating hydrocarbon such as toluene as the dispersed phase, hosting the metallocene catalyst, were utilized for the polymerization of gaseous ethylene and propylene. In order to stabilize these novel emulsions, appropriate emulsifiers needed to be designed. First attempts via a low molecular weight emulsifier with a lipophilic and a fluorophilic moiety (Figure 1) did not provide sufficiently stable emulsions, probably due to the aforementioned reasons.

In numerous examples high molecular weight emulsifiers provided good droplet stability due to enlarged specifically soluble moieties even in solvent mixtures with low polarity differences (*15*). Therefore, statistical copolymers based on poly(4-hydroxy styrene) with M_w = 20,000 g mol^{-1} were synthesized by two subsequent Williamson etherifications (Figure 2) (*23*). These amphiphilic structures, especially with $R_F = CH_2(CF_2)_{10}CF_3$, provided a satisfying emulsion stability and, therefore, allowed the polymerization of gaseous ethylene and propylene, resp., in toluene droplets dispersed in perfluoromethylcyclohexane. The resulting droplet diameters showed the expected dependency on the emulsifier amount and could be decreased down to 90 nm.

The fluorous oil-in-oil emulsions were exploited subsequently for the synthesis of polyolefin particles by metallocene catalysis from gaseous ethene or propene (Figure 3; Activity: 400 kg PP (mol Zr hr bar)$^{-1}$ at 1 bar and 60 °C with MBI catalyst ([dimethylsilanediylbis(3,3′-(2-methylbenz[e]indenyl))]-zirconium dichloride)). After formation of the micelles by stirring and ultrasonication treatment, the hydrocarbon and the cocatalyst (methylaluminoxane MAO) were added and the emulsion was saturated with the gaseous monomer. Subsequent addition of the metallocene-based precatalyst initiated the polymerization. As the metallocene/MAO catalyst system is only barely soluble in the continuous phase, it is assumed that the reaction takes place in the confined geometry of the dispersed hydrocarbon droplets ("nanoreactors"). The nonaqueous fluorous emulsions gave high molecular weight polyolefin particles (*e.g.* $M_{n,PE}$ = 450,000 g mol^{-1}, PDI = 3.1), and the diameters of the resulting spheres (Figure 4) were controlled by parameters like polymerization time as well as pressure. A decrease of the particle diameters was accomplished by decreasing them, which is typical for a waterborne emulsion. Thus, these new emulsions allow polymerization under pseudo-homogeneous conditions.

In order to overcome the diffusion limitation and to alter the conditions towards miniemulsion conditions, polymerization of liquid propylene in fluorous emulsions without any additional solvent as the dispersed phase was developed (*24*). Due to these more demanding conditions, stabilization of the emulsions needed to be improved by modification of the emulsifiers. High molecular weight block copolymer emulsifiers were synthesized in a nitroxide mediated controlled

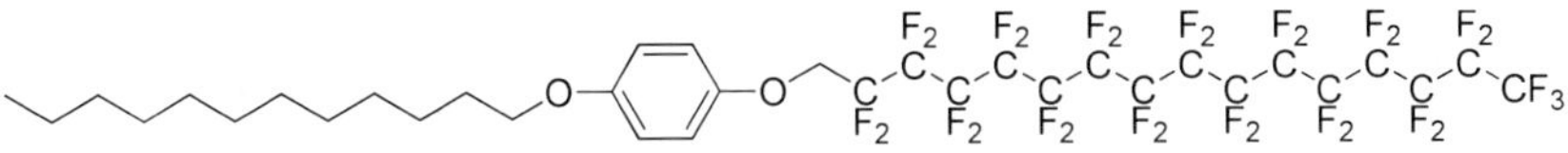

Figure 1. Low molecular weight lipophilic/fluorophilic emulsifier.

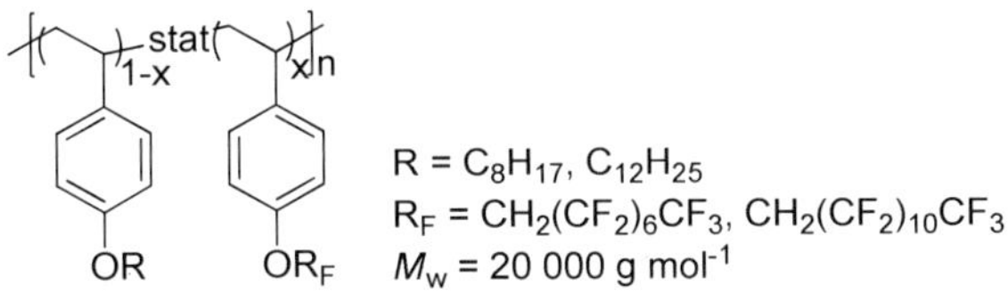

Figure 2. Statistical high molecular weight lipophilic/fluorophilic emulsifier.

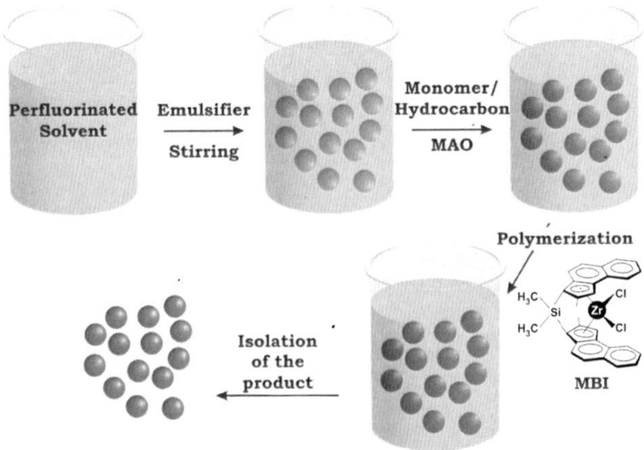

Figure 3. Schematic description of olefin polymerization in perfluorinated nonaqueous emulsions with MBI catalyst.

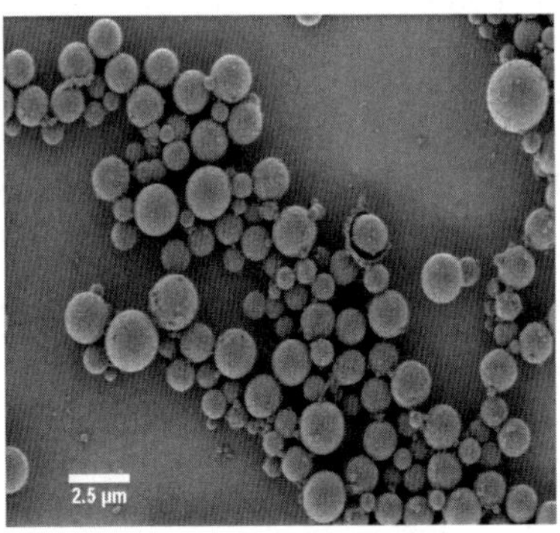

Figure 4. SEM micrograph of PE particles after 30 min at 60 °C and 40 bar reaction with MBI/MAO catalyst.

polymerization of styrene (S) and pentafluorostyrene (FS) (Figure 5). Two high molecular weight blocks were utilized which have a specific affinity to the hydrocarbon and the perfluorinated solvent, respectively. Increased affinity of the fluorinated block to the surrounding medium was accomplished by an additional perfluorinated alkyl chain in *p*-position.

These structures turned out to be the most suitable for the stabilization of the oil-in-oil emulsions. Light scattering measurements indicated the formation of stable droplets (*24*), and the classical dependence on the amount of emulsifier was observed, while there was no change in droplet size over a broad temperature range – even close to the boiling point of the continuous phase. With the amounts of up to 10 wt.-% emulsifier concerning the dispersed phase droplet sizes were decreased to 50 nm.

Switching completely to miniemulsion conditions resulted in a significant increase of the metallocene catalyzed polymerization of liquified propene (3500 kg PP (mol Zr hr)$^{-1}$ at 25 bar and 60 °C with MBI catalyst; see Figure 3). With this new method, monomer diffusion through the continuous phase was avoided whilst additional organic solvents were omitted. Due to the surrounding perfluorinated solvent, heat transfer was promoted from this highly exothermic process, and therefore reactor fouling, which is caused by molten polymer under homogeneous conditions, was supressed.

Besides pure olefin polymerization, these stable heterogeneous systems allowed the synthesis of core-shell structures. For example particles were obtained based on a core consisting of isotactic polypropylene with the shell of poly(*n*-butyl acrylate). Thus, the surrounding shell with its relatively low T_g showed good film forming properties of the stiff core material and provides an alternative to similar core-shell structures gained from polystyrene and poly(*n*-butyl acrylate).

Polar/Nonpolar Nonaqueous Emulsions – Step Growth Polymerization

In order to conduct polymerizations of *polar* monomers to form *e.g.* polyurethanes or polyesters, the aforementioned solvent mixtures could not be applied. In contrast, emulsions consisting of a polar dispersed phase and a nonpolar continuous phase were formed. Different solvent combinations were suitable for this approach; for example *N,N*-dimethyl formamide (DMF) dispersed in *n*-hexane or acetonitrile dispersed in cyclohexane. These mixtures were already used by Riess *et al.* (*11*) in order to prepare emulsions stabilized by copolymers of poly(styrene) (PS) and poly(isoprene) (PI) or poly(methyl methacrylate) (PMMA). On the basis of this concept new block copolymers consisting of PI and PMMA synthesized by anionic polymerization (*25*, *26*) were developed within our group (Figure 6). These turned out to be most suitable to disperse a polar organic solvent as DMF in a nonpolar phase like *n*-hexane.

The PMMA part of the copolymer is selectively soluble inside the polar dispersed phase and the PI sequence in the nonpolar sourrounding medium. The different polarities and, therefore, also the solubilities of the segments ensured good stabilization behavior in the two phases. Additionally, the investigation of the appropriate block lengths and the block length ratios played a crucial role in the specific designing of the surfactants (Table I).

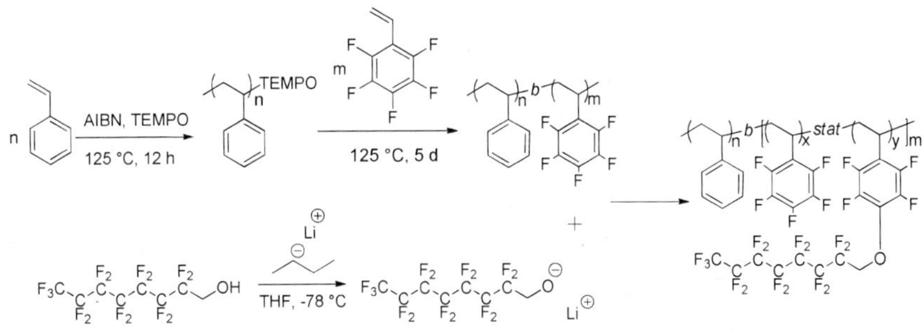

Figure 5. High molecular weight lipophilic/fluorophilic PS-b-PFS block copolymer.

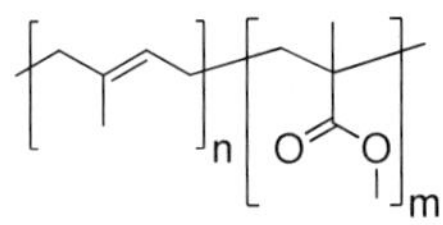

Figure 6. PI-b-PMMA block copolymer for polar/apolar oil-in-oil emulsions.

Table I. Block lengths and block ratios as well as phase stabilization behavior of different block copolymers (*27*)

Polymer	*PI-block[a]* M_n / *g mol*$^{-1}$	*PMMA-block[b]* M_n / *g mol*$^{-1}$	M_n/M_w	*Ratio PI/PMMA*	*Hydrodyn. diameter[c] / nm*
1	7,500	27,500	1.4	29:71	phase separation
2	3,000	4,000	1.3	52:48	phase separation
3	15,500	15,500	1.1	59:41	58
4	5,500	2,500	1.2	76:24	37
5	15,500	7,000	1.2	76:24	42
6	23,000	7,000	1.3	83:17	32

[a] by GPC in THF *vs.* PI standards; [b] by ratio calculation via ^{1}H NMR in $CDCl_3$ at 250 MHz and 298 K; [c] by DLS of 3 g MeCN and 0.5 g PI-*b*-PMMA in 24 g cyclohexane at scattering angle $\theta = 90\,°$.

Table I demonstrates that only copolymers with a PI fraction of at least 50% and a molecular weight of 8,000 g mol^{-1} resulted in stable emulsions. With respect to the droplet size and stability, the optimal ratio of the block copolymer consisted of approximately 75 mol-% PI and 25 mol-% PMMA. Concerning the molecular weight dependency, it is known that the stabilization efficiency of low molecular weight surfactants like SDS is unsatisfactory for emulsions with low polarity differences as it was already shown above for the perfluorinated systems. SDS molecules not only arrange on the droplet surface but are also soluble in the continuous phase and form micelles there. This results in a depletion of the surfactant from the particles destabilizing the emulsion. In contrast to that, high molecular weight emulsifiers possess much lower dynamics and this kinetical effect leads to a better stabilization of the droplets.

While the data concerning the emulsifier composition (Table I) were measured using a constant block copolymer concentration, Table II shows the effect on the particle size by changing the concentration of emulsifier in a constant mixture of 15 mL *n*-hexane and 1.5 mL of DMF.

The droplet size decreased with a decreasing amount of emulsifier while at a mass concentration of approximately 5 mg copolymer per gram hexane, phase separation occured and no droplet formation was detected. Additionally, it was observed that at a concentration of about 25 mg PI-*b*-PMMA / 1 g *n*-hexane a plateau value was reached and the droplet size did not decrease any more due to a equilibrium between umulsifier whish is bound in a micelle and free one.

In order to investigate the time dependency of the emulsion stability, the hydrodynamic radii of the droplets resulting of 0.2 g block copolymer (PI_{700}-*b*-$PMMA_{160}$), 0.5 g DMF in 12 g *n*-hexane were measured (Figure 7).

Over a period of 3 hours the hydrodynamic radii remained almost constant in a size range of approximately 50 nm and no aging effects were determined. Obviously, the mixture was stable enough to keep its droplet size without stirring and sonication for a long time. Figure 8 shows the development of the droplet radii during a slow increase of temperature to 45 °C.

Cooling the heated emulsion to room temperature led to an increase of the droplet radius to approximately 50 nm, which was in the size range of the droplets at the beginning of the measurement. Thus, raising the temperature caused a reversible decrease in droplet size and *vice versa*. A potential reason for this behavior might be the slight solubility of DMF in *n*-hexane which increased with rising temperature. It was assumed that the shrinking effect of the droplets was caused by a partial enrichment of DMF molecules of the dispersed phase in the continuous phase. This assumption was supported by ^1H NMR experiments. It was proven that 1 mL of *n*-hexane solved at 20 °C approximately 0.02 mL of DMF, whereas at 40 °C the dissolved amount increased to 0.05 mL and according to the literature both phases mix at 68 °C (*28*).

The synthesis of particles in these solvent mixtures can usually be considered as an emulsion polymerization based on a diffusion controlled process. Hitherto, there are also approaches available using compounds which are insoluble in the continuous phase (*29*). Thus, there is no diffusion control but only "nanovessels" are present (mini-emulsion process). In our standard procedure for particle formation by emulsion polymerization, the emulsifier (PI-*b*-PMMA) was dispersed homogeneously inside the continuous phase and, finally, micelle formation occurred (Figure 9). The addition of the polar phase containing compound A (*e.g.* a diol for the formation of polyurethanes or polyesters) and several minutes of ultrasonication led to a stable biphasic system. By dropwise addition of compound B (*e.g.* a diisocyanate for polyurethanes, a diacid dichloride for polyesters or a catalyst for ROMP) and allowing it to diffuse into the droplets, initiation of the the reaction took place. Since component A is barely soluble in the continuous phase, the reaction occured mainly inside the droplets. This was presented particularly for different types of reactions such as polyadditions, polycondensations and catalytic reactions (*15*).

Table II. ***Effect of the emulsifier concentration on the droplet size of mixtures of 15 mL* n*-hexane and 1.5 mL DMF***

m / g[a]	*PI-b-PMMA*[b] *ω10*$^{-2}$ */ wt%*	*Droplet diam.*[c] */ nm*
0.40	3.97	40 (±14)
0.29	2.88	41 (±19)
0.21	2.11	56 (±28)
0.15	1.49	79 (±34)
0.10	0.99	98 (±43)
0.05	0.50	phase separation
0.02	0.20	phase separation

[a] amount of emulsifier; [b] weight portion of emulsifier considering Hexane; [c] by DLS at a scattering angle $\theta = 90$ °.

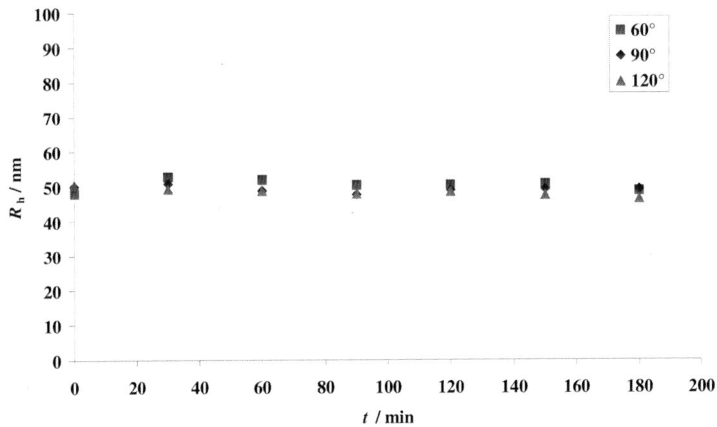

Figure 7. Time dependency of the droplet radii in a mixture of 12 g n-hexan, 0,5 g DMF and 0,2 g PI_{700}*-b-*$PMMA_{160}$.

The use of this nonaqueous emulsions opened access to versatile polymerization procedures and, therefore, different materials for the preparation of nanoparticles (*9*, *15*). One important example is the polyaddition reaction which, for example, was used to form polyurethane particles (*30*). In order to accomplish such a synthesis, a diol and a diisocyanate were chosen to be compound A and B. In most cases the polar phase was DMF dispersed in *n*-hexane. However, acetonitrile dispersed in cyclohexane or *N*-methylpyrrolidone in tetradecane were also used. It was possible to prepare polyurethane particles with molecular weights as high as $M_n = 40{,}000$ g mol^{-1} (PDI = 2.0) which equals – in regards to the Carothers equation – a conversion higher than 99%. It was assumed that two main factors were responsible for the high molecular weights and conversions: (i) the absence of water during the reaction, which diminished the amount of side reactions of the isocyanates such as urea formation; (ii) Schotten-Baumann conditions, meaning that the 1:1 stoichiometry which is necessary for the polyaddition was achieved by the diffusion of the monomers through the interface of the continuous and the dispersed phase. As the diols were predominantly soluble in the dispersed droplets, the stoichiometry of the reaction was controlled by the diffusion of the diisocyanate component into the micelles.

PU particles which were prepared by this method possessed a very narrow size distribution and a diameter of approximately 20 to 70 nm (Figure 10). The particles had a well defined spherical morphology and did not agglomerate (monomodal distribution according to DLS measurement). The slight aggregation visible in the SEM image was caused by the sample preparation during solvent evaporation.

Besides the synthesis of particles consisting solely of non-porous PU, more complex morphologies such as core-shell or porous structures were achieved by these types of emulsions (*31*). For instance, the preparation of porous particles by the slow and controlled addition of a well defined amount of water was possible. Hereby, some of the isocyanate groups underwent side reactions with water, forming urea derivatives. Simultaneously, carbon dioxide was released forming pores inside the viscous polymeric material. These particles with diameters from several hundreds of nanometers up to several micrometers had pore sizes in the range from 30 to 500 nm. Also, experiments concerning the use of porous PU particles as a support for the metallocene polymerization were accomplished. Another possible morphology for these PU particles is the preparation of core-shell particles. It was possible to create a polymeric shell from different methacrylate monomers by free radical polymerization around the existing PU core. These particles were prepared with small diameters of about 100 to 300 nm and high molecular weights (Figure 11).

Besides the preparation of core-shell particles via physical absorption of one polymer on the surface of another, the chemical attachment of surface functionalizing groups were investigated as well. Therefore, different model compounds were chosen in order to accomplish click type reactions in nonaqueous emulsions. Even though click reactions were performed in an aqueous environment, the possibility to use such a versatile tool in our system offers the possibility to attach manifold functional compounds to the polymer particles like dyes, polyethylene oxide chains or even DNA or peptides. Such a functionalization will result in water dispersible, fluorescent or biofunctional particles.

In order to investigate this application, poly(propargyl methacrylate)-*co*-poly(methyl methacrylate) (PgMA-*co*-PMMA) particles were prepared and converted with different azides (*e.g.* 1-amino-11-azido-3,6,9-trioxaundecane or 5-azidopentanoic acid). These azides were chosen due to their polarity which is necessary to ensure their diffusion into the dispersed phase. Depending on the amount of PgMA in the copolymer, the performed "click"-reaction gave a hardly soluble polymer which was most probably due to the attached side groups leading to strong H-bonds. However, a PgMA concentration of approximately 10 mol-% inside the PgMA-MMA-copolymer yielded a soluble polymer even after the polymer analogous "click"-reaction. Via ^{1}H NMR (for soluble samples) and IR spectroscopy (for insoluble polymers) it was proven that the signal intensity

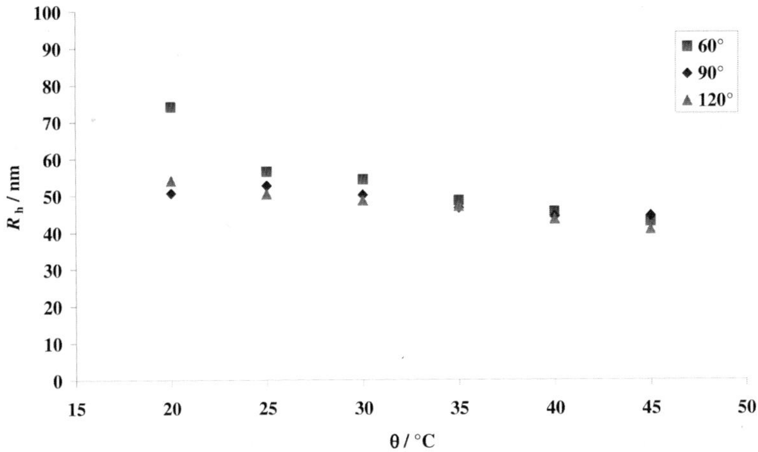

Figure 8. Temperature dependency of the droplet radii in a mixture of 12 g n-hexan, 0,5 g DMF and 0,2 g PI_{700}-b-$PMMA_{160}$.

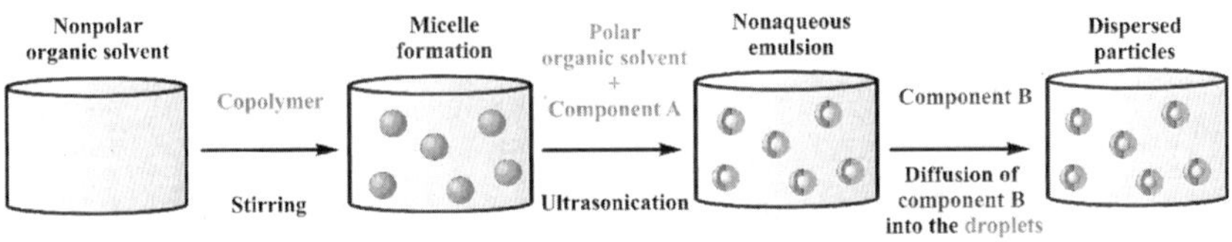

Figure 9. Schematic description of the polymerization in nonaqueous emulsion. (see color insert)

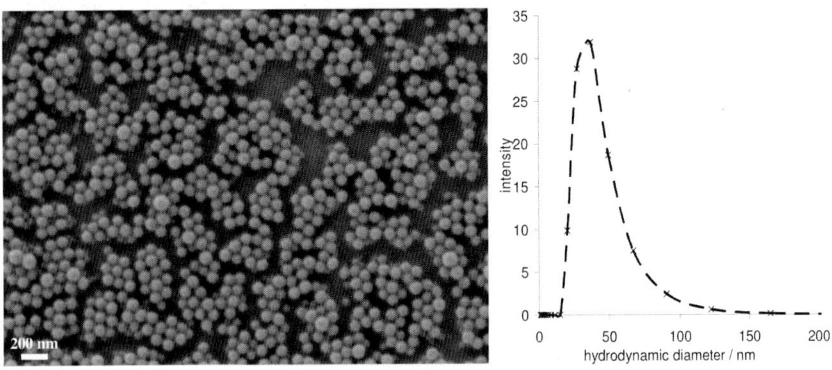

Figure 10. left) SEM image of PU particles at 120 V drop casted on Si waver; right) DLS curve from diluted primary PU particle dispersion at 90 ° scattering angle.

of the triple bonds was decreasing (^1H NMR at 250 MHz and 298 K in THF-d_8: 4.8 ppm (s, 2H); ATR-IR: 2100 cm^{-1}) whereas a signal caused by the triazole ring appeared (^1H NMR at 250 MHz and 298 K in THF-d_8: 7.8 ppm (s, 1H)).

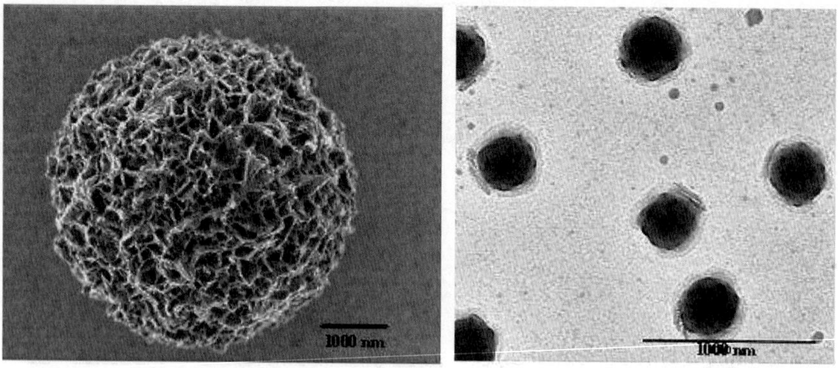

Figure 11. left) SEM image of porous PU particles at 750 V drop casted on Cu grid; right) TEM image of PU-core PMMA-shell particles at 200 kV.

Conclusion

It is demonstrated that for new types of hetereogeneous mixtures such nonaqueous emulsions, the design of appropriate stabilizing agents is essential. Block-structured surfactants provide the best stability due to their long segments. It was crucial to optimize the distinct solubilities and sterical shielding while the polarities and block lengths were adjusted carefully to the applied biphasic solvent mixtures. The biphasic systems which are presented herein have been stabilized by PS-*b*-PFS and PI-*b*-PMMA block copolymers, respectively. Emulsions consisting of *n*-hexane / DMF and toluene / perfluoromethyl cyclohexane were obtained which allow for the polymerization of moisture sensitive monomers or catalysts within the confined geometry of the stable droplets. While mainly solvent mixtures have been previously applied, these nonaqueous systems allow for the application of highly sensitive metallocene catalysts for the olefin polymerization as only aprotic solvents are used. Also polycondensations and polyaddition reactions of water sensitive monomers are accessible, yielding nanoparticles of polyester and polyurethane nanoparticles. In all cases the decisive point was the design of high molecular weight surfactants capable to form stable biphasic systems.

Acknowledgments

The authors acknowledge Thomas Wagner and Jürgen Thiel for the synthesis of the PI-*b*-PMMA copolymers, as well as Gunnar Glasser for the SEM measurements and Sandra Seywald for GPC analysis. Financial support of the Stiftung Stipendien-Fonds des Verbandes der Chemischen Industrie e.V. and providing of the metallocene catalyst by LyondellBasell is also gratefully acknowledged.

References

1. Whitby, G. S.; Katz, M. *Ind. Eng. Chem.* **1933**, *25*, 1338.

2. Riess, G.; Labbe, C. *Macromol. Rapid Commun.* **2004**, *25*, 401–435.
3. Odian, G. *Principles of Polymerization*; John Wiley & Sons: Hoboken, NJ, 2004.
4. Taylor, J. W.; Winnik, M. A. *J. Coat. Technol.* **2004**, *1*, 163–190.
5. Schuman, T.; Karlsson, A.; Larsson, J.; Wikstrom, J.; Rigdahl, M. *Prog. Org. Coat.* **2005**, *54*, 360–371.
6. Kawaguchi, T.; Tsugane, A.; Higashide, K.; Endoh, H.; Hasegawa, T.; Kanno, H.; Seki, T.; Juni, K.; Fukushima, S.; Nakano, M. *J. Pharm. Sci.* **1992**, *81*, 508–512.
7. Hatakeyama, M.; Iwato, S.; Hanashita, H.; Nakamura, K.; Fujimoto, K.; Kawaguchi, H. *Colloids Surf.*, **1999**, *153*, 445–451.
8. Naundorf, C.; Matsui, S.; Saito, J.; Fujita, T.; Klapper, M.; Müllen, K. *J. Polym. Sci., Part A: Polym. Chem.* **2006**, *44*, 3103–3113.
9. Klapper, M.; Clark, C. G.; Müllen, K. *Polym. Int.* **2008**, *57*, 181–202.
10. Landfester, K.; Willert, M.; Antonietti, M. *Macromolecules* **2000**, *33*, 2370–2376.
11. Periard, J.; Banderet, A.; Riess, G. *J. Polym. Sci., Part B: Polym. Lett.* **1970**, *8*, 109–112.
12. Durand, A.; Marie, E. *Adv. Colloid Interface Sci.* **2009**, *150*, 90–105.
13. Tauer, K.; Goebel, K. H.; Kosmella, S.; Staehler, K.; Neelsen, J. *Makromol. Chem., Macromol. Symp.* **1990**, *31*, 107–121.
14. Hansen, C. *Hansen Solubility Parameters: A User's Handbook*; CRC Press: Boca Raton, FL, 2000.
15. Klapper, M.; Nenov, S.; Haschick, R.; Müller, K.; Müllen, K. *Acc. Chem. Res.* **2008**, *41*, 1190–1201.
16. Riess, G. *Prog. Polym. Sci.* **2003**, *28*, 1107–1170.
17. Klapper, M.; Haimer, K. *Polym. Mater. Sci. Eng.* **1999**, *81*, 490.
18. Vass, S.; Haimer, K.; Meier, G.; Klapper, M.; Borbely, S. *Colloid Polym. Sci.* **2002**, *280*, 245–253.
19. Gao, Z.; Eisenberg, A. *Macromolecules* **1993**, *26*, 7353.
20. Kaminsky, W. *Macromol. Chem. Phys.* **1996**, *197*, 3907–3945.
21. Hlatky, G. G. *Chem. Rev.* **2000**, *100*, 1347–1376.
22. Mecking, S.; Held, S.; Bauers, F. M. *Angew. Chem., Int. Ed.* **2002**, *41*, 544–561.
23. Nenov, S.; Clark, C. G.; Klapper, M.; Müllen, K. *Macromol. Chem. Phys.* **2007**, *208*, 1362–1369.
24. Nenov, S.; Hoffmann, M. S.; Steffen, W.; Klapper, M.; Müllen, K. *J. Polym. Sci., Part A: Polym. Chem.* **2009**, *47*, 1724–1730.
25. Ni, S.; Zhang, P.; Wang, Y.; Winnik, M. A. *Macromolecules* **1994**, *27*, 5742–5750.
26. Tcherkasskaya, O.; Spiro, J. G.; Ni, S.; Winnik, M. A. *J. Phys. Chem.* **1996**, *100*, 7114–7121.
27. Müller, K.; Klapper, M.; Müllen, K. *Macromol. Rapid Commun.* **2006**, *27*, 586–593.
28. Francis, A. *Adv. Chem. Ser.* **1961**, *31*, 1–246.
29. Müller, K.; Klapper, M.; Müllen, K. *J. Polym. Sci., Part A: Polym. Chem.* **2006**, *45*, 1101–1108.

30. Müller, K.; Klapper, M.; Müllen, K. *Colloid Polym. Sci.* **2007**, *285*, 1157–1161.
31. Haschick, R.; Müller, K.; Klapper, M.; Müllen, K. *Macromolecules* **2008**, *41*, 5077–5081.

Chapter 7

Fluorinated Hydrogen Bonded Electrolytes of PVA/Nafion® and PAH/Nafion® Complexes

Electrochemical Applications

Tarek R. Farhat[*]

Chemistry, University of Memphis, Memphis, TN 38152
[*]tfarhat@memphis.edu

The newly discovered polyelectrolyte (PE) self-assembly by fluorinated hydrogen bonding (FHB) is harnessed to assemble FHB films and hydrogels. PE couples of poly(vinyl alcohol)/perfluorosulfonic acid PTFE copolymer (PVA/Nafion®) or poly(allyl amine) PAH/Nafion® would self assemble into layer-by-layer films or FHB hydrogels under high centrifuge using controlled pH and ion strength conditions. After centrifuge, FHB hydrogels can be isolated from their liquor, washed, and sliced. Selected physical properties such as stability, solubility, molecular partitioning, and ion conductivity are discussed therein aided by ATR-FTIR and potentiostatic/ac impedance analysis. Stability to a wide pH range and relatively high ionic conductivity (i.e. > 0.1 Scm^{-1}) makes these FHB gels potential candidates for electrolyte applications in solid-state electrochemical systems.

Introduction

Solid-state electrolyte systems used in power, detection, and energy storage applications are under continuous development especially when it comes to minimizing electrolyte thickness and fuel cross over, increasing ionic conductance and dielectric permitivity while preserving mechanical resiliency. Self-assembled supra-molecular structures (SAMS) that are used in the assembly of membranes, ultra-thin films, and gels or hydrogels are one of the leading "soft materials" in electrolyte design of solid-state systems. We introduce the SAMS polymer couples

© 2011 American Chemical Society

of poly(vinyl alcohol)/perfluorosulfonic acid PTFE copolymer (PVA/Nafion®) and poly(ally amine)/Nafion® (PAH/Nafion®) as potential soft materials in electrolyte design. First, we need to stress that the said systems are different from PVA hydrogels prepared by segmental cross linking using agents such as gultaraldehyde, thermal annealing, and freeze thawing techniques discussed by Peppas (*1*, *2*). Another close system that was used by Elabd is the PVA/Nafion® blend membranes which essentially mixes the polymer matrices of 25-90% viscous gels to form a polymer blend (*3*, *4*). Under controlled pH conditions Farhat harnessed physical cross links of long range *fluorinated hydrogen bonding* (*FHB*) forces between fluorinated Nafion® segments and their polyelectrolyte counter parts to initiate self assembly (*5*). The FHB forces occur between the CF units of Nafion® and the hydroxyl OH groups of PVA or the amine units NH_2 of PAH. The major difference in the polymer matrix structure between the three cases stated above is depicted in the schematics shown in Figure 1. FHB forces of the type (C-F:→H-N), and (C-F:→H-O) were already characterized between molecules by Howard and Strauss (*6*, *7*).

Interactions of fluorinated polymers with OH and NH functional surfaces were detected by McCarthy and Chen (*8*, *9*). The medium of FHB self assembly is water. Polyelectrolytes (PE) such as Nafion®, PVA, and PAH imply water-soluble polymers. However, when Nafion® physically cross-links with PVA (or PAH) by long-range FHB forces, under controlled pH and ion strength conditions, it would form PE insoluble complexes in water. In general, PE insoluble complexes formed by electrostatic self assembly or by hydrogen bonds as described by Decher and Osawa (*10*, *11*) is a white precipitate similar to the precipitates of inorganic salts. The white color is due to light scattering by the porous network of PE agglomerates. For example, the white fluffy precipitate of PVA/Nafion® is converted to semi-transparent hydrogels at 8000 rpm after 10 minutes centrifugation as shown in Figure 2. Transparency and hardness of the gel depends on the time and the rate of centrifuge. In general, the higher the rate of centrifuge the higher is the hardness. For example, the hardness of a PVA/Nafion® gel changed from paste-like to soft rubbery when the centrifuge rate is increased from 3000 to 8000 rpm.

Cross linking between the CF units of Nafion® and the OH or NH_2 units of the PVA and PAH segments would leave the sulfonate groups SO_3^- of Nafion® free of any segmental coupling imparting a strong cation exchanger character to the PVA/Nafion® and PAH/Nafion® gels. For example, ammonical complex of copper $Cu[NH_3]_4^{2+}{}_{(aq)}$ strongly exchanges with SO_3^- in PVA/Nafion® matrix such that:

$$[\text{R-OH} \dashrightarrow \text{CF}_2\text{-SO}_3^-] + \text{Cu[NH}_3]_4^{2+}{}_{(aq)} \rightarrow [\text{R-OH} \dashrightarrow \text{CF}_2\text{-SO}_3^- \ \text{Cu[NH}_3]_4^{2+}]$$

Author has already verified using FTIR spectroscopy and zeta potential analysis that the long range interactions between the PE segments is due to FHB forces using peak shifts, peak positions, and peak broadening of O-H *str* at 3410 to 3480 cm^{-1}, CF_2 *sym str* at 1235 cm^{-1} and the CF_2 *asym str* at 1152 cm^{-1}, N-H *str* at 3410 cm^{-1}, N-H *asym bend* at ~1580 cm^{-1}, and N-H *asym str* at 1320 cm^{-1} (*5*). The insoluble complexes of PVA/Nafion® are stable on the

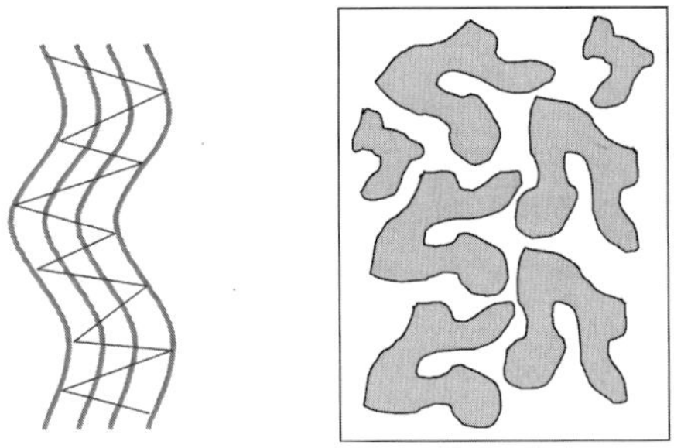

Figure 1. Schematics of polymer matrices (left) chemical cross-linking of poly(vinyl alcohol) PVA segments, (middle) a blend of PVA and Nafion®, (right) self assembly by physical cross linking of PVA and Nafion® segments due to fluorinated hydrogen bonds (FHB).

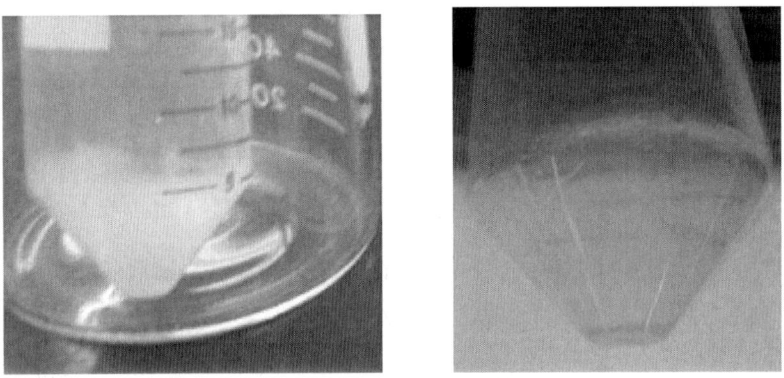

Figure 2. Pictures of a poly(vinyl alcohol)/Nafion® insoluble complex (left) agglomerate before centrifuge, (right) agglomerate after centrifuge for 10 minutes at 8000 rpm to form a FHB hydrogel.

pH scale 1.0 to 14 while the PAH/Nafion® maintains FHB cross-linking at pH $\geq$ 11. Self assembly of the PVA/Nafion® and PAH/Nafion® was also verified using layer-by-layer self assembly where these measured 300 A°/bilayer and 135 A°/bilayer respectively (*5*). The segmental interaction in layer-by-layer can be depicted by the schematic of Figure 1 (right) but in the corresponding hydrogels and depending on the centrifuge rate water is going to be the dispersion medium of the coupled PVA/Nafion® or PAH/Nafion® segments.

To this end, we will highlight some properties of these FHB hydrogels that includes gel structural analysis, ionic conductance, and faradaic properties. That is, this book chapter would address rigidity, texture, stability towards pH changes, ion-exchange interactions and partitioning, water loss, FTIR analysis, surface roughness and film thickness. Conductance and faradaic analysis are conducted to show the potential application of these "soft materials" as solid electrolytes for battery, capacitor, and electrochemical applications thus potential impact in many areas of industry (*12*).

FHB Gel Physical Analysis

Technologists who search for newly introduced soft materials are interested in their characteristics that fit a particular application. In this section some selected physical properties that characterize the fluorinated hydrogen bonded (FHB) gels of poly(vinyl alcohol)/perfluorosulfonic acid PTFE copolymer (PVA/Nafion®) and poly(ally amine) PAH/Nafion® are presented and discussed. It would highlight the FHB gel mechanical, partitioning, pH and ion-exchange behavior, ionic conductance, and faradaic properties. In this chapter we opted to study the PVA/Nafion® and PAH/Nafion® gels that maintain their hydrogel nature under the influence of FHB forces and its water dispersion without introducing any modifiers. That is, no cross-linking agents such as glutaraldehyde or dicarboxylics were used to chemically bond PVA or PAH segments within the gel. No chemicals such as ethylene glycol were used to suppress water evaporation from the gel.

FHB Gel Structural Analysis

The PVA/Nafion® and PAH/Nafion® FHB hydrogels are assembled under centrifugation after carefully decanting them from their respective equimolar 10mM:10mM mixtures. Subjecting pure 10.0 mM solutions of PVA, PAH, and Nafion® to 5000 to 8000 rpm did not produce any gels. The gels can be isolated from their solutions as semi-transparent lumps, washed in water, and dried on a filter or blotting paper, Figure 3. In terms of rigidity their durashore OO varies based on the conditions used. For example, a PVA/Nafion® lump soaked in 0.1M HCl has a shore ~20 to 25 and a shore ~40 to 45 when soaked in 0.1M $[Cu(NH_3)_4]^{2+}$ solution, Figure 3.

The PAH/Nafion® FHB hydrogels prepared at pH 11 would show a semi-transparent soft texture. A white rubbery texture indicated that preparation was not conducted at pH ≥ 11 for both PE solutions. A PAH/Nafion® gel soaked in 0.1M $[Cu(NH_3)_4]^{2+}$ solution changed slowly in color from deep blue to dark brown (not shown) which is getting investigated for the formation of metallic copper or copper oxide clusters, while dichlorofluoresceine is favorably absorbed by PVA/Nafion® gel, Figure 3. Insoluble complexes of polystyrene sulfonate PSS and poly(diallyl dimethyl ammonium chloride) PDAC that self assemble by electrostatic forces form an opaque rubbery matrix at 8000 rpm and a tough but clear pale yellow hydrogel at > 40,000 rpm.

FTIR and ATR-FTIR spectra of PVA/Nafion® layer-by-layer films having a thickness of ~ 6000 A°, their insoluble complexes (*5*), and their insoluble gels provided direct evidence of self assembly by fluorinated hydrogen bonding (FHB), Using a Thermoelectron/Specac FTIR-ATR spectro-optical box we observed that the OH *str* blue shifted by ~50 cm^{-1} and the CF *str* by ~20 cm^{-1} as a fingerprint of hydrogen bonding thus FHB. The OH *str* of a dry PVA/Nafion® hydrogel film applied by smearing to an ATR window shifted from 3343 cm^{-1} to 3470 cm^{-1} with a corresponding 15 cm^{-1} red shift in the CF *str*, Figure 4. The NH *str* was difficult to monitor but a similar shift of 12 cm^{-1} in the CF *str* down to 1205 cm^{-1} is observed in the PAH/Nafion® system. We have tested a pH 11 PAH/Nafion® hydrogel by treating a clear lump with pH = 1 acidic solution that changed to white color.

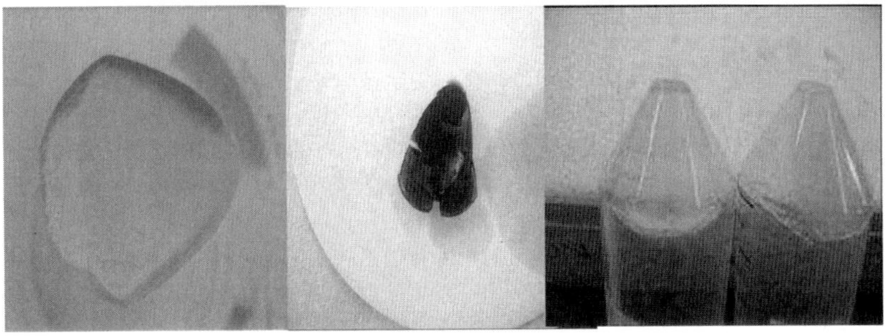

Figure 3. A typical PVA/Nafion® FHB hydrogel lump isolated from its liquor, (left) as made semi-transparent, (middle) soaked in 0.1M $[Cu(NH_3)_4]^{2+}$ solution, (right) strongly retained the fluorescent yellow-green dichlorofluoresciene molecule (yellow) while the PAH/Nafion® FHB hydrogel (colorless) did not. (see color insert)

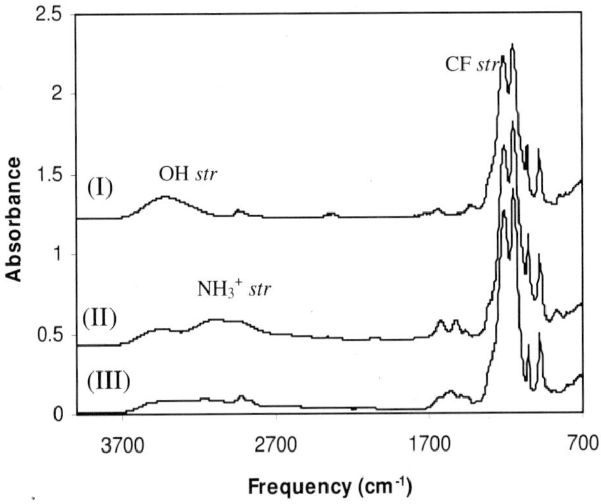

Figure 4. ATR-FTIR spectra of a dry FHB hydrogels of (I) PVA/Nafion®, (II) PAH/Nafion® treated with pH = 1 solution, (III) PAH/Nafion® prepared in pH 11 polyelectrolyte equimolar mixture.

ATR/FTIR showed clearly that the NH_2 groups in PAH get protonated revealing the spectral features of the protonated form NH_3^+ at 3100 cm^{-1}, Figure 4. At low pH it is expected that the PAH/Nafion® assembly transforms from FHB interaction into electrostatic [R-SO_3^- •••NH_3^+-R] (*13*) causing a red shift in the SO_3^- *str* which was not detected in this work. Previously, author detected degradation in PAH/Nafion® Layer-by-layer films when the pH is decreased (*5*). At this juncture, we are testing two methods to obtain micron thin *FHB hydrofilms* from these insoluble complexes that would compete with the slow conventional technique of layer-by-layer deposition. It is likely that the PAH/Nafion® matrix maintains FHB at lower pH even though the matrix degrades due to repulsive forces among the R-NH_3^+ segments. That is, the PAH/Nafion® matrix might behave like an amphiphilic PE. The SO_3^- *str* of PVA/Nafion® at pH = 7 is non interacting because it occurs at 1065 cm^{-1} that overlaps with the SO_3^- *str* assignment of free Nafion® segments at pH = 7. Note that a 0.5 micron silica particle coated with PAH at pH ~ 2.5 changed zeta potential from +40 mV to approximately neutral at pH ≥ 11 similar to the neutral charge of PVA. When PAH assembles with Nafion® segments at pH ≥ 11 the zeta surface charge changes to -52 mV.

Since both types of gels are hydrogels the PVA/Nafion® FHB hydrogel is 72% water determined by monitoring the change in weight at 28 to 33% humidity. When sealed these FHB hydrogels do not lose their water dispersion, however, when exposed to dry conditions a PVA/Nafion® matrix loses its water dispersion at a rate of 0.15g/hr per 1.0 g of hydrogel at ~30% humidity. The PVA/Nafion® hydrogel did not degrade when treated with 0.4 M HCl, pH = 12 NaOH or NH_3 solution, and ammonical metallic complexes such as copper, Figure 3, and cobalt. For methanol fuel cell applications the PAH/Nafion® gel did not degrade in pure methanol and ethanol solutions but the PVA/Nafion® gel was not stable in both fuel solvents.

It was mentioned earlier that both the PVA/Nafion® and PAH/Nafion® FHB hydrogels behave as strong sulfonate cation exchangers. However, only PVA/Nafion® gels strongly retained multivalent metal ions such as Cu^{2+}, $[Cu(NH_3)_4]^{2+}$, Co^{2+}, $[Co(NH_3)_4]^{2+}$, and Cr^{3+} by its fluoro-sulfonated matrix with no leaching of ions when stored indefinitely under water, Figure 3. Similar experiment with a PAH/Nafion® gel strongly sequestered the $[Cu(NH_3)_4]^{2+}$ ions that slowly transformed from deep blue to deep brown color. X-ray analysis confirmed strong presence of copper but yet to identify metallic copper or copper salt clusters. Monovalent salts, acids, and bases such as NaCl, HCl, and NaN_3 partition into these gels at the solution-gel interface. However, these salts leached out when the hydrogel is soaked in water. For example, the rate of leaching of HCl acid out of 1.0 g of PVA/Nafion® slice is ~26×10^{-7} $molL^{-1}s^{-1}$ determined by pH and conductivity vs. time measurements. Evidence for the partitioning of ions and entrapment inside the hydrogel matrix was obtained by ATR-FTIR peak area analysis, Figure 5. Concentration of the azide $N_3^-{}_{(aq)}$ ion at 2104 to 1913 cm^{-1} inside the gel was determined using the peaks at 1650 cm^{-1} and 985 cm^{-1} as references to account for refractive index changes. Entrapment means the said gels are washed and not stored in pure water before any testing. The cation exchange nature of the FHB gels would repel any anion if the FHB hydrogel is stored in pure water. After storing FHB PVA/Nafion® and PAH/Nafion® hydrogels in ferricyanide solution for 24 hours followed by soaking in pure water for 5 hours both FHB hydrogels turned from light yellow to colorless. Overnight soak of 0.06 g slice of FHB PVA/Nafion® in 0.1 M NaN_3 solution to equilibrate the $N_3^-{}_{(aq)}$ between azide solution and the gel matrix yielded a partition coefficient of $K_{(N3/PVA\text{-}Nafion)}$ = 0.16, Figure 5, with a similar result for a PAH/Nafion® sample. In general, PVA/Nafion® hydrogel acted as a sponge to neutral molecules such as fluorescein while the PAH/Nafion® matrix was reluctant to permeation.

For example, dichloro-fluoresceine strongly partitions into a FHB PVA/Nafion® matrix imparting a strong yellow fluorescent color while it hardly permeates inside a PAH/Nafion® matrix at pH = 6.5 where the gel remained nearly colorless, Figure 3 (right). Strong ion exchange and selective partitioning of ions and molecules between the FHB hydrogel and the solution interface should have potential applications in electro analytical and opto-analytical chemistry. That is, these FHB gels can be tailored to have high ionic conduction and dielectric permitivity.

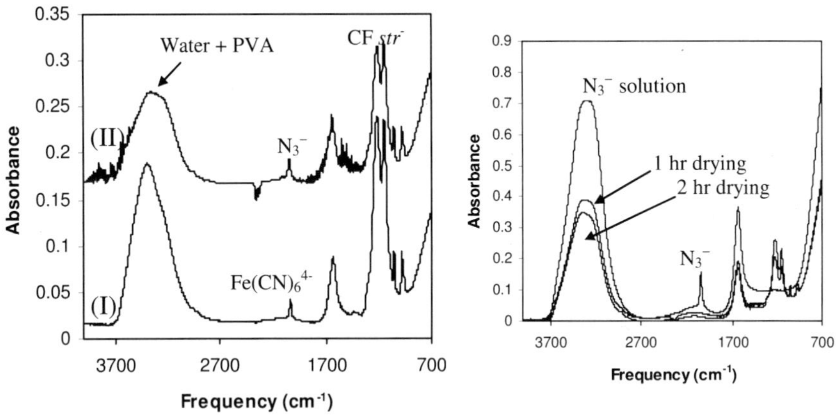

Figure 5. Ion and molecular partitioning between poly(vinyl alcohol) PVA/Nafion® FHB hydrogels and salt solutions, (left) ATR-FTIR spectra of PVA/Nafion® FHB hydrogel showing anion uptake of (I) ferrocyanide ion $Fe(CN)_6^{4-}$ at 2038 cm^{-1} and (II) the azide ion N_3^- at 2038 cm^{-1}, (right) ATR-FTIR spectra of 0.1M azide N_3^- solution, blotted PVA/Nafion® FHB hydrogel after 1 hour drying and 2 hours drying where the water peak at 3500 cm^{-1} decreases in height with time. Note: The broad OH str of water dispersion masks the OH str of PVA.

FHB Gel Conductance and Faradaic Properties

Ability of ion uptake would allow these gels to be used as electrolyte membranes, micron gel films, or layer-by-layer nano thin films in power and detections systems. The FHB matrix is easily dosed with strong acids to maintain high ionic conductance for battery and fuel cell technology. The FHB matrix strongly retains salt cations for devices based on ion-exchange technology. Testing for ion conductance and faradaic currents were performed using a Gamry Ref-600 potentiostat/AC impedance analyzer. After soaking the gel in an electrolyte solution overnight approximately 0.06 to 0.07 g of the gel was blotted and dried such that its water dispersion medium decreased from 70-75% to 45-50%. The gel sample is placed in a solid-state conductivity (SSC) cell. The electrodes of the cell were titanium with Teflon spacers. The sample gel is molded within the 0.3 cm^2/1.0 mm Teflon spacers. We will denote the RC series connection in all experiments as R_tSSC cell. In general the shape of the Bode plots (i.e. impedance /Z/ vs. frequency (ω)) of these gel-containing R_tSSC cells reflects an RC series circuit, Figure 6 (left).

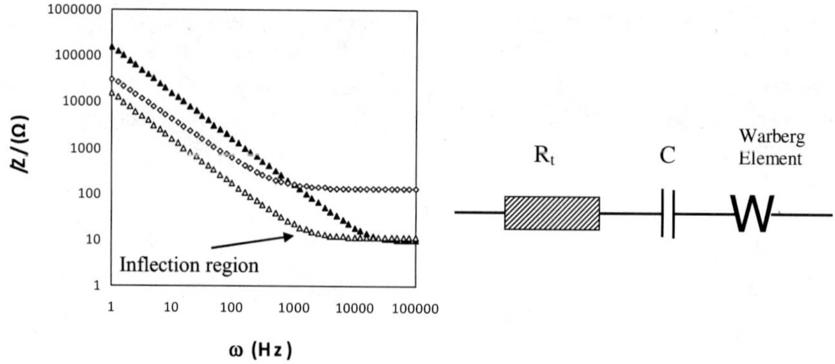

Figure 6. (left) Bode plots of RC series circuits comprising a 10 Ω resistor and 1μF (▲), 10 μF (△) electrolytic capacitors, and for the R_tSSC cell (◇) using PAH/Nafion gel prepared at pH = 11 of R_t = 1240 Ωcm^{-1} and C = 25 μFcm^{-2}; (right) A schematic of RCW series circuit used in finding the values of R_t, C, and W.

Any Bode plot that indicated a Randle cell meant a wet SSC cell and the experiment was discarded. That is, the SSC cell should behave similar to a commercial electrolytic capacitor with no charge transfer resistance. The Bode plots of RC series circuits comprising a 10 Ω resistor and 1μF and 10 μF electrolytic capacitors is shown in Figure 6 (left). The Bode plot of an R_tSSC cell using PAH/Nafion® gel prepared at pH = 11 yielded R_t = 1240 Ωcm^{-1} and C = 25 μFcm^{-2} similar in shape to a RC series circuit, Figure 6 (left).The main difference in the Bode plots is the inflection region where changes in the Warberg Element would apply the best fit. The inflection point shifts to lower frequencies as the capacitance increases and the curve shifts to higher impedance if R_t increases, Figure 6 (left). The best theoretical solution that fits the Bode plot of impedance /Z/ vs. frequency (ω) profile is a simulated $(RCW)^{\varphi}$ series circuit where W accounts for Warberg coefficient τ and φ is the constant phase element, Figure 6 (right). Theoretical fits were either computer generated using the Simplex method for a RCW software circuit or calculated using the impedance relation such that (*14*):

$$Z = R_t + \frac{\tau}{\sqrt{\omega}} - \left(\frac{1}{\omega C} + \frac{\tau}{\sqrt{\omega}}\right)^{\varphi} \cos\frac{\pi}{2}\varphi - j\left(\frac{1}{\omega C} + \frac{\tau}{\sqrt{\omega}}\right)^{\varphi} \sin\frac{\pi}{2}\varphi$$

$$\tau = \frac{RT}{F^2 A}\left(\frac{1}{c\sqrt{2D}}\right)$$

C is the gel capacitance, R_t is the total resistance, φ constant phase element, τ Warberg coefficient that accounts for ion mobility, *c* concentration of ions in gel, *D* effective diffusion coefficient of ions in gel, *A* area of electrode, and RT/F^2 = 2.66×10^{-7}.

Infra Red active salts of the azides N_3^- and cyanides $Fe(CN)_6^{3-}$ demonstrated that the FHB gels could partition salt ions into their matrix, Figure 5 (left). However, our objective is to check if these gels act as effective ionic conductors. The word gel does not necessary mean high ionic conductivity (σ). The ionic conductivity of PE complex gels that self assemble by electrostatic forces is extremely low of $\sigma < 1x10^{-7}$ Scm^{-1}. The reason is that salt ions in electrostatic assembled gels are tightly locked in a reluctant ion-pair effect with water clusters confined to the exchanger sites (*15*, *16*). Such locking lowers the ionic mobilities or diffusion coefficients of ions. In general, the FHB gels are characterized by high ionic conductivity that approaches or exceeds many of the commercial ion conducting membranes of $\sigma \sim 0.1$ Scm^{-1}. In this work conductivity analysis is confined to changes in electrolyte type, concentration or pH, polymer type, and water retention. Starting with the electrolyte type, gels that were soaked in different electrolytes gave different conductivities. For example, a PVA/Nafion® gel in solid-state conductivity (SSC) cell that used (a) 0.1M HCl gave $\sigma \sim 0.17$ Scm^{-1}, (b) 0.1M NaN_3 gave $\sigma \sim 0.025$ Scm^{-1}, (c) pH = 11 NaOH gave $\sigma \sim 0.13$ Scm^{-1}, and (d) 0.1M $CuSO_4$ gave $\sigma \sim 0.015$ Scm^{-1}. For acidic solutions as the pH is increased towards pH = 7 the ionic conductivity changed from $\sigma \sim 0.17$ Scm^{-1} to $\sigma \sim 0.037$ Scm^{-1}. Minimum water retention of $\geq 10\%$ is needed to maintain constant conductance. Since water is the dispersion medium then it would affect ionic conductance because conductivity waves would traverse through the dispersion medium. In all experiments ionic conductance was maintained constant for at least 5 hours however the home made SSC cell is not a tightly sealed cell and during overnight storage the gel was seen to collapse to a thin film. When the SSC cell was tightly sealed in a parafilm pack it maintained its original conductance for days. After multiple electrochemical runs an ATR-FTIR spectra, Figure 5 (left), of PVA/Nafion® gel treated with $Fe(CN)_6^{4-}$/NaCl electrolyte with a water content of ~ 15% but sealed in the SSC-Ti cell gave an ionic impedance of 9.5 Ω corresponding to ionic conductivity of $\sigma \sim 0.07$ Scm^{-1}. For polymer type, a PAH/Nafion® SSC cell yielded ionic conductivities lower than the PVA/Nafion® SSC cell in all comparable experiments. For example, a PAH/Nafion® SSC cell that used 0.1M NaN_3/pH = 11 gave $\sigma \sim 0.016$ Scm^{-1} while a pH = 11 NaOH SSC gave $\sigma \sim 0.003$ Scm^{-1}. At this stage of research it is difficult to tell whether a lower ionic conductance by the PAH/Nafion® matrix is attributed to lower ionic content, lower water content, or a particularity in the PAH segmental structure. In all conductivity experiments the capacitance values ranged from $C \sim 10$ μFcm^{-2} to $C \sim 40$ μFcm^{-2} while the Warberg instrumental factor (not τ) averaged at 0.5 to 1.5×10^{-3} using Gamry ac impedance software. Simulations that use the above impedance equation produce a better fit than instrumental software for values of $\varphi = 0.75$, $\tau \sim 4$, using $c = 10^{-4}$ molcm^{-3}, and $D = 2 \times 10^{-6}$ cm^{2}s^{-1}.

The previous section illustrated some ion uptake by the PVA/Nafion® matrix and a high ionic conductance where both factors qualify the gel for electrochemical experiments. Both ion uptake and high ionic conductance were confirmed by fabricating a miniature galvanic cell and an electrolytic cell. The galvanic cell comprised Cu/Zn electrodes and a 2 mm layer of $Cu^{2+}_{(g)}$ in contact with a 2 mm layer of $H^+_{(g)}$ gel electrolyte where both layers (i.e. $Cu^{2+}_{(g)}/H^+_{(g)}$) were sandwiched between the Cu/Zn electrodes having a surface area = 0.3 cm^2.

The open circuit potential of the $Cu^0_{(s)}/Cu^{2+}_{(g)}//H^+_{(g)}/Zn^0_{(s)}$ galvanic cell recorded 1.4 V. By applying voltage drop analysis that uses different resistors the average internal resistance $R_i \sim 205\ \Omega cm^{-1}$ corresponding to an ionic conductivity of $\sigma \sim 0.016\ Scm^{-1}$ (note: $R_i = R_t$). This is close to the ionic conductivity of $Cu^{2+}_{(g)}$ that appeared in the previous section (i.e. $\sigma \sim 0.015\ Scm^{-1}$). The faradaic currents in the galvanic cell deposited a metallic copper film at the gel/Cu interface. AC impedance analysis on the $Cu^0_{(s)}/Cu^{2+}_{(g)}//H^+_{(g)}/Zn^0_{(s)}$ galvanic cell measured an $R_t = 250\ \Omega cm^{-1}$ and relatively high capacitance of 30 $mFcm^{-2}$. At this juncture, these gels can be part of making materials of high dielectric constants (*17*). For example, a PAH/Nafion® matrix treated with $[Cu(NH_3)_4]^{2+}$ produced capacitance close to 1.3 $mFcm^{-2}$ using a symmetric Titanium cell. For an electrolytic cell that was not designed to take a reference electrode and uses a platinum working electrode we tested the electroactive $Fe(CN)_6^{3-}$ and $Fe(CN)_6^{4-}$ ions (30 mM) in 0.1M NaCl electrolyte. The cyclic voltammograms shown in Figure 7 are produced by the PVA/Nafion® gel soaked in either $Fe(CN)_6^{3-}$ or $Fe(CN)_6^{4-}$ electrolytes. The shape of the CVs is similar to their solution counterparts but tended to show an irreversible behavior within a potential window of -0.9 to +0.9V. The electrochemical potential of the same couple is displaced by ~0.3 V higher on the potential axis because no reference electrode was used. Evidence for the consumption of the electro active probes is verified by ATR-FTIR where a single $Fe(CN)_6^{4-}$ peak at 2050 cm^{-1} or $Fe(CN)_6^{3-}$ peak at 2120 cm^{-1} transforms to two peaks indicating a mixture of $Fe(CN)_6^{3-}$ or $Fe(CN)_6^{4-}$ as shown in Figure 7 (right) after multiple CV sweeps. Capability of the PVA/Nafion® gel to perform electrochemical experiments under a quasi-dry state would qualify the FHB gel for solid-state electrochemical detection systems.

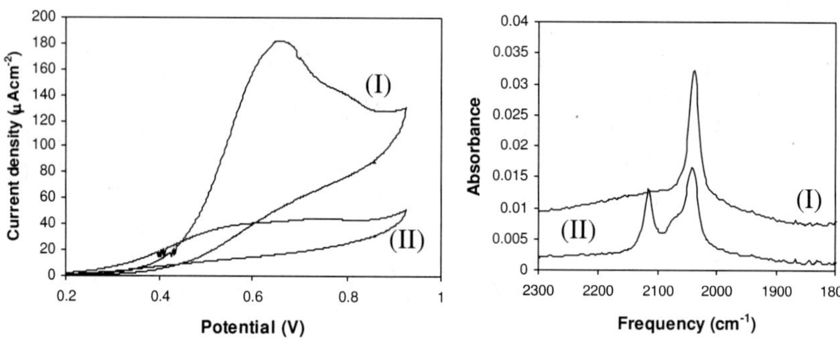

Figure 7. (left) Cyclic voltammograms (CV) produced by a PVA/Nafion® gel treated with (I) 30 mM $Fe(CN)_6^{3-}$/0.1M NaCl, (II) 30 mM $Fe(CN)_6^{4-}$/0.1M NaCl; (right) ATR-FTIR spectra of (I) $Fe(CN)_6^{4-}$ peak at 2050 cm^{-1} before starting multiple CV sweeps, (II) two peaks of a mixture of $Fe(CN)_6^{3-}$ at 2120 cm^{-1} and $Fe(CN)_6^{4-}$ after multiple CV sweeps; Area = 0.3 cm^2, T = 23 °C, v = 10 $mV\ s^{-1}$

To this end, author has demonstrated the mechanism of self-assembly of fluorinated hydrogen bonded (FHB) hydrogels and films. Various techniques were used to touch base on the various physical properties that characterize the FHB hydrogels. The ease of preparation, handling, stability to pH changes, and promising high ionic conductance and dielectric properties strongly points to applications in the electrochemical field. Although detailed physical analysis will be discussed in separate papers the author objective is to demonstrate that water-soluble fluoro-polymers (e.g. Nafion®, FluoroLink®) would couple to poly(alcohols) and poly(amines) (e.g. poly(vinyl alcohol), poly(ethylene imine)) by *fluorinated hydrogen bonding* to assemble *FHB electrolytes* of different physical properties. No chemical modifiers were used to manipulate the chemical and mechanical stability of these FHB films or hydrogels as well as their water content. The reason is to study the FHB matrix under the effect of FHB long-range interactions without using chemical cross linkers and water repellents. Ultimately, this study aims to introduce *FHB electrolytes* into power, detection, and energy storage solid-state electrochemical systems.

Conclusion

The newly discovered self-assembly by fluorinated hydrogen bonds (FHB) that occurs between fluorinated polyelectrolytes (PE) such as Perfluorosulfonic acid PTFE copolymer (i.e. Nafion®) and Poly(vinyl alcohol) (PVA) or Poly(allyl amine) (PAH) is a precursor to FHB hydrogels. The FHB hydrogels can be extra soft to rubbery depending on the centrifugal rate during preparation and chemical treatment. In a dry medium the FHB hydrogels emit their water in few hours but in sealed or humid environment dispersion water is retained and the gels act as strong cation exchangers for multivalent cations. Ions and molecules partition between the gel/solution interface (e.g. $K_{(N3/PVA\text{-}Nafion)} = 0.16$) affecting their hardness and ionic conductivity. In a PVA/Nafion® hydrogel high ionic conductance is maintained between 72% to 15% water content. PVA/Nafion® gel did not degrade when treated with 0.4 M HCl, pH = 12 NaOH or NH_3 solution, and even when exposed to ammonical metallic complexes. The PAH/Nafion® gel did not degrade in pure methanol and ethanol solutions but the PVA/Nafion® gel was not stable in both solvents. When treated with electrolytes both gels are high ionic conductors and behave similar to electrolytic capacitors. A $(RCW)^{\varphi}$ series circuit was used to describe the performance of a conducting solid-state cell comprised of these gels. Depending on the electrolyte used capacitance ranged from a low of 2 to 12 μFcm^{-2} up to relatively high 1.3 $mFcm^{-2}$ while ionic conductivity ranged from 0.05 Scm^{-1} to 0.15 Scm^{-1}. These gels can effectively replace liquid electrolyte solutions for the purpose of making miniature solid-state electrochemical systems.

References

1. Peppas, N. A.; Merrill, E. W. *J. Appl. Polym. Sci.* **1977**, *21*, 1763–1770.
2. Stauffer, S. R.; Peppas, N. A. *Polymer* **1992**, *33*, 39392.

3. DeLuca, N. W.; Elabd, Y. A. *J. Membr. Sci.* **2006**, *282* (1, 2), 217–224.
4. Shao, Z.-G.; Wang, X.; Hsing, I.-M. *J. Membr. Sci.* **2002**, *210* (1), 147–153.
5. Abebe, D. G.; Farhat, T. R. *Soft Matter* **2010**, *6* (6), 1325.
6. Barbarich, T. J.; Rithner, C. D.; Miller, S. C.; Anderson, O. P.; Strauss, S. H. *J. Am. Chem. Soc.* **1999**, *121* (17), 4280–4281.
7. Howard, J. A. K.; Hoy, V. J.; O'Hagan, D.; Smith, G. T. *Tetrahedron* **1996**, *52* (38), 12613–12622.
8. Anac, I.; Hoven, V. P.; McCarthy, T. J. *Langmuir* **2008**, *24*, 5740–5745.
9. Coupe, B.; Chen, W. *Macromolecules* **2001**, *34*, 1533.
10. Decher, G.; Schlenoff, J. B. *Multilayer Thin Films*; Wiley-VCH: 2003.
11. Oosawa, F. *Polyelectrolytes*; Marcel Dekker, Inc.: New York, 1971; Ch. 4, pp 41−49.
12. Bai, L. *Macromol.: Indian J.* **2006**, *2* (1), 40–43.
13. Daiko, Y.; Katagiri, K.; Matsuda, A. *Chem. Mater.* **2008**, *20* (20), 6405–6409.
14. Bard, A. J.; Faulkner, L. R. *Electrochemical Methods Fundamentals and Applications*, 2nd ed.; J. Wiley & Sons, Inc.: 2001.
15. Farhat, T. R.; Schlenoff, J. B. *Langmuir* **2001**, *17*, 1184–1192.
16. Michaels, A. S. *Ind. Eng. Chem.* **1965**, *57*, 32–40.
17. Choudhury, N. A.; Sampath, S.; Shukla, A. K. *Energy Environ. Sci.* **2009**, *2* (1), 55–67.

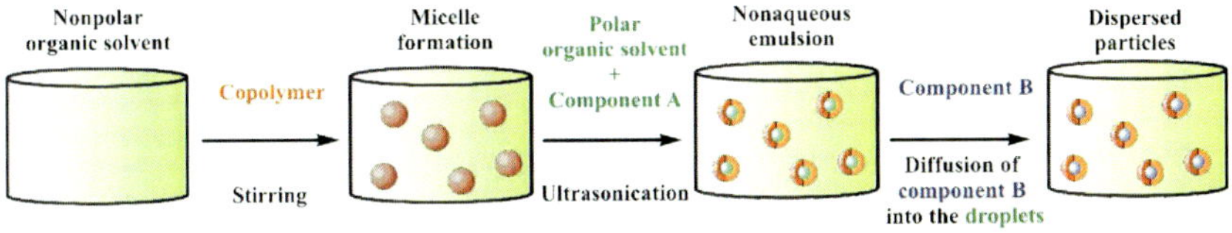

Figure 9. Schematic description of the polymerization in nonaqueous emulsion.

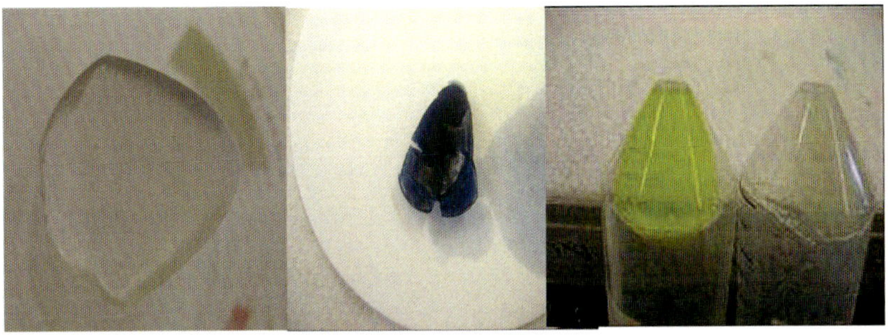

Figure 3. A typical PVA/Nafion® FHB hydrogel lump isolated from its liquor, (left) as made semi-transparent, (middle) soaked in 0.1M $[Cu(NH_3)_4]^{2+}$ solution, (right) strongly retained the fluorescent yellow-green dichlorofluoresciene molecule (yellow) while the PAH/Nafion® FHB hydrogel (colorless) did not.

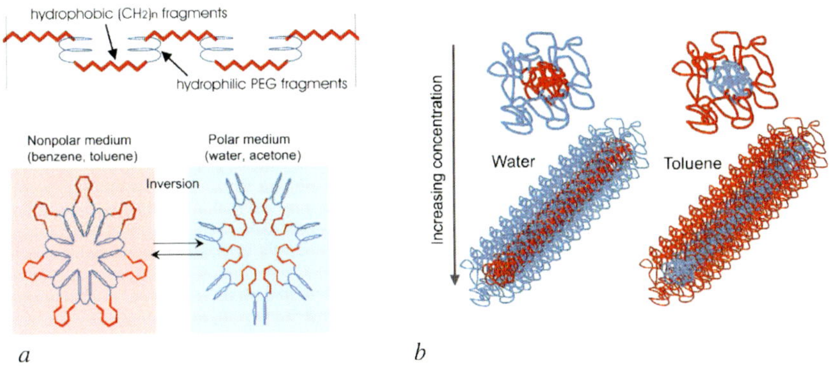

Figure 1. AIPs structure and scheme of AIPs micelle in polar and non-polar media (a). AIPs self-assembly by increasing concentration in aqueous and organic solutions (b).

Chapter 8

Structure Formation of Adsorption Layers of Ionic-Amphiphilic Copolymers on Inorganic and Organic Pigment Surfaces As Studied by ESA

Claus D. Eisenbach,[1,*] Nikolay Bulychev,[1,2] Klaus Dirnberger,[1] Bart Dervaux,[3] Filip E. DuPrez,[3] and Vitali Zubov[4]

[1]Institute for Polymer Chemistry, Universitaet Stuttgart, Pfaffenwaldring 55, D-70569 Stuttgart, Germany
[2]N.S. Kurnakov Institute of General and Inorganic Chemistry of Russian Academy of Sciences, 119991, Leninsky avenue, 31, Moscow, Russia
[3]Department of Organic Chemistry, Polymer Chemistry Research Group, Ghent University, Krijgslaan 281 S4, B-9000 Ghent, Belgium
[4]Lomonosov Moscow State Academy of Fine Chemical Technology, pr. Vernadskogo, 86, 117571, Moscow, Russia
[*]Corresponding author: Prof. Dr. Claus D. Eisenbach, tel.: +49 711 68564441, fax.: +49 711 68564396, *E-mail*: cde@ipoc.uni-stuttgart.de

The solid-liquid interface behaviour and the structure of adsorption layers of ionic-amphiphilic acrylic acid (AA)/ isobornyl acrylate (iBA) based copolymers on hydrophilic titanium dioxide (TiO_2) and hydrophobic copper phthalocyanine (CuPc) has been studied by the electrokinetic sonic amplitude (ESA) method. It was shown that the polymer gel layer theory can be applied to polyelectrolytes, giving detailed information not only about the polymer-particle interaction but also the thickness and structure of the coating layer around the particles.

CuPc was found to be covered by a relatively thin copolymer layer irrespective of the copolymer architecture, i.e., for both PAA-b-PiBA block copolymers with sharp block boundary and PAA-b-P(AA-co-iBA) block-like copolymers with tapered block-transition and isolated AA units in the hydrophobic block. In terms of the gel layer theory, PiBA blocks form the inner dense layer, and the PAA blocks represent

© 2011 American Chemical Society

the outer, permeable region of the adsorbed layer. Contrarily, different adsorption mechanisms of block and block-like copolymers were observed for the TiO_2. In case of block copolymers, a multilayer coating consisting of PAA anchor blocks, an interphases layer of PiBA blocks, and an outer PAA shell is obtained, whereas a solloid monolayer is formed for block-like copolymers.

Introduction

Aqueous colloidal dispersions can be effectively stabilized by the addition of amphiphilic polymers (*1–4*). In this context, polyelectrolyte-type copolymers are of particular interest because of their electrosteric stabilization properties (*5–7*). For ionic-hydrophobic poly(styrene)-poly(acrylic acid) block and graft copolymers, specific polymer – titanium dioxide (TiO_2) pigment surface interactions in combination with self-assembly processes result in a special pigment coating characterized by a polyelectrolyte shell (*7*). With the example of a series of well-defined amphiphilic block and block like-copolymers of isobornyl acrylate (iBA) and acrylic acid (AA), it was shown (*8*) that the interaction and layer formation of polyelectrolyte copolymers on, e.g., TiO_2 (hydrophilic) and copper phtalocyanine CuPc (hydrophobic) particles dispersed in aqueous media strongly depend on the composition and structure of the copolymers.

As to the adsorption of poly(isobornyl acrylate)-*block*-poly(acrylic acid) block copolymers $PiBA_x$-*b*-PAA_y and poly(isobornyl acrylate)-block-poly(isobornyl acrylate-co-acrylic acid) block-like copolymers $PiBA_x$-*b*-($PiBA_y$-*co*-PAA_z) on TiO_2 or CuPc particles, in-situ analysis of the adsorption process by applying the electrokinetic sonic amplitude (ESA) method under constant low frequency (cf. (*9*)) allowed to establish models of the interaction of the copolymers with the pigment surface, and to determine the thickness of the resulting polymer adsorption layer as a whole (*8*). However, insights into the dimensions of the inner structure of the adsorbed polymer layer as related to the copolymer architecture were not accessible in these experiments.

Detailed information about the inner structure and the thickness of the polymer adsorption layer can only be obtained by means of ESA measurements under variation of the applied alternating electrical field as demonstrated for non-ionic polymers (*10*, *11*). Here we report on the investigation of the adsorption of ionic amphiphilic block and block-like copolymers $PiBA_x$-*b*-PAA_y and $PiBA_x$-*b*-($PiBA_y$-*co*-PAA_z), respectively, onto TiO_2 and CuPc particles in aqueous dispersion by applying frequency dependant ESA measurements (cf. (*12*)).

Experimental

Materials and Measurements

Titanium dioxide TiO_2 rutile pigment Kronos 2310 with a particle size of 0.3 µm and breadth of particle size distribution of 60 nm, and copper phthalocyanine

(CuPc, BASF AG, Ludwigshafen) with a primary particle size of 0.1 μm and a particle size distribution breadth of 15 nm were employed as received.

Block copolymers $PiBA_x$-*b*-PAA_y were synthesized via the macroinitiator strategy whereas block-like copolymers $PiBA_x$-*b*-($PiBA_y$-*co*-PAA_z) were prepared via a sequential monomer addition (Figure 1). More details about the synthesis and advantages of either route and the characterization of these tailored copolymers can be found in the literature (*13*).

Electrokinetic sonic amplitude (ESA) measurements were carried out with an Acoustosizer 2 Instrument (Colloidal Dynamics, Sydney, Australia). 1 wt.-% aqueous dispersions of TiO_2 of CuPc were employed in the copolymer adsorption studies. Frequency dependent ESA measurements (frequency range 1.0 - 18 MHz were carried out. In these measurements, the saturation concentration (SC (*7*)) which had been determined in separate experiments at fixed low frequency (1 MHz) was chosen as concentration of the added copolymers. At SC, all the added polymer is adsorbed to the pigment surface, i.e., practically none is remaining in solution.

For the evaluation of the ESA data according to the polymer gel layer theory (*10*) (cf. (*14*)), the following theoretical parameters were used: dynamic viscosity $\eta = 0.95$ N·s·m^{-2} (viscosity of water at 22 °C), Debye-Hueckel parameter $\kappa = 0.2$ nm^{-1}, drag coefficient $\alpha = 0.02$ N·s·m^{-2}·nm^{-2}, and relaxation frequency $\omega_0 = 0.85$ s^{-1}. For details of the experimental protocol and data analysis it is referred to earlier publications (*8*, *11*).

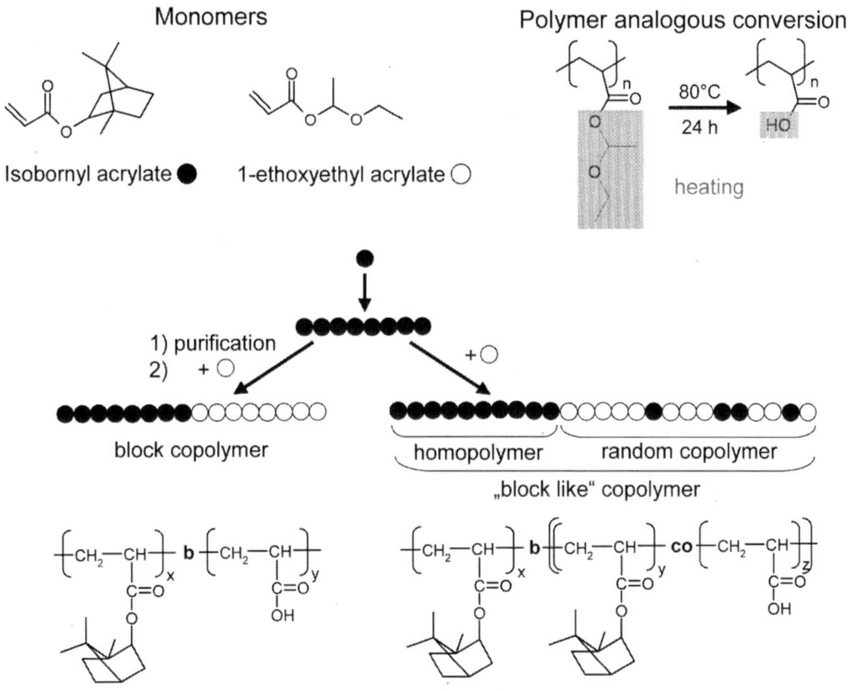

Figure 1. Scheme of the synthesis (13) of isobornyl acrylate (iBA) and acrylic acid (AA) block and block-like copolymers, and copolymer structures.

Results and Discussion

Methodology

Electrokinetic sonic amplitude (ESA) measurements have been demonstrated to be a powerful method for providing information about the process of polymer adsorption onto particles dispersed in aqueous media (*9, 10*). In the ESA technique, an alternating electrical field of frequency ω is applied across a colloidal system. The alternating field exerts an electrical force on the charged colloidal particles which causes them to move back and forth with a sinusoidal velocity. It is this backwards and forwards motion that generates pressure waves which propagate as sound waves and are recorded as ESA signal. The ESA signal is directly related to the dynamic mobility μ_D, i.e. electrophoretic mobility of a particle in the alternating field (cf. (*10*)).

The suitability of the ESA measurement to detect changes in the surface charge density of particles upon addition of polyelectrolytes which interact with the particle surface follows from the considerations of the limiting conditions from a theoretical point of view. This is shortly addressed below.

For a dilute suspension (particle volume fraction $\varphi<0.02$), it has been shown (*14*) that the ESA signal is given by eq 1:

$$\mathrm{ESA}(\omega) = A(\omega)\cdot\varphi\cdot\frac{\Delta\rho}{\rho}\cdot\langle\mu_D\rangle \tag{1}$$

Here ω is the angular frequency of the applied electric field, $A(\omega)$ is an apparatus constant, ρ is the density of the liquid, $\Delta\rho$ the density difference between the particles and the liquid, and μ_D is the particle-averaged dynamic mobility (*15*). In the case of spherical particles with a thin double layer, μ_D is related to the zeta potential ζ by the following formula eq 2 (*14*):

$$\mu_D = \frac{2\cdot\varepsilon\cdot\zeta}{3\cdot\eta}\cdot G\left(\frac{\omega\cdot a^2}{\nu}\right)\cdot\left[1+f(\lambda,\omega')\right] \tag{2}$$

Here a is the particle radius, η is the viscosity of the liquid, ε is the permittivity (dielectric constant) of the liquid, and ν ($=\eta/\rho$) is the kinematic viscosity.

The function $1+f(\lambda,\omega')$ in eq 2 is proportional to the tangential component of the electric field at the particle surface (*16*). For spherical particles with thin double layer, $f(\lambda,\omega)$ is a correction term to account for the enhanced surface conductance of the particle due to the double layer at the particle surface in relation to the conductivity of the aqueous phase (K^∞). The function $f(\lambda,\omega')$ with $\omega' = \omega\varepsilon/K^\infty$ is defined by eq 3a (*10, 14*):

$$f(\lambda,\omega') = \frac{1 + i\omega' - (2\lambda + i\omega'(\varepsilon_p/\varepsilon))}{2(1 + i\omega') + (2\lambda + i\omega'(\varepsilon_p/\varepsilon))} \tag{3a}$$

The (dimensionless) quantity λ is the surface conductance parameter which measures the ratio of the particle's surface conductivity to that of the liquid electrolyte (*15*) as already addressed above. In the case of water-based suspensions, where the dielectric constant ε_p of the particle is much smaller than ε of the liquid, the ratio $\varepsilon_p/\varepsilon$ and λ are usually small, i.e. for many colloids eq 3a reduces to f = 0.5 (cf. (*16*)).

The $G(\alpha)$ factor ($\alpha=(\omega a^2/\nu)$) represents the effect of the inertia forces which reduces the magnitude of the dynamic mobility μ_D, and which increases the phase lag as the frequency ω increases. This factor is given by the formula eq 3 (*14*):

$$G(\alpha) = \frac{1+(1+i)\sqrt{\alpha/2}}{1+(1+i)\sqrt{\alpha/2}+i(\alpha/9)\cdot(3+2(\Delta\rho/\rho))} \tag{3b}$$

The change of μ_D upon addition of a polymer to the pigment dispersion reflects the formation of thin polymer layers on the particle surface and allows calculating the zeta-potential as well as the particle size.

It is because of the inertia effect at higher frequencies that the particle size can be determined from the dynamic mobility by the ESA technique. At low frequencies (ω < 1 MHz) and small particle size (a < 1 μm), α ($=(\omega a^2/\nu)$) is much smaller than 1, and thus $G(\alpha)$ reduces to 1; this means that the inertia forces of the particle can be neglected. Furthermore, for most water based colloidal systems $f(\lambda,\omega)$ becomes ½, and eq 2 reduces to the well known Smoluchowski equation, eq 4 (cf. (*17*)):

$$\mu_D = \frac{\varepsilon \cdot \zeta}{\eta} \tag{4}$$

From eq 4 it is obvious that the dynamic mobility μ_D at low frequencies is only affected by the surface charge density of particles (as expressed by the zeta potential ζ). This means that the dynamic mobility μ_D as obtained by ESA measurement at 1.0 MHz is suited for the investigation of the particle/polymer interaction. At this low frequency the particle size and mass as well as the small mass increase due to adsorbed polymers (as compared to the mass of the naked particle) does not influence the dynamic mobility μ_D which is controlled by particle surface charge density.

For normalization and better elucidation of the pigment/polymer interactions, it is often advantageous to introduce the reduced dynamic mobility μ/μ_0 here μ stands for the mobility of particles with adsorbed polymer layer, and μ_0 represents the mobility of the non-treated particles (cf. (*9*, *10*)).

According to the gel layer theory (*10*, *14*), the frequency depending decrease of the magnitude of the dynamic mobility can be correlated with the thickness as well as the elasticity of the adsorbed layer of non-ionic polymers. In this theory it is assumed that the polymer adsorption layer consists of an inner dense layer of thickness Δ adsorbed on the particle surface, and an outer permeable region of thickness δ; the inner layer has a higher, density and its formation is a direct result of pigment-polymer interactions.

The procedure given in literature (*10*) for the calculation of the thicknesses of the inner and outer polymer adsorption layers after the polymer gel layer theory was applied. The formula for calculating the relative dynamic mobility μ/μ_0 is given in formula eq 5:

$$\frac{\mu}{\mu_0} = \ell^{-\kappa\Delta}\left\{-\frac{i\omega\alpha/\gamma-k^2}{k^2-\kappa^2}*\left[\ell^{-\kappa\delta}(1+\frac{\kappa}{k}\tanh k\delta)-\frac{\kappa^2}{k^2\cosh k\delta}\right]-\frac{i\omega\alpha}{\gamma k^2}\right\} \quad (5)$$

with

$$k^2=\frac{\alpha}{\eta}\left(1+\frac{i\omega\eta}{\gamma}\right)$$

and

$$\gamma=\omega_0\eta$$

Here Δ and δ is the thickness of the inner and outer adsorption layer, respectively, ω is the frequency of the applied electric field, η is the dynamic viscosity, κ is the Debye-Hueckel parameter, α is the drag coefficient, and ω_0 is the relaxation frequency.

Experimental Results

The employed $PiBA_x$-*b*-PAA_y block and $PiBA_x$-*b*-($PiBA_y$-*co*-PAA_z) block-like copolymers differ in the architecture of the hydrophilic block as illustrated in the schematics Figure 3. In case of $PiBA_x$-*b*-PAA_y, the hydrophilic PAA block consists of acrylic acid (AA) repeat units only, whereas the hydrophilic $PiBA_y$-*co*-$PAAz$ block of the block-like copolymer $PiBA_x$-*b*-($PiBA_y$-*co*-PAA_z) contains a relatively low mole fraction of the hydrophobic iBA comonomer in a random sequence.

As explained above, an observed change in the ESA signal upon addition of a (co)polymer directly reflects the altering of the surface charge density as related to the (co)polymer adsorption. In agreement with earlier results obtained with similarly structured iBA and AA based copolymers (*8*), the copolyelectrolytes employed in this work revealed characteristic adsorption features as reflected from the ESA measurements under constant low frequency. The corresponding dynamic mobility μ vs. polymer concentration curves for aqueous TiO_2 and CuPc dispersions are depicted in Figure 4.

For hydrophilic TiO_2 particles, $PiBA_{17}$-PAA_{72} block copolymer unimers first adsorb via the ionic carboxyl(ate) groups of PAA, resulting in a layer of more or less immobilized PAA chains in a train-like conformation, with islands of collapsed PiBA blocks on top (Figure 5a: low concentration regime; cf. (*8*)). This goes along with partial screening of the particle's surface charges: the ionic-hydrophilic character of the now polymer-coated particle is reduced, as reflected by the initial decrease of the ESA signal (dynamic mobility μ) upon addition of copolymer, finally reaching a minimum value (Figure 4, curve 1). This monolayer covered TiO_2 is subsequently stabilized by further adsorption of

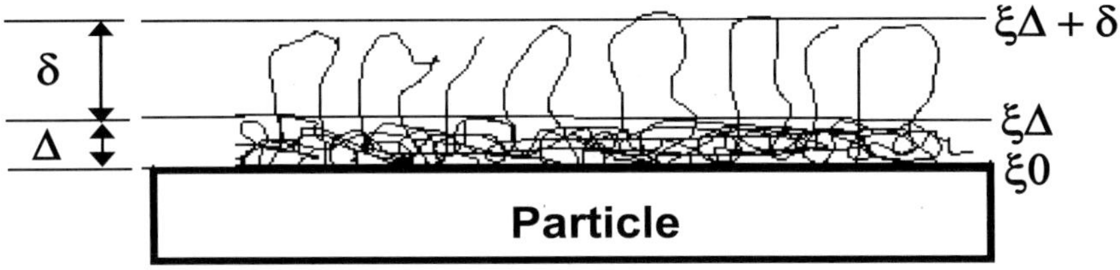

Figure 2. Schematic of the adsorbed layer of a polymer on the particle surface, and shear plane displacement caused by the polymer coating (after (10)).
***Δ** – thickness of the inner, dense adsorption layer.*
***δ** – thickness of the outer, more permeable layer.*
***ξ0** – zeta potential of the pristine surface.*
***ξΔ** – zeta potential at the border between the inner and outer layer.*
***ξΔ+δ** – zeta potential at the polymer layer surface.*

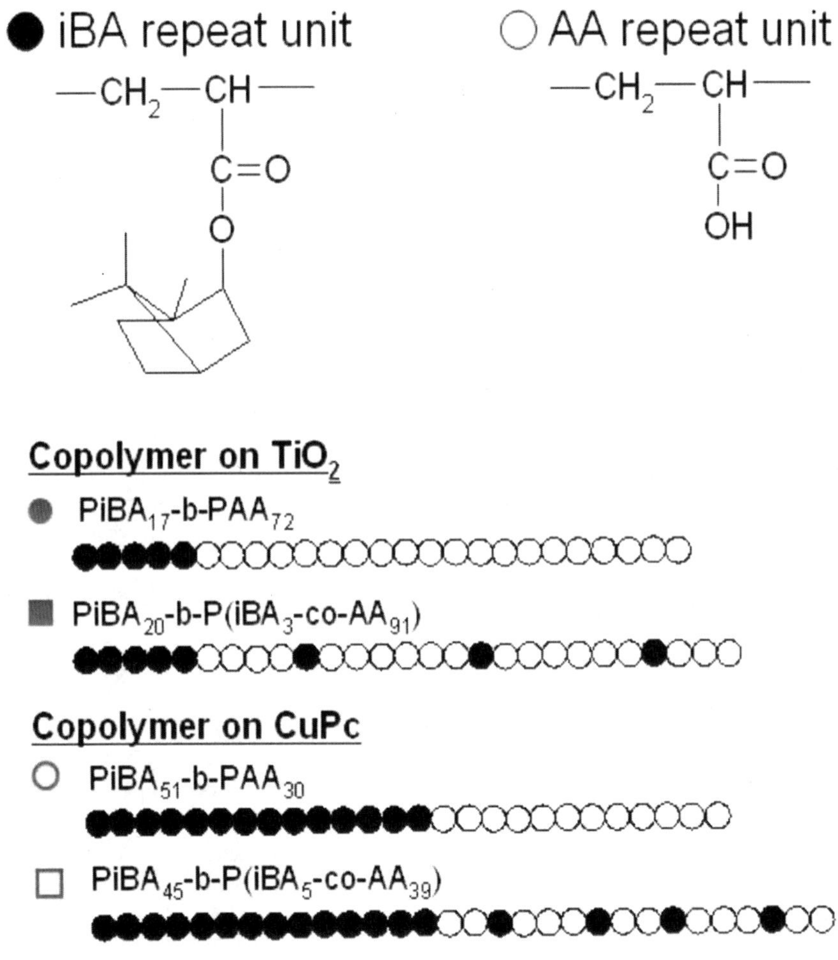

Figure 3. Schematic of the architecture of the copolymers employed in this study.

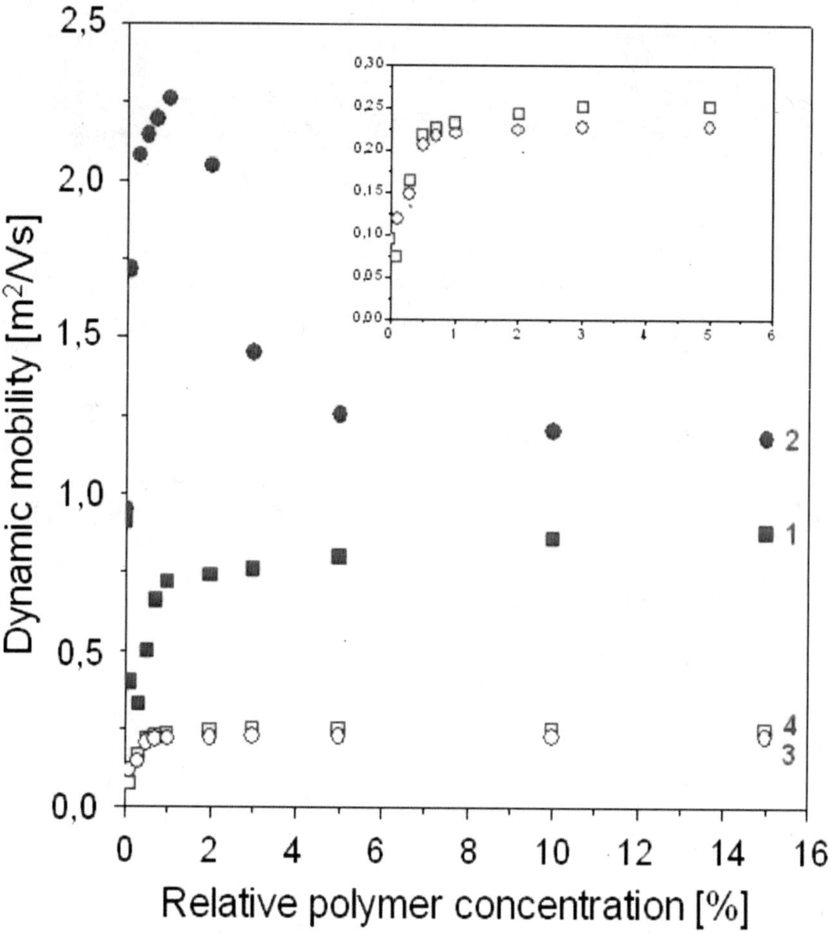

Figure 4. Dependence of the dynamic mobility μ on the relative (to pigment weight in the 1 wt.-% pigment aqueous dispersion) concentration of block and block-like copolymers $PiBA_x$-b-PAA_y and $PiBA_x$-b-($PiBA_y$-co-PAA_z), respectively. TiO_2 aqueous dispersion: block copolymer $PiBA_{17}$-b-PAA_{72} (●) and block-like copolymer $PiBA_{20}$-b-P(iBA_3-co-PAA_{91}) (■). Copper phthalocyanine (CuPc) aqueous dispersion: block copolymer $PiBA_{51}$-b-PAA_{30} (○) and block-like copolymer $PiBA_{45}$-b-P(iBA_5-co-PAA_{39}) (□)

copolymer via interactions between the existing surface-immobilized (collapsed) PiBA blocks and PiBA blocks of dissolved block copolymer (Figure 5a: intermediate concentration regime; cf. (*8*)). As a result, a thermodynamically more favourable ionic hydrophilic outer surface is obtained, as reflected by the recovery (increase) of the ESA signal after a minimum value has been passed (see Figure 4, curve 1). Schematically (see Figure 5), the sequential copolymer adsorption process leads to a 3-layer architecture normal to the pigment surface, i.e., a PAA anchoring layer, a PiBA interphase, and an outer layer of PAA polyelectrolyte chains reaching into the aqueous environment (cf. (*7*, *8*)).

Contrary to the TiO_2/block copolymer system, when employing the $PiBA_{20}$-*b*-($PiBA_3$-*co*-PAA_{91}) block-like copolymer, the ESA signal increases from the very beginning of the copolymer addition (Figure 4, curve 2). This distinct difference is related to the fact that block-like copolymer molecules form single molecule aggregates in aqueous solution consisting of a core of the collapsed PiBA block covered by loops and tails of PAA segments. The PAA loops and dangling ends are a result of the intramolecular interaction of isolated iBA units of the hydrophilic block with the hydrophobic PiBA core (Figure 5b). Such spheres are attached to the TiO_2 surface via a fraction of the carboxylate groups of the PAA segments of the shell of the micelle-like aggregate (Figure 5b: low concentration regime). With increasing polymer concentration, a so-called solloid (cf. (*2*)) layer is formed (Fig. 5b: intermediate concentration regime). This adsorption process does not cause any decrease in the TiO_2 surface charge density (screening of charges) as in the case of the initial adsorption of block copolymers (see above), but to the contrary, the zeta potential of the coated particle increases as compared to the bare particle.

As to the hydrophobic CuPc pigment, the initial adsorption of copolymer always proceeds via the hydrophobic PiBA block, presumably again in a train-like conformation, with the hydrophilic block reaching into to the aqueous phase (Figure 5a and 5b, bottom). In case of block copolymer adsorption ($PiBA_{51}$-PAA_{30}), the dispersed particle is stabilized by the dangling PAA block (Figure 5a), whereas in the case of the block-like copolymer $PiBA_{45}$-*b*-($PiBA_5$-*co*-PAA_{39}), primarily by loops of PAA segments that are attached to the adsorbed PiBA block by isolated iBA units (Figure 5b). The copolymer adsorption immediately results in a pigment hydrophilization. Consequently, and irrespective of the copolymer architecture, the ESA signal increases from the very beginning of the copolymer addition (Figure 4, curves 3 and 4).

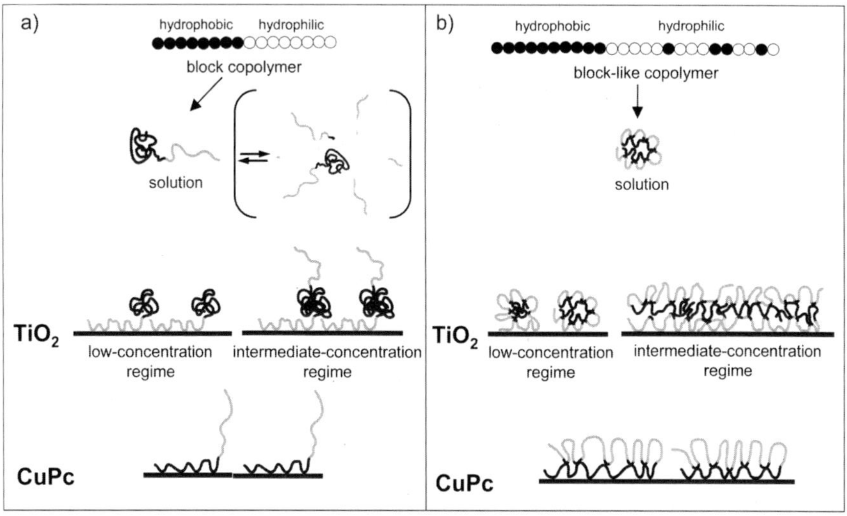

Figure 5. Schematic of the adsorption process of ionic-amphiphilic block and block-like copolymers onto TiO_2 or copper phthalocyanine (CuPc) pigment in aqueous dispersion.

Irrespective of the pigment type or copolymer, for all systems investigated, an upper limit of the dynamic mobility is observed upon polymer addition: a so-called polymer saturation concentration (SC) is defined beyond which only comparatively small, i.e., negligible changes of the dynamic mobility μ is observed (*7–9*). Further details and the way the observed phenomena can be explained by differences in the conformation of the adsorbed copolymers as related to the copolymer architecture have been already discussed elsewhere (*8*).

The amount of polymer adsorbed to the pigment surface, and thus the overall thickness of an adsorption layer can be calculated for collapsed coils by relating the SC values (as determined from the curves depicted in Figure 4) to the surface area of the uncoated pigment (cf. (*8*, *11*)). Here the SC is given by the concentration of added copolymer above which the dynamic mobility of the dispersed particle remains more or less constant, i.e. becomes independent of polymer concentration. The significant differences in the thickness of the adsorption layer on TiO_2 and CuPc (Table 1) is in full agreement with the previously proposed (and outlined above) adsorption mechanisms: monolayer formation with hydrophobic PiBA anchor blocks in case of CuPc, and multilayer formation for TiO_2 (*8*).

In order to further elucidate the structure of the polymer adsorption layer, ESA measurements under varying frequency of the applied alternating electrical field have been carried out. In principle, such data can be used to get detailed information about the structure of adsorption layers, as first shown and theoretically treated for the adsorption of poly(vinyl alcohol) to colloidal silica (*10*). This method was confirmed for the adsorption of cellulose based polymers to titanium dioxide and ferrous oxide (*11*). According to the polymer gel layer theory (*10*), it is assumed that the structure of the adsorbed polymer layer consists of an inner layer of polymer molecules adsorbed on the particle surface, and an outer, more permeable layer toward the solution (see Figure 2). The inner layer is characterized by a comparatively higher density, and its formation is a direct result of pigment surface - polymer interactions (*10*, *11*).

The procedure and formula (see eq 5) given in literature (*10*, *11*) for the calculation of the thicknesses of the inner (Δ) and outer (δ) polymer layer (see Figure 2) from frequency dependent measurements of the reduced dynamic mobility μ/μ_o was applied for block as well as block-like PiBA/PAA copolymers adsorbed on both TiO_2 and CuPc particles. According to the boundary conditions of the polymer gel layer theory (*10*), requiring that no free polymer is left in the solution after the process of polymer adsorption is completed, the ESA measurements were carried out for the SC of added copolymers. At SC, all the added polymer is adsorbed to the pigment surface, i.e., practically none is remaining in solution.

The changes of the reduced dynamic mobility (ratio μ/μ_o of the dynamic mobility of polymer-coated to uncoated particle) measured for the frequency range 1.0 - 18 MHz are given by the data points in Figure 6a and Figure 6b. The averaged thicknesses of the inner (Δ) and outer (δ) adsorption layers (see Figure 2) as obtained from the analysis of these experimental data according to literature protocols (*16*) are compiled in Table 1.

The experimental μ/μ_o – magnitudes are reasonably well described by the theoretical fit-curve (solid lines in Figures 6a and 6b) as calculated after the

Table 1. Saturation concentration (SC) of $PiBA_x$-*b*-PAA_y block or block-like $PiBA_x$-*b*-($PiBA_y$-*co*-PAA_z) copolymers adsorbed to the surface of TiO_2 or copper phthalocyanine (CuPc) particles, total thickness d of the copolymer adsorption layer, and thickness Δ of the inner as well as thickness δ of the outer adsorption layer of the copolymers

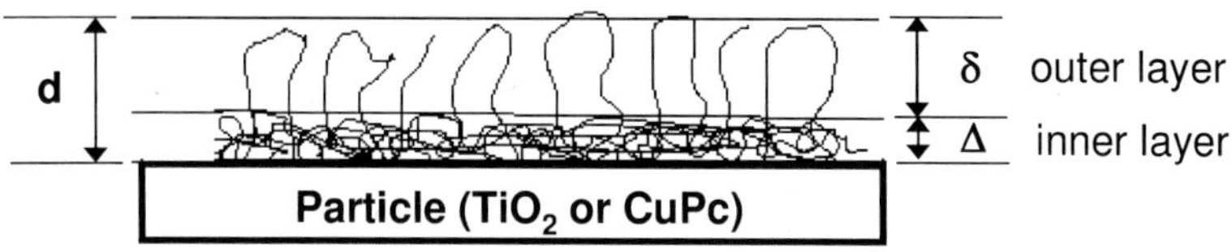

System	$SC^{a)}$ wt.-%	$d_{SC}^{b)}$ nm	$\Delta^{c)}$ nm	$\delta^{c)}$ nm
TiO_2 + $PiBA_{17}$-*b*-PAA_{72}	1,0	5,5	1,6	4,2
TiO_2 +$PiBA_{20}$-*b*-P(iBA_3-*co*-PAA_{91})	3,0	3,6	2,3	1,2
CuPc + $PiBA_{51}$-*b*-PAA_{30}	0,5	1	0,6	0,5
CuPc +$PiBA_{45}$-*b*-P(iBA_5-*co*-PAA_{39})	0,7	0,8	0,6	0,3

[a] Obtained from reduced dynamic mobility μ/μ_0 vs. polymer concentration curves Figure 4.
[b] Calculated on the basis of the saturation concentration SC (cf. (7)).
[c] Calculated from the frequency dependent ESA data (Figure 6a and 6b) according to the polymer gel layer theory (cf. (10, 11)).

polymer gel layer theory by using the averaged layer thickness data of the inner (Δ) and outer (δ) adsorption layers (given in Table 1). This means that the polymer gel layer theory that was originally developed for neutral, i.e., non-ionic polymers can be applied to the adsorption of polyelectrolyte copolymers on hydrophilic as well as hydrophobic surfaces.

The combined layer thicknesses Δ and δ (Table 1, columns 4 and 5), as individually obtained by the analysis of the frequency dependent reduced dynamic mobility data according to the polymer gel layer theory, is in good agreement with the thickness d of the whole adsorption layer (Table 1, column 3) which has been calculated from the SC as established independently from the concentration dependency of the ESA signals at constant, low frequency of the applied electrical field (see also curves in Figure 4). This is clear evidence for the conclusiveness of the complementary experimental approaches, i.e., frequency independent and frequency dependent ESA measurements.

Following the already discussed model of the block copolymer adsorption onto TiO_2 (Figure 5a), in case of $PiBA_x$-*b*-PAA_y block copolymer, the inner, dense adsorption layer as defined by the polymer gel layer theory consists of flatly adsorbed PAA blocks and overlaying islands of collapsed PiBA blocks. Because

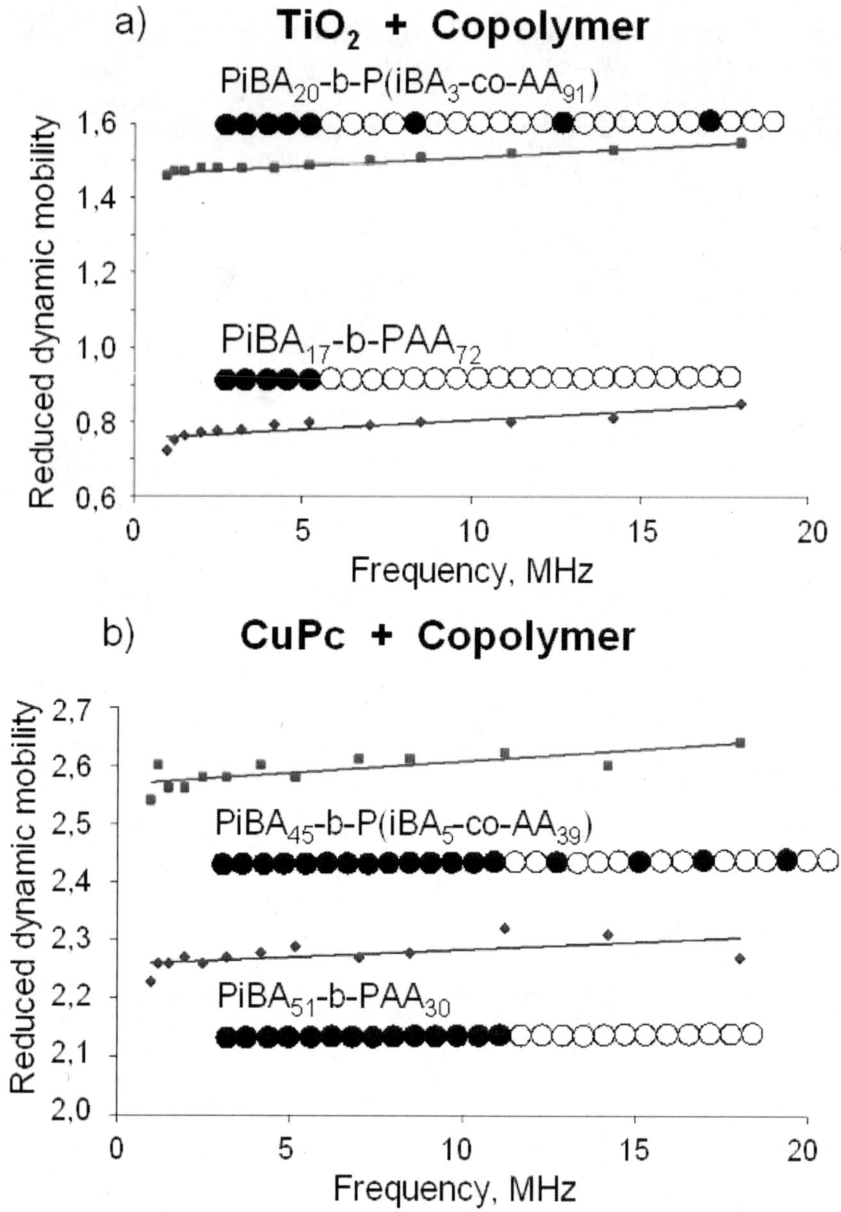

Figure 6. Dependence of the reduced dynamic mobility μ/μ_0 on the frequency of the applied alternating electric field for aqueous TiO_2 (a) and copper phthalocyanine (CuPc) dispersions stabilized by copolymers at saturation concentration (SC). Points: experimental data; solid line: calculated theoretical curve. TiO_2: block copolymer $PiBA_{17}$-b-PAA_{72}; block-like copolymer $PiBA_{20}$-b-P(iBA_3-co-PAA_{91}). CuPc: block copolymer $PiBA_{51}$-b-PAA_{30}; block-like copolymer $PiBA_{45}$-b-P(iBA_5-co-PAA_{39}).

of the differences in chain length between the long PAA_{72} and comparatively short $PiBA_{17}$ block, the PAA adsorption layer will not be completely covered by collapsed PiBA. Therefore the zeta potential $\xi\Delta$ at the border between the inner and outer layer (see Figure 2) will be still partially affected by PAA that is not covered by PiBA. The thickness Δ of this inner adsorption layer (1.6 nm) correlates with the molecular (chain cross-sectional) dimension of the PAA and PiBA chain. The thickness (4.2 nm) of the significantly thicker outer adsorption layer is related to the random flight coil dimensions of dangling PAA blocks as estimated for a PAA chain of about 70 repeat units assuming a characteristic ratio $C_n = 11.3$ (*18*). The corresponding profile of the block copolymer adsorbed onto TiO_2 pigment particles is sketched in Figure 7a.

Contrary to block copolymers, the adsorption model of block-like copolymers PiBA-*b*-P(iBA-*co*-AA) to TiO_2 is based on the adsorption of micelle-like aggregates consisting of the collapsed PiBA block as a core (see Figure 5b). The shell of such single molecule aggregates predominantly consist of loop-like PAA attached to the PiBA core by isolated iBA units of the P(iBA-*co*-AA) block. Considering the iBA mole content of the P(iBA_3-*co*-AA_{91}) block, and also the random incorporation of iBA as given by the copolymerization parameters (*9*), the PAA loops or tails comprise in average about 23 AA units. The aggregates adsorb to the TiO_2 via ionic interactions of the PAA loops and dangling ends. The corresponding profile of the block-like copolymer adsorbed onto TiO_2 pigment particles is sketched in Figure 7b.

The resulting solloid monolayer can be divided into 1) a relatively dense inner adsorption layer of thickness Δ consisting of immobilized PAA that is ionically bond to the TiO_2 surface, including the overlaying blobs of PiBA, and 2) an outer layer of PAA loops and tails of thickness δ. The thicker inner layer as compared to the case of block copolymer adsorption may be associated to both the comparatively large volume of the aggregate's core consisting of a collapsed PiBA block, and steric restrictions in the interaction of relatively short PAA loops of the P(iBA_3-*co*-AA_{91}) block of the block-like copolymer in comparison to the much longer PAA block of the $PiBA_{17}$-*b*-PAA_{72} block copolymer. Accordingly, the much thinner outer adsorption layer in case of the block-like copolymers is related to restrictions in the permeability of the comparatively shorter PAA loops (average of 23 AA repeat units) of the P(iBA-*co*-AA) block that are less mobile than the only single tethered, and much longer PAA block (72 AA repeat units) of PiBA-*b*-PAA. This view is supported by a rough estimate of the differences in the end-to-end distance between a PAA_{73} block and a PAA_{23} loop, using the same characteristic ratio as above.

As already elucidated above, the adsorption model of block and block-like copolymers onto the hydrophobic CuPc significantly differs from the adsorption models of both block and block-like copolymers onto TiO_2 (see Figure 5a and 5b, bottom). However, the adsorption of block and block-like copolymers onto CuPc is quite similar for both block and block-like copolymers in that no sequential polymer adsorption is observed, moreover, single molecule adsorption occurs.

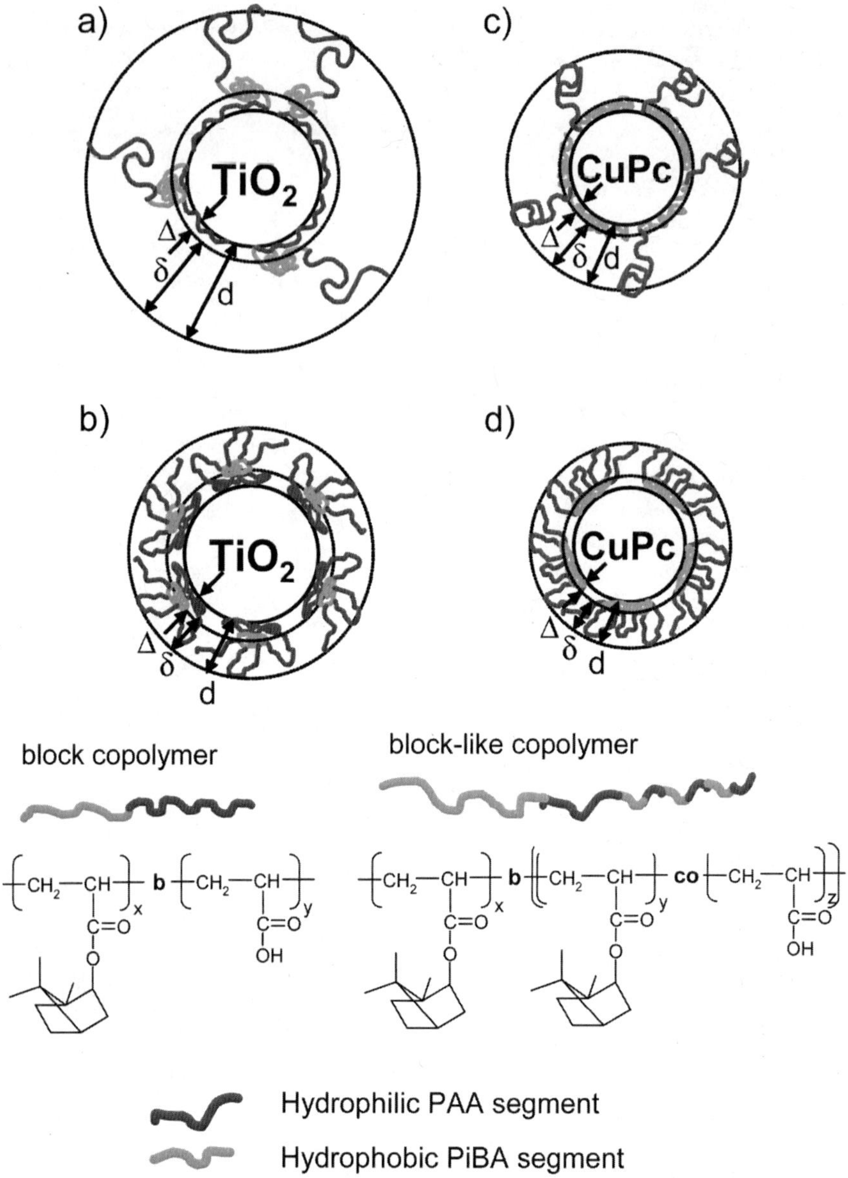

Figure 7. Schematic representation of the assumed structure of adsorption layers formed by isobornyl acrylate (iBA) / acrylic acid (AA) based block and block-like copolymers:
***A:** Adsorption model for block copolymers $PiBA_x$-b-PAA_y.*
***B:** Adsorption model for block-like copolymers $PiBA_x$-b-($PiBA_y$-co-PAA_z).*
***Δ:** thickness of inner adsorption layer; **δ:** thickness of outer adsorption layer;*
***d:** total thickness of the adsorption layer (see Table 1).*

The anchoring of the copolymers to the CuPc surface is given by the hydrophobic PiBA block in a presumably train-like fashion. This means that the adsorption process of PiBA chains to the CuPc surface involves a decoiling of the globular, collapsed conformation of the PiBA block of the copolymers in aqueous solution. The formation and structure of the adsorption layer according to the polymer gel layer theory is sketched in Figure 7c and 7d.

The inner, dense adsorption layer is comparatively thin and consists of flatly adsorbed PiBA blocks. This is in accordance with the determined thickness Δ (Table 1, column 4) which corresponds to the geometrical dimension of the isobornyl residue. The even thinner outer layer (Table 1, column 5: thickness δ), which consists of dangling PAA ends in case of block copolymers (see Figure 7c), and of PAA loops in case of block-like copolymers (see Figure 7d), is explained by the fact that the PiBA block is not only much longer but also larger in volume than the PAA or P(iBA-*co*-AA) block, respectively. As a consequence, only a fraction of the inner PiBA adsorption layer (see Figure 2: zeta potential $\xi\Delta$) is shielded by the mobile PAA, meaning that the shear plane (see Figure 2, zeta potential $\xi\Delta+\delta$ at the polymer layer surface), in average, is not that much shifted.

This explanation is in agreement with the differences in the thickness of the outer adsorption layer of the block and block-like copolymer coating on the CuPc particle: the permeable "outer layer" is expected to be thicker for the 30 AA repeat units long PAA block of the $PiBA_{51}$-*b*-PAA_{30} block copolymer as compared to the loops (in average 6 AA repeat units) of the P(iBA_5-*co*-AA_{39}) block of the block-like copolymer.

Conclusion

Electrokinetic sonic amplitude (ESA) studies of the adsorption of specially tailored amphipolar copolyelectrolytes on hydrophilic (TiO_2) or hydrophobic (copper phthalocyanine; CuPc) particles dispersed in aqueous media have given new insights into both the adsorption mechanism and the layer structure formation. It was established that the polymer gel layer theory can be applied to polyelectrolytes, and it was found that both the surface characteristics of the dispersed (pigment) particles and the architecture of the iBA/AA comonomer based copolymers distinctly affected not only the copolymer – particle interaction but also the thickness and structure of the coating layer formed around the particles.

The applicability of the polymer gel layer theory originally developed to describe the electroacoustic behaviour of an elastic gel layer of uncharged adsorbed polymer (*10*) to the analysis of ESA data obtained for the adsorption of charged copolyelectrolytes onto hydrophilic as well as hydrophobic particles is explained by the fact that the basic process of the adsorption of a neutral, non-ionic macromolecule (or non-ionic surfactant) and a polyelectrolyte onto a colloidal particle is similar. Irrespective of the molecular structure of the polymer – neutral or charged, adsorption to the particle's surface means bond formation and thus immobilization of part of the macromolecule. This always results in a distance dependence of the density and mobility of the adsorbed polymer normal to the particle's surface. Consequently and according to the gel layer theory, a comparatively dense inner adsorption layer can be discriminated from a more permeable, elastic outer layer.

As elucidated above, the data analysis of the frequency dependent ESA measurements of copolyelectrolyte adsorption on hydrophobic CuPc or hydrophilic TiO_2 allowed insights into the structure formation beyond the effects observed with neutral polymers (cf. (*10*)). This means that, if the boundary conditions of absence of free polymer after the process of polymer adsorption is completed are fulfilled, the polymer gel layer theory is a powerful tool to study polyelectrolyte – colloidal particle interactions on a molecular scale.

The electroacoustic data and their analysis on the basis of the gel layer theory revealed that particles with hydrophobic surface such as the CuPc pigment are covered for both the block and block-like copolymers by a relatively thin copolymer monolayer, with the hydrophobic PiBA blocks acting as anchor, and the PAA blocks as electrosteric stabilizing moieties. In terms of the polymer gel layer theory, the PiBA blocks form the inner dense layer, and the PAA blocks represent the outer, permeable region of the adsorbed layer.

For the copolymer adsorption on TiO_2, the data revealed not only distinctly different adsorption mechanisms but also different polymer layer structures. In case of block copolymers, the inorganic pigment is covered by a multilayer: PAA of a first adsorption layer of block copolymer act as anchor blocks, an interphase layer consisting of interpenetrating PiBA blocks of this first and a second copolymer absorption layer follows, and PAA polyelectrolyte blocks of the second adsorption layer form an outer shell layer. For block-like copolymers, a solloid monolayer is formed, with some of the PAA loops acting as anchor, and some as electrosteric stabilizing moieties. Here, according to the polymer gel layer theory, the PAA anchor block layer together with the PiBA core of the solloids correspond to the dense inner layer, and the remaining PAA loops and short dangling ends represent the permeable outer layer.

Acknowledgments

The authors would like to thank the Alexander von Humboldt Foundation for granting a fellowship within the re-invitation program for former Humboldt Research Fellows (V.Z.). The *Belgian Program on Interuniversity Attraction Poles* [Program P6/27] is acknowledged.

References

1. *Polymeric Stabilization of Colloidal dispersions*; Napper, D. H., Ed.; Academic Press: New York, 1983.
2. Somasundaran, P.; Krishnakumar, S. *Colloids Surf., A* **1997**, *123*, 491.
3. *Surfactants and Polymers in Aqueous Solution*, 2nd ed; Holmberg, K., Jöhnsson, B., Kronberg, B., Lindman, B., Eds.; J. Wiley & Sons: Chichester, 2003.
4. Cartridge, D. J. *Polym. Paint Color J.* **1994**, *184*, 10.

5. Kaczmarski, J. P.; Tang, M.-R.; Glass, J. E.; Buchacek, R. J. *Prog. Org. Coat.* **1997**, *30*, 15.
6. Hoogeveen, N. G.; Cohen Stuart, M.; Fleer, G. J. *J. Colloid Interface Sci.* **1996**, *182*, 133.
7. Schaller, C.; Schauer, T.; Dirnberger, K.; Eisenbach, C. D. *Eur. Phys. J. E* **2001**, *6*, 365.
8. Bulychev, N.; Van Camp, W.; Dervaux, B.; Kirilina, Y.; Dirnberger, K.; Schauer, T.; Zubov, V.; Du Prez, F. E.; Eisenbach, C. D. *Macromol. Chem. Phys.* **2009**, *210*, 287.
9. Eisenbach, C. D.; Schaller, C.; Schauer, T.; Dirnberger, K. In *Particle Sizing and Characterization*; Provder, T., Texter, J., Eds.; American Chemical Society: Washington, DC, 2004; p 215.
10. Carasso, M. L.; Rowlands, W. N.; O'Brien, R. W. *J. Colloid Interface Sci.* **1997**, *193*, 200.
11. Bulychev, N.; Dirnberger, K.; Arutunov, I. A.; Kopold, P.; Schauer, T.; Zubov, V.; Eisenbach, C. D. *Prog. Org. Coat.* **2008**, *62*, 299.
12. Bulychev, N.; Dervaux, B.; Dirnberger, K.; Zubov, V.; DuPrez, F. E.; Eisenbach, C. D. *Macromol. Chem. Phys.* **2010**, *211*, 971.
13. Dervaux, B.; Van Camp, W.; Van Renterghem, L.; DuPrez, F. E. *J. Polym. Sci., Part A: Polym. Chem.* **2008**, *46*, 1649.
14. O'Brien, R. W. *J. Fluid Mech.* **1988**, *190*, 71.
15. Hunter, R. J.; O'Brien, R. W. *Colloids Surf., A* **1997**, *126*, 123.
16. O'Brien, R. W.; Cannon, D. W.; Rowlands, W. N. *J. Colloid Interface Sci.* **1995**, *173*, 406.
17. Hunter, R. J. *Colloids Surf., A* **1998**, *141*, 37.
18. Kurata, M.; Tsunashima, Y. In *Polymer Handbook*, 3rd ed.; Brandrup, J., Immergut, E. H., Eds.; Wiley-Interscience: New York, 1989; p VII/34.

Chapter 9

Salt Concentration Dependence of Swelling States for Poly(sulfobetaine) Brush at Aqueous Solution Interface

Yuki Terayama,[1] Moriya Kikuchi,[2] Koji Mitamura,[2] Motoyasu Kobayashi,[2] Norifumi L. Yamada,[4] and Atsushi Takahara[1,2,3,*]

[1]Graduate School of Engineering, [2]JST, ERATO, [3]Institute for Materials Chemistry and Engineering, Kyushu University, 744 Motooka, Nishi-ku, Fukuoka 819-0395, Japan
[4]Neutron Science Laboratory, High Energy Accelerator Research Organization, 1-1 Oho, Tsukuba, Ibaraki 305-0801
[*]takahara@cstf.kyushu-u.ac.jp

Swelling states of poly(3-(*N*-2-methacryloyloxyethyl-*N*,*N*-dimethyl) ammonatopropanesulfonate) (MAPS) brush in aqueous solutions were investigated by atomic force microscopy (AFM) and neutron reflectivity measurement. The thickness of swollen poly(MAPS) brush evaluated by AFM was increased from 61 to 87 nm with an increase in NaCl concentration from 0 to 0.05 M, and was dramatically increased up to 153 nm at 0.5 M NaCl aqueous solution. At low salt concentration, poly(MAPS) brush chains formed shrunk structure due to the inter- and intrachain attractive electrostatic interactions, while the brush chains stretched up with increasing salt concentration due to the screening effect by hydrated salt ions.

Introduction

Polysulfobetaine, such as poly(3-(*N*-2-methacryloyloxyethyl-*N*,*N*-dimethyl) ammonatopropanesulfonate) (MAPS), is one of the zwitterionic polymers, and has attracted much attention due to its unique solution properties (*1*–*3*). For instance, poly(MAPS) is insoluble in pure water due to strong inter- or intramolecular attractive electrostatic interaction between ammonium cations and

© 2011 American Chemical Society

sulfonate anions, while the poly(MAPS) is soluble in a salt aqueous solution because the attractive electrostatic interactions among side chains are screened by hydrated salt ions. Consequently, the chain dimension of poly(MAPS) expands with increasing with salt concentration in aqueous solution (*4*). In the case of surface-tethered poly(MAPS), these solution properties affect the swollen thickness and the internal structure of the poly(MAPS), which would give various useful functionalities such as superhydrophilic- (*5*), protein antifouling- (*6*), and water lubrication surfaces of polysulfobetaine.

The solution properties of surface-tethered poly(MAPS) in aqueous solution have been investigated. Kato et al. estimated the thickness L of physically adsorbed layer of poly(MAPS) (weight-average molecular weight (M_w) = 1.5×10^6) on silica surface in NaCl aqueous solution by ellipsometry (*7*). The L linearly increased from 8.4 nm to 202 nm by increasing salt concentration from 0.06 M to 1.0 M, which indicated that the conformation change of adsorbed poly(MAPS) occurred among collapsed-, normal-, and elongated states depend on the salt concentration due to the polymeric osmotic pressure and deformation pressure of adsorbed layer. Ducker et al. have also investigated the swollen structure and the diffuse electric double layer of the adsorbed poly(MAPS) on the silica surface by neutron reflectivity (NR) measurement, atomic force microscopy (AFM), and zeta potential measurement (*8*). The diffuse electric double layer of adsorbed poly(MAPS) with 3.7 nm thickness was decreased from 1.35 nm to 0.43 nm with increasing salt concentration from 0.05 M to 0.5 M imposing that the salt ions screened the attractive electrostatic interaction among sulfobetaine group, which affects the colloidal suspension of the silica particle adsorbed poly(MAPS) in aqueous solution. However, these reports investigated only physically adsorbed poly(MAPS), of which grafting density must be changed by salt concentration. In order to investigate the solution properties of surface-tethered poly(MAPS) on the substrate systematically, the chain ends of poly(MAPS) were necessary to be immobilized by covalent bonds.

Polymerization from an initiator immobilized with a solid surface by covalent bond is useful way to prepare the surface-tethered polymer on the substrate because the swollen thickness and graft density of the tethered polymers can be controlled precisely by the molecular weight of the polymer and the graft density of surface-initiator, respectively. In this study, we prepared the poly(MAPS) brush by surface-initiated atom transfer radical polymerization (ATRP) and investigated the swollen thickness and the internal structure of the poly(MAPS) brush in aqueous solution with various salt concentration by AFM and NR measurement. Compared with a previous report (*7*), the dependency of the swollen structure change of the surface-tethered poly(MAPS) on the salt concentration can be observed more clearly to investigate the poly(MAPS) brush with the precise architecture.

Experimental

Materials

Copper(I) bromide (CuBr, Wako Pure Chemicals) was purified by successive washing with acetic acid and ethanol and was dried under vacuum. Ethyl 2-bromoisobutylate (EB, Tokyo Chemical Inc.) was distilled from calcium hydride before use. 2,2′-Dipyridyl (bpy, Wako pure chemicals) and 2,2,2-trifluoroethanol (TFE, Tokyo Chemical Inc.) were used as received. Surface-initiator, (2-bromo-2-methyl)propionyloxyhexyltriethoxysilane (BHE) (*9*, *10*), and MAPS monomer (*11*) were synthesized using previous reported procedures, respectively. Silicon wafer (10 × 10 × 0.5 mm³) and quartz substrate (65 × 65 × 10 mm³) were washed with piranha solution at 428 K for 1 h and subsequently exposed to the vacuum ultraviolet-ray (λ = 172 nm) for 10 min under 30 Pa. The BHE monolayer was immobilized on the substrates using the chemical vapor adsorption (CVA) method (*12*). Deionized water was prepared with arium 611UV (Sartorius Stedim Biotech GmbH, resistivity = 18.2 MΩ cm, 298 K).

Surface-Initiated ATRP of MAPS (13)

Silicon substrate (or quartz plate) and MAPS monomer diluted with TFE were placed into a glass vessel with stopcock connecting to vacuum and argon lines, and degassed by repeated freeze-pump-thaw cycles to remove oxygen. A mixture of CuBr, bpy, and EB in TFE solution was also degassed by a freeze-pump-thaw process and was quickly transferred into monomer solution by a syringe. After the reaction mixture was degassed again by repeated freeze-pump-thaw cycles to remove oxygen, the polymerization was carried out at 333 K for 18 h to generate a poly(MAPS) brush from the silicon wafer (Scheme 1) and a free poly(MAPS) from EB. The reaction mixture was diluted with TFE and poured into methanol to precipitate the free polymer. The silicon wafer was washed with TFE using a Soxhlet apparatus for 12 h to remove the free polymer adsorbed on their surface, and was dried under reduced pressure at 323 K for 6 h.

Measurements

Size exclusion chromatography of the free soluble poly(MAPS) was applied to determine number average molecular weight (M_n), M_w, z-average mean-square radius of gyration, and molecular weight distribution (MWD) using a Shimadzu HPLC system connected to three polymethacrylate-based TSKgel columns (Tosoh Bioscience): G6000PWXL (pore size > 100 nm, bead size 20 μm) + G5000WXL (pore size 100 nm, bead size 10 μm) + G3000WXL (pore size 20 nm, bead size 7 μm) and equipped with a multiangle light scattering detector (MALS; Wyatt Technology DAWN-EOS, 30 mW GaAs linearly polarized laser, wavelength: λ = 690 nm) and refractive index detector (Shimadzu RID-10A, tungseten lamp (wavelength 470 – 950 nm), 40 °C) using 200 mM NaCl aqueous solution as an eleuent at a rate of 0.8 mL min^{-1}. The Rayleigh ratio at a scattering angle of 90° was based on that of pure toluene at a wavelength of 632.8 nm at 25

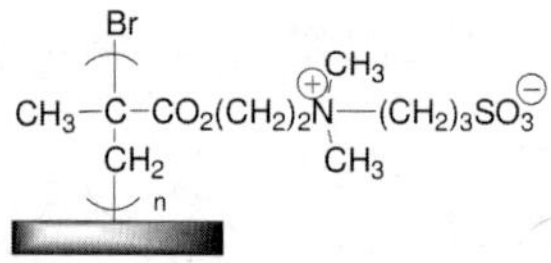

Scheme 1. Chemical structure of poly(MAPS) brush.

°C. The sensitivities of 17 detectors at angles other than 90° and the dead volume for each detector were determined by the scattering intensities of 0.30 wt% aqueous solution of poly(ethylene oxide) standard with 2.22×10^4 g mol^{-1} and $M_w/M_n = 1.08$. The specific refractive index increment (dn/dc) of poly(MAPS) in 200 mM NaCl aqueous solution was determined as 0.120 mL g^{-1} by differential refractometer (Otsuka Electronics DRM-3000, wavelength $\lambda = 632.8$ nm). The thickness of the brush was determined by spectroscopic ellipsometer (Five Lab Co. MASS-102) with an Xenon arc lamp (wavelength 380-890 nm) at a fixed incident angle of 70°. The assumed refractive index and density values were 1.50 (*14*) and 1.34 g cm^{-3} (*8*), respectively, as referenced from literatures. Graft density of the poly(MAPS) brush σ was calculated using the equation followed as $\sigma = d\ L\ N_A \times 10^{-21} / M_n$, where d, L, and N_A are bulk density of the polymer, the dry thickness of the polymer brush, and Avogadoro number, respectively. Swollen thickness of the polymer brush was evaluated by AFM (Agilent Technologies Inc., Agilent5000). The radius of a SiO_2 tip of a cantilever (Nanoworld, Sphere Tips) was 1000 nm. The bending spring constant of the cantilever tip was 0.2 N m^{-1}. The cantilever surface was covered with *n*-propyltriethoxysilane (Sigma Aldrich Co.) monolayer by CVA method. A liquid cell was used for the measurement in deionized water and 0.01, 0.05, 0.5, 1, 5 M NaCl aqueous solutions at 298 K. NR measurement was performed on an ARISA II reflectometer using white neutrons with wavelengths of 0.20 – 0.88 nm in MLF at J-PARC pulsed-neutron source. The neutron beam was irradiated from quartz glass to the interface between D_2O with different concentration of NaCl and the immobilized poly(MAPS) brush. Incident angles were fixed at 0.30°, 0.75°, and 1.2° which covered a q_z range of 0.07–0.33, 0.19–0.82 and 0.30-1.31 nm^{-1}, respectively. The incident slits were adjusted to keep a 55 mm footprint size on the sample surface as well as the angular resolution, $\Delta\theta/\theta$, at 5%. The NR profiles were analyzed by fitting calculated reflectivities from model scattering length density profiles to the data, using Parratt32 software (*15*).

Results and Discussion

Swollen Thickness of Poly(MAPS) Brush Evaluated by AFM

The swollen thickness of poly(MAPS) brush ($M_n = 2.65 \times 10^5$ g mol^{-1}, $M_w/M_n = 1.80$, $\sigma = 0.085$ chains nm^{-1}) in aqueous solution was evaluated by AFM (*16*). Figure 1(a) shows a typical approaching force curves between the tip modified by *n*-propyltriethoxysilane and bare silicon wafer. The horizontal axis in Figure 1 is relative distance between the tip and brush surface estimated from the piezo displacement distance. The force F was increased linearly as a function of the distance when the tip was attached to the bare silicon wafer. The slope of force curve tells us the deflection sensitivity (mV nm^{-1}) of the cantilever based on the deflection voltage (mV) of the refracted beam and relative displacement (nm). Since the deflection sensitivity varies with salt concentration in aqueous solution, we estimated the deflection sensitivity every force curve measurements to normalize the observed force.

Figure 1 (b) shows a typical approaching force curve between the modified tip and poly(MAPS) brush. When the tip was approached to the brush, the F was exponentially increased due to the repulsively steric interaction between the swollen poly(MAPS) brush and the tip. The length of the exponential region was defined as L'.

When the F was reached at 2 nN, the F was linearly increased with decreasing the distance indicating that the brush was not compressed any more even though the larger F was loaded on the brush surface. The thickness of compressed brush layer at 2 nN was evaluated by contact mode AFM. Figure 2 shows the typical topographic image of the scratched poly(MAPS) brush. The boundary layer between the scratched and the unscratched area was scanned to determine the thickness of the compressed brush in aqueous solution, which was defined as L_0. The total value of the height of compressed brush L_0 and the exponential region L' in the force curve was defined as the swollen thickness L_e. In dry state, the L' was not detectable and dry thickness L_d was determined by L_0. The thickness of poly(MAPS) brush in the air (relative humidity = 19%) was estimated at 31 nm, which was agreed with the value estimated by ellipsometry.

Figure 3 shows the revised approaching force curves of swollen poly(MAPS) brush in pure water and NaCl aqueous solutions at 0.01, 0.05, 0.5, 1, and 5 M. The axis of L in Figure 3 was absolute distance from the silicon wafer based on the values of L' and L_0. The swollen thickness L_e was corresponded to the place which the F began to increase exponentially. Actually, the values of L_e measured at 0, 0.01, 0.05, 0.5. 1 and 5 M were estimated at 69, 71, 87, 148, 153, and 176 nm, respectively. The L_e in pure water was 69 nm, which was larger than that in the dry state (31 nm). With an increase in salt concentration from 0 to 0.05 M, the thickness of the brush was slightly increased from 69 to 87 nm. However, the thickness of poly(MAPS) brush significantly increased to be 153 nm at 0.5 M. The thickness of the polymer brush at 5 M reached to be 179 nm. In particular, a large difference in brush thickness was observed between 0.05 and 0.5 M of NaCl aqueous solution. In the case of lower salt concentration than 0.05 M, the ammonium cation and sulfonate anion in poly(MAPS) brush chains were attractively interacted with each other in the neighboring brush chains. Therefore, the brush chains were shrunk due to the attractive electrostatic interaction between ammonium cation and sulfonate anion. On the other hand, the L_e was dramatically increased at 0.5 M because the attractive electrostatic interaction of ionic groups was screened by hydrated salt ions. The brush chains formed relatively extended chain structure. In general, polyelectrolytes in aqueous solution have an unique critical salt concentration at which the polymer chains begin to change its chain dimension drastically, so called θ concentration. Kato et al. evaluated θ salt concentration for isolated free poly(MAPS) to be around 0.060 M (*4*). Although the evaluation of L_e should be carried out in other salt concentrations, the drastic change in thickness of poly(MAPS) brush between 0.05 and 0.5 M might be

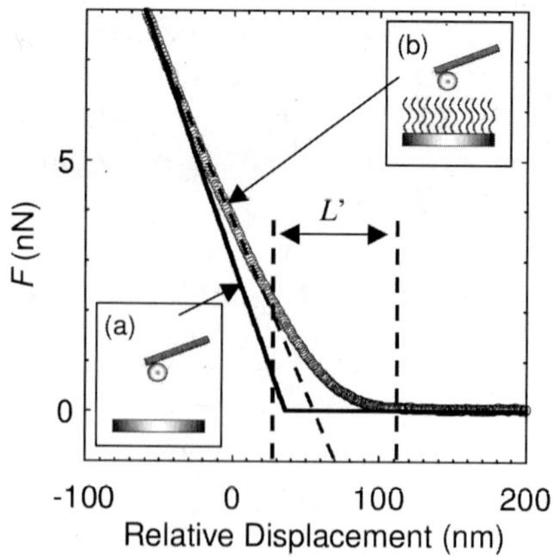

Figure 1. Typical approach curves of (a) bare silicon wafer and (b) swollen poly(MAPS) brush in aqueous solution. The gap L' shows exponential region derived from the swollen brush.

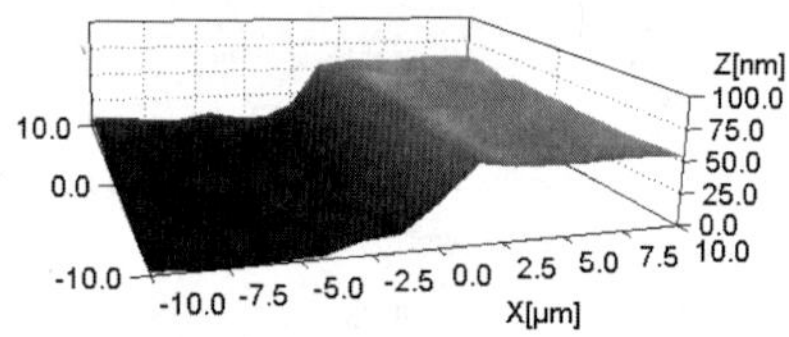

Figure 2. Typical topographic AFM image across the boundary between the scratched and unscratched parts of poly(MAPS) brush. The image was taken under constant force 2 nN between sphere probe and brush surface in aqueous solution.

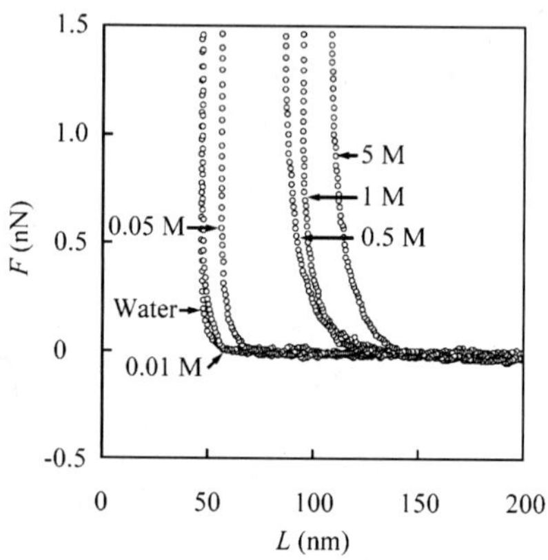

Figure 3. Force vs L curves of the poly(MAPS) brush in water, 0.01 M, 0.05 M, 0.5 M, 1 M, and 5 M NaCl aqueous solutions. The origin of L corresponds to the substrate surface.

dominated by θ concentration of poly(MAPS). Actually, the L_e estimated in this method cannot describe the true thickness of swollen brush because the electrostatic interaction between the probe head and the outermost polyelectrolyte brush should be considered, especially in the case of a low ionic strength solution. Then, we tried to estimate the swollen thickness of poly(MAPS) brush by NR measurement.

NR Study of Swollen Poly(MAPS) Brush in Aqueous Solution

The poly(MAPS) brush prepared by quartz substrate (M_n =1.20 × 10^5 g mol^{-1}, M_w/M_n = 1.55, Thickness in the dry state = 19.0 nm, σ = 0.072 chains nm^{-1}) was measured by NR measurement. D_2O was used as a solvent to enhance the contrast of scattering length density (SLD) between the polymer and the solvent. Figure 4 shows the NR curves of poly(MAPS) brush in D_2O, 0.05, 1, and 5 M NaCl/D_2O and their corresponding fitting curves calculated by assumed SLD profiles consisting of quartz layer (3.40×10^{-4} nm^{-2}), initiator layer (1.90×10^{-4} nm^{-2}), gradient layer of swollen poly(MAPS) in D_2O, and D_2O layer (6.28×10^{-4} nm^{-2}).

Figure 4 (right) shows the SLD profiles of poly(MAPS) brush in D_2O, 0.05, 1, and 5 M NaCl/D_2O. The gradient increase of SLD with increasing the distance from the quartz substrate was observed indicating that the brush chains were extended to D_2O interface against the perpendicular direction from the quartz substrate. The SLD of poly(MAPS) brush in D_2O was dramatically increased from 4.44 × 10^{-4} to 6.28 × 10^{-4} nm^{-2} at 88 nm along with the distance from the substrate. Similar SLD profile was observed in 0.05 M solution. With increasing salt concentration to 1 M, the SLD was considerably extended in 234 nm region from the substrate. The SLD profile of poly(MAPS) brush at 5 M NaCl/D_2O increased from 5.11 × 10^{-4} to 6.11 × 10^{-4} nm^{-2} at 240 nm from the substrate.

In D_2O or 0.05 M NaCl/D_2O, the chain ends of the brush were reached to be 88 or 96 nm, respectively, which were larger than the dry thickness (19 nm) estimated by ellipsometry. However the brush chains were shrunk due to the attractive electrostatic interaction of sulfobetaine groups. D_2O could not be entered to inside shrunk brush chains. On the other hand, the brush chains were further extended from the quartz surface at 1 and 5 M. The brush chains were swollen and formed extended structure due to the screening effect by both positive and negative ions in aqueous solution. The drastic change from the shrunk- to the extended structure was observed between 0.05 and 1 M. Consequently, the θ concentration for poly(MAPS) would be between 0.05 and 1 M of NaCl solution. These values are well corresponding to the critical transition concentration in brush thickness observed by AFM.

Both AFM and NR analysis showed that interesting change from shrunk- to extended structure of poly(MAPS) brush depend on the salt concentration. This dependency of salt concentration on swollen structure of poly(MAPS) brush contracted with poly(2-methacryloyloxyethyl phosphorylcholine) (MPC) brush, which is also zwitterionic polymer. NR of swollen poly(MPC) brush revealed that the extended chain structure formed in pure water as well as in aqueous NaCl solution with 1 M due to its unique weak interaction of phosphorylcholine with water molecules (*17*, *18*). Hydrated poly(MPC) brush showed negligible

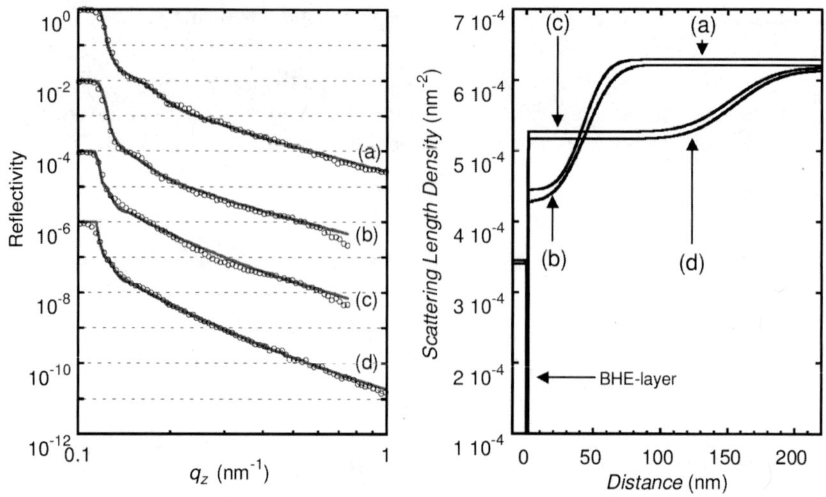

Figure 4. Experimental NR curves (open circle) of poly(MAPS) brush in (a) D_2O, (b) 0.05 M, (c) 1 M, and (d) 5M NaCl/D_2O and their corresponding fits (solid line). SLD profiles along with a distance from the quartz surface were shown in the right side.

dependency on salt concentration, which is significantly different from the dependency of salt concentration on poly(MAPS) brush probably because of the strong attractive electrostatic interaction between ammonium and sulfonate units.

Conclusion

The swollen poly(MAPS) brush structure in aqueous solution with different salt concentrations was investigated by AFM and NR measurement. The swollen thickness L_e of poly(MAPS) brush in aqueous solution was evaluated by AFM. The L_e was slightly increased from 69 to 87 nm with an increase of salt concentration from 0 to 0.05 M. At 0.5 M NaCl aqueous solution, the L_e was dramatically increased to 153 nm. Drastic change in brush thickness was observed between 0.05 and 0.5 M of NaCl aqueous solution. The similar change was also observed between 0.05 and 1 M by NR measurement. At lower salt concentration than 0.05 M, the brush chains are collapsed state due to the inter- and intrachain attractive electrostatic interaction between ammonium cations and sulfonate anions. On the other hand, the brush chains were swollen due to the screening effect of sodium and chloride ions at higher salt concentration than 0.5 M. This interesting thickness change was dominated by θ concentration. In particular, the high density polymer brush chains were attractively interacted with neighboring brush chains to form shrunk structure at lower than θ concentration. At higher than θ concentration, the brush chains were extended to the perpendicular direction from the substrate due to high osmotic pressure of the densely grafted brush chains.

Acknowledgments

The present work is supported by a Grant-in-Aid for the Global COE Program, "Science for Future Molecular Systems", and partially supported by JST, ERATO, Project "Takahara Soft Interfaces". Y.T. acknowledges the financial support of a Grant-in-Aid for JSPS Fellows.

References

1. Kantor, Y.; Kardar, M.; Li, H. *Phys. Rev. Lett.* **1994**, *49*, 1383–1392.
2. Liaw, D.-J.; Lee, W.-F.; Whung, Y.-C.; Lin, M.-C. *J. Appl. Polym. Sci.* **1987**, *34*, 999–1011.
3. Schulz, D. N.; Peiffer, D. G.; Agarwal, P. K.; Larabee, J.; Kaladas, J. J.; Soni, L.; Handwerker, B.; Garner, R. T. *Polymer* **1986**, *27*, 1734–1742.
4. Kato, T.; Takahashi, A. *Ber. Bunsen. Phys. Chem.* **1996**, *100*, 784–787.
5. Cheng, N.; Brown, A. A.; Azzaroni, O.; Huck, W. T. S. *Macromolecules* **2008**, *41*, 6317–6321.
6. Yang, W.; Chen, S.; Cheng, G.; Vaisocherova, H.; Xue, H.; Li, W.; Zhang, J.; Jiang, S. *Langmuir* **2008**, *24*, 9211–9214.
7. Kato, T.; Kawaguchi, M.; Takahashi, A. *Langmuir* **1999**, *15*, 4302–4305.
8. Starck, P.; Mosse, W. K. J.; Nicholas, N. J.; Spiniello, M.; Tyrrell, J.; Nelson, A.; Qiao, G. G.; Ducker, W. A. *Langmuir* **2007**, *23*, 7587–7593.
9. Ohno, K.; Morinaga, T.; Koh, K.; Tsujii, Y.; Fukuda, T. *Macromolecules* **2005**, *38*, 2137–2142.
10. Kobayashi, M.; Takahara, A. *Chem. Lett.* **2005**, *34*, 1582–1583.
11. Duann, Y. F.; Chen, Y. C.; Shen, J. T.; Lin, Y. H. *Polymer* **2004**, *45*, 6839–6843.
12. Koga, T.; Morita, M.; Ishida, H.; Yakabe, H.; Sasaki, S.; Sakata, O.; Otsuka, H.; Takahara, A. *Langmuir* **2005**, *21*, 905–910.
13. Terayama, Y.; Kikuchi, M.; Motoyasu, K.; Takahara, A. *Macromolecules* **2010**, *44*, 104–111.
14. Zhai, G; Yu, W. H.; Kang, E. T.; Neoh, K. G.; Huang, C. C.; Liaw, D. J. *Ind. Eng. Chem. Res.* **2004**, *43*, 1673–1680.
15. Berlin Neutron Scattering Center (BENSC) Parratt32 Program. http://www.hmi.de/bensc/instrumentation/instrumente/v6/refl/parratt_en.htm.
16. Yamamoto, S.; Ejaz, M.; Tsujii, Y.; Matsumoto, M.; Fukuda, T. *Macromolecules* **2000**, *33*, 5602–5607.
17. Kobayashi, M.; Terayama, Y.; Hino, M.; Ishihara, K.; Takahara, A. *J. Phys: Conf. Ser.* **2009**, *184*, 012010.
18. Kitano, H; Takaha, K.; Gemmei-Ide, M. *Phys. Chem. Chem. Phys.* **2006**, *8*, 1178–1185.

Chapter 10

Ultrasonic Rheology of Mixed-Phase Systems

Using a QCM as an Effective Rheological Device – Theory and Applications for Combined Surface and Bulk Rheology

Thomas J. Lane[*,†]

Pomona College Department of Chemistry, Claremont, CA
[*]tjlane@stanford.edu
[†]Currently at Stanford University, Palo Alto, CA

Rheology provides a powerful way to access molecular scale data from a macroscopic measurement. Ultrasonic rheology, performed with a quartz-crystal microbalance (QCM) instrument, provides a straightforward route to sensitive (< 0.1 mPa·s STD) rheology of either bulk or surface phases. Presented is the necessary theory (Johannsmann and Voinova models) to perform such measurements, and some applications of that theory - including rheology on polymer brush layers, lipid vesicle adsorption, proteins in the bulk phase, and microemulsions. In each of these applications, an advantage of QCM-based rheology is described, such as small sample size, precision and the ability to perform either bulk or surface measurements. Additionally, corrections to the standard Sauerbrey approximation, which often results in an underestimation of adsorbed mass, are presented. In conclusion, the QCM is shown to provide a facile route to rheological measurements in the MHz regime.

Rheology, or the measurement of viscoelastic properties of complex fluids, has provided a powerful tool for scientists to characterize the behavior of a wide variety of materials, most notably polymer gels. Complex fluids, such as polymer gels (*1*), both store and dissipate kinetic energy under applied stress and therefore exhibit both viscous and elastic properties, characteristic of solids and

© 2011 American Chemical Society

liquids respectively. Additionally, this response often changes with the frequency of applied stress, and variation in frequency dependence yields additional information about the molecular nature of the system (*2*, *3*). Complex fluids are ubiquitous; surfactant emulsions, egg whites, and silly putty are all examples of complex systems from everyday life. The unusual response of these materials to stress is a result of their specific compositions. Composition and orientation on a molecular scale within a material can drastically affect the way that material responds to physical stress; measurement of material strain, therefore, provides a simple yet powerful route to the characterization of molecular structures.

This type of molecular technique has gained importance as the lines between chemistry, chemical engineering, and nanotechnology blur together. Recently, scientists have rushed to build and utilize analytical techniques for the effective and efficient characterization of nano-scale ordered structures. Rheology, desirable for its sensitivity to the interactions of molecular structures has been adopted into the arsenal of tools useful for characterizing small, ordered molecular structures (*4*).

Traditionally, rheology is considered a technique for the characterization of large quantities of bulk material, most notably polymer gels or composites. Rheological measurements are attractive for bulk polymer applications for the same reasons as molecular scale applications – obtaining viscoelastic information about a polymer gel gives a clear picture about how polymer molecules arrange themselves in the material. Consider a hydrogel; polymer branching, cross-linking, molecular weight, strand alignment, concentration of trapped solvent, and molecular nature of the polymer(s) will all affect how the gel responds to stress (*1*). Characterizing the strain response gives the polymer engineer a way to monitor these molecular features that ultimately determine the properties of the material.

Recently chemists and biologists have realized the potential for detailed information rheology provides, and have adapted rheology as an analytical tool capable of revealing information about the nature of unknown materials and processes. This evolution has taken different forms depending on the application, but two major new fields have emerged – microrheology and ultrasonic surface rheology. While by no means the only applications of rheology to nano-scale structures, these two techniques in particular are of interest since they differ from traditional rheology in significant ways.

For biological applications, the focus has been on microrheology, performed by monitoring one or two colloidal particles suspended in a small region of interest (*5*). Microrheology is typically performed by inserting a probe, usually a magnetic fluorescent colloid, into a specific medium and monitoring its movement with a camera or laser tweezers device. The probes can either be actively driven by a magnetic field or laser to measure specific frequencies, or passively monitored, relying on thermal fluctuations as a driving force.

By limiting these particles to the micron scale, it is possible to obtain measurements of the rheological parameters of a very small region, such as the cytoplasm of one individual living cell (*6*). These measurements can then be compared to values for other cells, other regions within the same cell, or the same cell under some kind of new condition, e.g. while exposed to a relevant active drug (*7*). Such measurements allow for the quantification of the material properties of either very small volumes (microliter scale) or good spatial specificity (*e.g.* a specific organelle in a cell), both of which are attractive for biological experimentation.

Over the past two decades, the field of microrheology has developed rapidly into a fruitful area of study. This chapter will not focus on the details of microrheology, as many excellent reviews are available outlining the progress of this field (*5*, *6*, *8–11*). Instead, it will focus on the less common field of surface-based ultrasonic rheology.

The Development of Ultrasonic Rheology

In contrast to the well-guided, linear development of microrheology, the development and application of ultrasonic rheology has grown organically from other technologies. Ultrasonic rheometers are typically adapted quartz crystal microbalances (QCMs), devices used to measure molecular-scale adsorption phenomena (*12*). The foundation of the QCM is a piezoelectric quartz crystal that oscillates in the megahertz frequency range, and measures changes in that oscillation to quantify adsorbed masses. Acoustic rheology takes advantage of that oscillation as a source of stress to probe analyte material, and represents an expansion of the QCM beyond its traditional gravimetric application, to a more advanced interpretation of the data provided. Alternatively, rheological knowledge can be used to ensure that gravimetric measurements are accurate, and correctly calculate adsorbed masses.

Origins

QCMs were originally developed in the mid 20th century in order to monitor vapor deposition processes, and have played a key role in vapor deposition apparatuses ever since (*12*, *13*, *15*). The QCM consists of a thin, AT-cut slab of quartz crystal wedged between two electrodes (Fig. 1). When an AC current is applied to the electrodes, the piezoelectric quartz oscillates at a well-defined resonant frequency depending on its thickness. Typically, crystals are cut to have fundamental resonances between 1-10 MHz. These crystals oscillate in the shear plane along the sensor surface at about 100 nm amplitude, with accelerations on the order of 10^8 m s^{-2} (*14*), propagating shear acoustic waves into the medium adjacent to the sensor surface.

When a thin film is deposited on the surface of one of the electrodes, the resonant frequency of the crystal decreases. Lock-in electronics make it possible to monitor this feedback, and quantify the decrease. Sauerbrey, in a classic paper (*15*), showed that for a thin, fully elastic film, under no-slip conditions, this frequency change, Δf, was directly proportional to the areal mass (total mass divided by the area over which it is distributed) deposited, m, by:

$$\Delta f = -\frac{2 f_0^2}{\sqrt{\rho_q \mu_q}} \Delta m \qquad [1]$$

where f_0 is the fundamental resonant frequency of the quartz slab (unloaded), $\rho q = 2.648$ g cm^{-3} and $\mu_q = 2.947 \times 10^{11}$ g cm^{-1} s^{-2} are the density and elastic modulus of quartz, respectively.

While QCMs became standard in many vapor deposition apparatuses, their use was predominantly limited to this application, due to the widely held belief that under liquid loading the quartz crystal would be overdamped, and therefore non-functional. In 1982, however, Nomura and Okuhara published a groundbreaking paper showing that this was not the case (*16*). They demonstrated, to the surprise of theorists, that the sensor reported a frequency change upon loading with organic solvents, and reported an empirical expression relating the density and viscosity of the solvents to the frequency shift.

With the potential to use the QCM in a liquid environment, applications for the technology quickly multiplied. The technology provides a facile route to measure the kinetics and magnitude of adsorption of almost any solute to a large variety of surfaces. Additionally, these measurements are quite sensitive – modern QCMs are precise to within tenths of Hertz, corresponding to a Sauerbrey mass of hundredths of ng cm^{-2}. There are many excellent reviews of the applications of QCM to various problems in immunosensing (*17*), polymer films (*18*), lipid fusion and rupture (*19*), etc., that the interested reader will find useful for demonstrating the modern gravimetric uses of the QCM.

Technological Advances

One of the two major technical advances in QCM technology over the past decade has been the development of instruments capable of monitoring not only the fundamental resonant frequency of the quartz crystal, but also its harmonics. With better frequency isolation chambers, electrode design, and electronics, the ability to measure odd overtones up to the 13th or 15th multiple of the fundamental has become possible. Since higher frequencies dissipate energy faster in a viscous medium, these higher overtones decay faster as the shear waves propagate into the bulk above the sensor and are more contained to the surface region. Additionally, if adsorbed material is viscous, it will result in a smaller (less negative) Δf for the higher frequencies than expected for a fully elastic film. Therefore, a viscoelastic material will show different Sauerbrey masses for each harmonic; in this situation, a more detailed analysis is necessary to account for "missing mass effects" (*20*) caused by sensor damping. A more detailed account of the theory behind harmonic overtones can be found in the *Microgravimetry Corrections* section below.

The second major advance in QCM technology was the introduction of dissipation monitoring (*12*). The dissipation, D is a measure of the ratio of the energetic storage and loss of a system in contact with the QCM per cycle. Specifically

$$D = \frac{G''}{2\pi G'} \qquad [2]$$

where G' and G'' are the energetic storage and loss moduli, respectively. These standard rheological parameters are functions of the frequency of oscillation, ω. The dissipation is also intrinsically related to the phase lag between stress and strain, δ, and is actually a measure of $\tan(\delta)$ (inverse Q-factor, or quality factor of a resonator) (*2*). The dissipation measurement is highly sensitive to experimental considerations, such as the orientation of molecules at the surface and quantity of surface-coupled water held by an adsorbed film. Dissipation is unitless, and typical values range from 10^{-6} for adsorption of a thin, soft film to 10^{-4} for fluid loading of a viscous substance.

Two different methods of dissipation monitoring have emerged, the power cycling ("ping") method and the impedance method. In power cycling, at regular intervals the AC current driving the QCM's oscillations is cut, and the attenuation of the drive is monitored (*21*). This attenuation can be fit to an exponential decay, yielding the dissipation (Eq. 3).

$$x(t) = A_0 \exp(-t/\tau)\sin(\omega t + \phi) \qquad [3]$$

Where $x(t)$ is the displacement of the crystal, A_0 is the amplitude at $t = 0$, τ is the characteristic decay time, and ω and ϕ are the angular frequency and phase. The dissipation can then be calculated by:

$$D = \frac{1}{\pi \tau f} \qquad [4]$$

Faster attenuation indicates viscoelastic characteristics, while a slow attenuation is indicative of a rigid, elastic film.

Alternatively, the dissipation factor can be measured from the electrical impedance (Z_{el}) of the circuit driving the quartz resonance. Monitoring the impedance of the electrical signal caused by the quartz resonance yields a measure of the phase lag between stress applied by the sensor and the strain induced in any coupled material. Higher impedance is indicitive of a higher dissipation factor, and vice versa. For a sensor of active area A, they are related by

$$D = \frac{4Z_m}{\pi^2 A\sqrt{\mu_q \rho_q}} n^{1/2} \qquad [5]$$

where Z_m is the mechanical impedance, and is directly proportional to Z_{el} (*22–24*). Both the impedance and power cut methods are effective means of determining the dissipation factor of a coupled film.

The modern QCM, with the capacity to monitor the frequency and dissipation at multiple harmonics in real time (on the millisecond time scale) has drastically grown in use over the past two decades. The most common uses of the QCM are to quantify the deposition of protein and polymer films (*18*), immunosensing (*17*), and monitoring vesicle fusion and rupture (*19*).

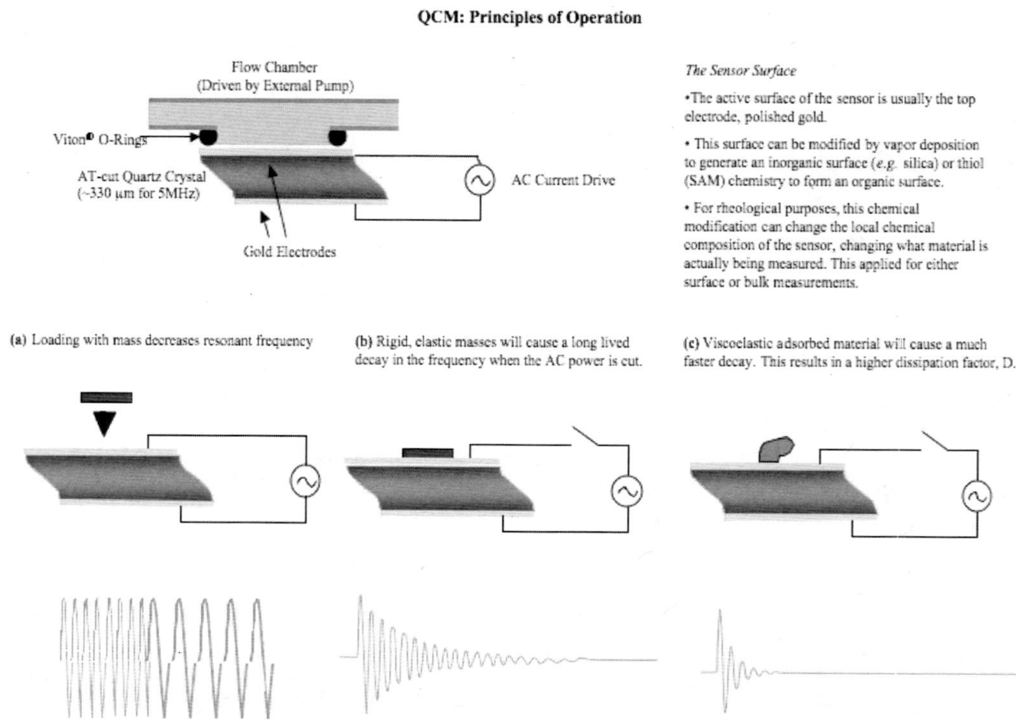

Figure 1. Schematic of a QCM/ultrasonic rheometer. Panels (a), (b), (c), show the instrument under different loading conditions.

Translation of QCM Data to Rheological Parameters

Rheological Foundations

The QCM, while usually used as a gravimetric device, can also be function a rheometer. Frequency changes are typically translated immediately to gravimetric measurements, through means such as the Sauerbrey equation. However, these translational equations, such as the Sauerbrey equation, are derived from an analysis of the way in which systems in contact with the QCM sensor dissipate energy. Since the QCM is founded on probing analytes with direct oscillatory motion, all QCM measurements are fundamentally rheological. Additionally, the surface-specific nature of QCM measurements mean that surface and bulk measurements can be differentiated if a rheological view is adopted.

Apparent in Equation 2 is the connection of QCM measurements to the rheological parameters of coupled material. The dissipation factor, directly dependent on G' and G'', provides a wealth of information in many experiments. One of the easiest applications of dissipation monitoring is to immunosensing. In a typical QCM immunosensing experiment, an antigen will first be immobilized on the sensor surface. An antibody will then be exposed to the surface. If the molecules interact specifically – *i.e.* the antibody is complimentary for that antigen, the molecules will orient themselves in an "upright" orientation that traps a large amount of water (Fig. 2). If the molecules are not complementary, any adsorption will be driven by non-specific interactions, and the bound molecules will result in little to no frequency change, trap less solvent, and have a lower dissipation factor. Figure 2 shows an example of this for a *Bacillus anthracis* (anthrax) membrane protein and its antibody (*25*).

In this example, the QCM data shows that rheological measurements are sensitive to determination of specific interactions in immunosensing experiments. The orientation of molecules at the surface determines the rheological parameters of the film, which in turn affect the dissipation factor. As we will see, an advanced treatment of the frequency and dissipation data will yield interesting rheological data for both adsorbed and bulk material. The QCM apparatus provides a foundation for the measurement of these rheological parameters.

The development of physical theories of the QCM sensor has been a collaborative effort distributed over many researchers, including Sauerbrey (*15*), Johannsmann (*26*), Kanazawa and Gordon (*27*), and Rodahl and Kasemo (*28*), among others. From these studies, a well-developed physical description of the QCM sensor operating under various conditions has arisen.

Rheological Theory of a Loaded Bulk Fluid

As an ultrasonic rheometer, the QCM is sensitive to two separate phases – any surface-associated materials, and a bulk fluid extending beyond the surface region. The operation of QCM under fluid loading was first described in physical detail by Kanazawa and Gordon, who showed from first principles that the change in frequency for a quartz sensor taken from vacuum to fluid with viscosity η_f and density ρ_f is given by:

$$\Delta f = -f_0^{3/2}\left(\frac{\rho_f \eta_f}{\pi \rho_q \mu_q}\right)^{1/2} \quad [6]$$

where ρ_q and μ_q are the density and elastic modulus of quartz, respectively. This equation was consistent with the previous transmission-line analysis of Stockbridge, who was working with sensors exposed to gases under various pressures (*29*).

Later, this model was expanded by Rodahl and Kasemo (*28*) to include an adsorbed film, the expected dissipation response, and to account for multiple overtones. Again, from a first principles analysis, they derived the following more general expressions for fluid loading of a film with mass m with a coefficient of friction with the surface of χ:

$$\Delta f = -\mathrm{Im}\left[\frac{2 f_0 \chi m \beta}{2\pi\sqrt{\rho_q \mu_q}(\chi m + \beta)}\right] \quad [7]$$

$$\Delta D = \mathrm{Re}\left[-\frac{2 f_0 \chi m \beta}{\pi f\sqrt{\rho_q \mu_q}(\chi m + \beta)}\right] \quad [8]$$

where $\beta = (1 - i)\eta_f\delta^{-1} \tan[(1 - i)]t_f/\delta]$ is an arbitrary parameter, $\delta = (2\eta_f/\omega\rho_f)^{1/2}$ is the exponential decay length of the fluid shear wave, and t_f is the thickness of the fluid. The subscript q represents a physical properties of quartz, while f represents the fluid loaded onto the QCM.

In the limiting case where the fluid extends beyond the effective sensing distance (about 250 nm in water, where $t_f = \infty$), and there is no film at the surface (m=0), we obtain:

$$\Delta f = -f_0^{3/2} n^{1/2}\sqrt{\frac{\rho_f \eta_f}{\pi \rho_q \mu_q}} \quad [9]$$

$$\Delta D = 2 f_0^{1/2} n^{-1/2}\sqrt{\frac{\rho_f \eta_f}{\pi \rho_q \mu_q}} \quad [10]$$

where n is the harmonic overtone of the quartz crystal. These equations form the beginnings of a rheological analysis of a pure fluid loaded onto the sensor – note the fluid viscosity is given as a function of QCM outputs, but only for Newtonian fluids. Equation 9 is identical to Equation 6, but now includes values for all measurable harmonics, n.

By substituting a complex term for viscosity into the derivation at an earlier point, it is possible to obtain straightforward expressions translating QCM output

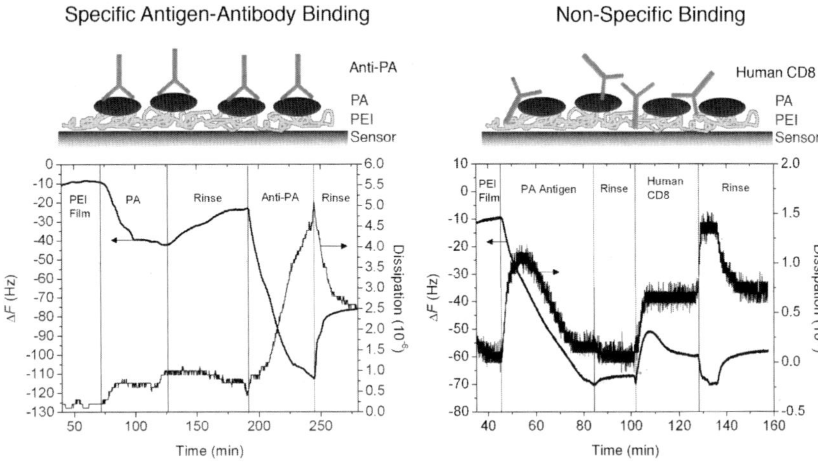

Figure 2. Bacillus anthracis (anthrax) protective antigen (PA) immunosensing monitored by QCM. The left panel contains a QCM data for the adsorption of Bacillus PA on a generic polymer (polyethylenimine, PEI) surface. This layer is then exposed to a specific antigen, which shows strong binding in a large increase in Δf and ΔD due to an upright orientation of the antibody at the surface, which traps a large amount of solvent (illustrated). The left panel shows the same experiment with a nonspecific control antibody, Human CD8 - the Δf and ΔD values are much smaller. Arrows indicate axes (25).

into rheological parameters (G' and G''). By using a complex viscosity, η^* (henceforth, an astric, *, will denote a complex quantity), defined as

$$\eta^* = \frac{iG^*}{\omega} = \frac{(iG' - G'')}{2\pi f} \qquad [11]$$

in Equations 7 and 8, and following the same assumptions used to derive Equations 9 and 10, we can eliminate systems of equations using Equation 2, to obtain

$$G' = \frac{\pi^2 \rho_q \mu_q}{f_0^2 \rho_f}\left(\frac{n^2 f_0^2}{4}\Delta D^2 - \Delta f^2\right) \qquad [12]$$

$$G'' = -\frac{\pi^2 n \rho_q \mu_q}{f_0 \rho_f}\Delta f \Delta D \qquad [13]$$

These are the same equations as those obtained by Patel *et. al.* (*30*) who worked from Johannsmann's impedance analysis in order to first derive this set of equations. From Equations 12 and 13, at a given solution density, we can calculate G' and G'' values as a function of frequency, as the QCM measures multiple frequencies, one for each harmonic of the quartz resonator.

This kind of analysis has been used to measure the rheological parameters of such materials as concentrated antibody solutions (*31*), foodstuffs (*32*) and additional applications, which will be further discussed in the *Applications* section later in the chapter. While these kinds of measurements are possible on other instrumentation, *e.g.* standard rheometers (*2*), the use of the QCM provides measurements of unparalleled sensitivity at much higher frequencies. These high-frequency measurements allow for the detection and characterization of rapid relaxation processes, typical for molecular structures (*33*). Typical rheology, done on in the kHz range, cannot access molecular processes, which typically occur in on the μs to fs timescale (*33*).

Surface Rheology with the QCM

Interest in the QCM typically focuses on its powerful application to the study of thin films. As previously mentioned, the QCM has been used as a balance, simply measuring the mass of absorbed material (*34*). This mass causes a decrease in the resonance frequency of the oscillating quartz, which can be related to mass adsorbed as in Sauerbrey's equation (Eq. 1).

It was quickly observed, however, that when loaded with a viscous material, Sauerbrey's equation would often under-report the mass adsorbed, leading to a 'missing mass' effect (*35*). Since Sauerbrey's analysis assumes a rigid, non-viscous film, it does not account for any energy dissipated within the film itself. Such losses can result in significant deviation from Equation 1, when the QCM is loaded with a viscous soft film in a liquid environment (see *Microgravimetry Corrections* section below for a specific example).

Since then, theory has been generalized to account for the viscoelastic properties of adsorbed films. Two main analytical interests arise in these analyses. The most common has been to obtain accurate mass measurements for dissipative adsorbed structures, such as the protein example mentioned above. The second has been to study the rheological properties of an adsorbed thin film to investigate its structure or composition.

Additionally, two different theoretical approaches have arisen to tackle the problem. Both are focused on relating Δf and ΔD to accurate masses via rheological parameters, but arrive at slightly different expressions to relate these quantities. Presented below are the origins and foundations associated with each model, followed by a section comparing them.

Johannsmann Mass – Correction from Impedance Analysis

First, we will examine approaches for obtaining accurate mass measurements for viscous thin films adsorbed to the QCM. Johannsmann's analysis (*26*) of adsorbed films provides an excellent framework for understanding missing mass effects on the QCM. It treats the QCM as an equivalent circuit, and then predicts changes in the resonant features of the circuit upon perturbation from a loaded viscoelastic film. From an equivalent circuit/impedance analysis, he and coworkers derived the following expression for the change in frequency, Δf, for a QCM loaded with a viscoelastic film and a bulk Newtonian ($G'' \gg G'$) fluid

$$\Delta f^* = -nf_0 \frac{\sqrt{\rho_s / J^*}}{\pi\sqrt{\rho_q \mu_q}} \tan(\omega h \sqrt{\rho_s J^*}) \quad [14]$$

where h is the film thickness and J^* is the complex compliance of the film (equal to $1/G^*$). A subscript s denotes a property of the surface-bound film. Considering a Taylor expansion of the tangent term, for a linear approximation $\tan x \approx x$ we obtain

$$\Delta f = -nf_0 \frac{\omega \rho_s h}{\pi\sqrt{\rho_q \mu_q}} \quad [15]$$

which is the Sauerbrey equation (Eq. 1). Note that the compliance term has been cancelled. This demonstrates why the Sauerbrey equation is often a good first-order approximation, and can be used as a starting point for further analysis. If we continue the expansion, including the second term (such that $\tan x \approx x + x^3/3$) we obtain

$$\Delta f^* \approx -nf_0 \frac{1}{\pi\sqrt{\rho_q \mu_q}} \left[\omega \rho_s h + J^*(\omega) \frac{\omega^3 \rho_s^2 h^3}{3} \right]$$

or, in terms of G^*, taking the imaginary part of the modulus,

$$\Delta f \approx -nf_0 \frac{1}{\pi\sqrt{\rho_q \mu_q}} \left[\omega \rho_s h - \frac{\omega^3 \rho_s^2 h^3}{3} \frac{G''}{G'^2 + G''^2} \right] \quad [16]$$

following Johannsmann, if we define an equivalent, or Sauerbrey, mass, m_{eq}, such that

$$m_{eq} = -\frac{\sqrt{\rho_q \mu_q}}{2 f_0} \frac{\Delta f}{f} \quad [17]$$

which will be the mass reported by the QCM via the Sauerbrey equation, then the effect of the viscous compliance, J^* on the reported mass is apparent. Substituting Equation 17 into Equation 16, and defining the real mass of the adsorbed film as m_0 yields

$$m_{eq}* = m_0\left[1 + J*(\omega)\frac{\rho_f \omega^2 h^2}{3}\right]$$

$$m_{eq} = m_0\left[1 - \frac{\rho_f \omega^2 h^2}{3}\frac{G''}{G'^2 + G''^2}\right] \quad [18]$$

From Equation 18, we can see that a significant compliance, the real component of which will be negative, will lead the Sauerbrey mass, m_{eq}, to underreport the true mass, m^0.

In order to obtain a more accurate mass for a viscous film it is possible to obtain Sauerbrey mass measurements at a number of frequencies by simply monitoring many overtones of the QCM response. If the real part of the compliance, J', is known, then Equation 18 can yield the real film mass m_0. Alternatively, if the true mass is known from an orthogonal technique (*e.g.* dual polarization interferometry, DPI (*36*), an optical technique), one can determine J', and its related value, G'.

Often, neither the true mass nor any viscoelastic properties of the film will be known. However, the frequencies being monitored may lie over a narrow enough frequency range such the adsorbed material will not have any characteristic decay features in that range. In that case, J' will be linear with respect to ω. Then plotting the Sauerbrey mass for each overtone as a function of frequency squared for that overtone will yield a line, with intercept m_0 and slope J' (*37*). Deviation from linearity in this plot is indicative that there are significant dissipative processes occurring in the monitored frequency range, and that a more thorough analysis is needed (see Fig. 4, discussion *vide infra*).

The final parameter to be extracted is the imaginary part of the compliance, J''. Solving Equation 18 for $J*(\omega)$ and taking only the imaginary part yields

$$J''(\omega) = \frac{3\sqrt{\rho_q \mu_q}}{\rho_f^2 \omega^3 h^3}\frac{\Delta D}{f_0} \quad [19]$$

after substituting in ΔD as a measure of energy loss in the system. Note that impedance-analysis theorists typically use a measure of the bandwidth of the resonance, the half-bandwidth at half height, Γ. Both are measurements of the energy dissipated by the system, and are related by $D = 2\Gamma/f$. This leads to a relatively straightforward determination of J''.

Voinova Mass – Correction from Energy Balance

Another appropriate mass correction has been derived from first principles by Voinova *et. al.* using slightly different tactics. While Johannsmann's work views the sensor as an electrical circuit, and analyzes energy dissipation as the impedance effect of waves on resonance phenomena, the Voinova approach treats the problem of as an energy balance issue. Note that, in contrast to the equations presented by Johannsmann, the Voinova analysis includes the contribution of the bulk fluid to the frequency change, and shows the effect of bulk fluid damping on the contribution of a thin film. This will be discussed in detail below.

We consider, as before, a thin viscoelastic film loaded onto the QCM surface, with a bulk Newtonian fluid extending infinitely beyond the film. This enables a number of simplifying assumptions. First, we assume no-slip boundary conditions, *i.e.*, where the force of friction causing energy losses or slip between media is ignored and there is complete transfer of momentum. This is a common assumption, and for the molecular-scale interface, results in negligible deviation from reality. Second, assume that the film is very thin, or $h \gg \delta_s$, where δ_s is the exponential decay length of the acoustic wave in the surface-bound film, $\delta_i = (2\eta_i/\omega\rho_i)^{1/2}$ (we will use a similar value corresponding to the bulk, δ_f). Finally, we assume that the thickness of the bulk is infinite, appropriate if the fluid extends beyond the QCM's sensing range, about 250 nm for water.

Applying these assumptions and calculating the expected shift in resonant frequency due to energy dissipation results in the following expression

$$\Delta f = -\frac{\eta_f}{2\pi\delta_f\sqrt{\rho_q\mu_q}} - \frac{h\rho_s\omega}{2\pi\sqrt{\rho_q\mu_q}}\left[1 - \frac{2}{\rho_s}\left(\frac{\eta_f}{\delta_f}\right)^2 \frac{G''}{G'^2 + G''^2}\right] \quad [20]$$

where subscript *s* denotes properties of the surface-bound film, and *f* represents the bulk fluid. The rheological parameters are specific to the film, since the bulk is assumed to be Newtonian, and therefore has only a viscous contribution to the equation.

Then in the Sauerbrey regime, $G'' \ll G'$, Equation 20 becomes

$$\Delta f = -f^{1/2}\sqrt{\frac{\rho_f\eta_f}{\pi\rho_q\mu_q}} - f\frac{h\rho_s}{\sqrt{\rho_q\mu_q}} \quad [21]$$

where the first term on the RHS is the Kanazawa equation for the loading of a Newtonian fluid (Eq. 6) and the second term is the Sauerbrey mass (Eq. 1). Equation 21 represents the frequency shift for loading of a Newtonian fluid added to the shift for loading of a Sauerbrey film.

In Equation 20, then, the term in brackets is a mass correction term for the Sauerbrey equation, and we may write

$$m_{eq} = m_0\left[1 - \frac{\eta_f\rho_f\omega}{\rho_s}\frac{G''}{G'^2 + G''^2}\right] \quad [22]$$

where, as before, m_{eq} is an equivalent (Sauerbrey) mass, and m_0 is the true mass of the film. Note that in this form, Equation 22 confers information only about the film, but assumes that there is a bulk fluid above that film. The contribution of viscous losses on the mass, measured by G'', is immediately apparent.

An equivalent analysis yields

$$\Delta D = \frac{1}{\pi f\sqrt{\rho_q \mu_q}}\left[\frac{\eta_f}{\delta_f} + 2\left(\frac{\eta_f}{\delta_f}\right)^2 \frac{h\omega G'}{G'^2 + G''^2}\right] \qquad [23]$$

for the dissipation value in the same situation. With this additional relation, given information about the film, specifically its thickness, and knowledge of the fluid properties of the bulk, we can use our system of equations (20 and 23) in order to calculate a unique solution for the rheological parameters G' and G''.

Notice that the mass correction, Equation 22, is much different that the previously reported Johannsmann correction, Equation 18 (reproduced below):

$$m_{eq} = m_0\left[1 - \frac{\rho_f \omega^2 h^2}{3}\frac{G''}{G'^2 + G''^2}\right]$$

This is actually the identical result of the Voinova analysis if we ignore any contribution from the bulk fluid (*20*). Notice that this contribution is much smaller, since it scales with h^2, as opposed to h, in a regime where h is very small (~1 to 100 nm).

As reported by Voinova *et. al.*, in situations where there is a viscoelastic thin film present, bulk fluid loading has more than just an additive effect on the resultant Δf and ΔD values. Due to dissipative energy transfer between the film and the bulk, adding a viscous bulk media *accentuates* the effect of G' and G'' on the QCM response. As made explicit in Equations 20 and 23, the contribution of G' and G'' scale with η_f, so the more viscous the bulk solution, the greater the contribution of the rheological features of the material to the sensor response.

Despite coming from two different perspectives, the Voinova and Johannsmann analyses reach identical conclusions. The Voinova analysis, however, is more general, accounting for bulk fluid in a more explicit manner. From the Voinova analysis, the rheological parameters for a thin adsorbed film can be determined. Both analyses faithfully reproduce the Sauerbrey equation in limiting conditions, and show the rheological origins of deviations from the Sauerbrey equation.

While sufficient theory has been covered in order to perform both bulk and surface rheological measurements with a QCM, a brief discussion of how to translate the generic rheological parameters G' and G'' to real physical properties is in order.

Viscoelastic Models for Thin Films and Bulk Solutions

Once the generic complex shear modulus, G^*, is acquired, we desire a route to real physical properties. Comparison of G' and G'' within one material can be quite insightful, and lead to meaningful advances in understanding of that material. To effectively compare these values to absolutes, and perform quantitative calculations, however, they are typically translated into other properties. G', the storage modulus, is typically converted into an elastic modulus, while G'' is related to a specific kind of viscosity depending on the material. This process takes the unitless moduli and translates them into physical properties with quantitative units associated with them.

Many models exist for these kinds of conversions, and the theory surrounding this process is well established, so only a brief discussion will be conducted here. Two models, however, are so ubiquitous in QCM analyses that they warrant a brief discussion. The Maxwell model is typically used to model a viscoelastic bulk fluid, while the Voigt model is used to model an adsorbed film (*2*, *3*, *12*).

Maxwell Model

The Maxwell model of a viscoelastic fluid considers the material to be a spring and dashpot in series (Fig 3a). This means that there is one energy storage term, equivalent to Hook's law, added to an energy loss term, which is simply a linear loss with respect to flow. When added, these terms for the Maxwell model, relating applied stress σ, to induced strain ε, in the material

$$\frac{\sigma(t)}{dt} = E\frac{d\varepsilon(t)}{dt} + \eta\varepsilon(t) \qquad [24]$$

where E is a storage modulus, equivalent to the spring constant in Hook's law, and η is the viscosity of the material. Note that in this expression, stress and strain are spatial units, expressing distance traveled by the material. For a more general discussion, see Ref. (*2*).

If we apply a sinusoidal stress, and substitute a wave equation into the LHS of Equation 24, we can obtain the relation between the complex shear modulus G^* and the Maxwell parameters E and η

$$G^* = G' + iG'' = \frac{1}{1/E - i/(\omega\eta)} = \frac{E\omega^2\eta^2 + iE^2\omega\eta}{\omega^2\eta^2 + E^2} \qquad [25]$$

which can easily be separated into real and imaginary components, corresponding to the storage (G') and loss (G'') moduli, respectively.

Notice that the Maxwell material will not creep – if a very small stress is applied over a long period of time, it will be dissipated by the dashpot term. The material will flow over time under an applied stress, and is therefore a fluid. Energy stored in the first (elastic) term shows how the material will spring if the applied stress is rapid, while the second term demonstrates how flow will occur on a slower time scale. Energy partitioning between these features will be determined by the relative ratios of E and η. Note, if an induced stress is constant $d\sigma/dt = 0$, the material will relax, releasing its elastic energy (strain) over time proportional to E, causing flow proportional to η. Therefore, the Maxwell model faithfully reproduces the properties of a fluid, and is used to model complex bulk fluids, such as vesicle solutions or microemulsions.

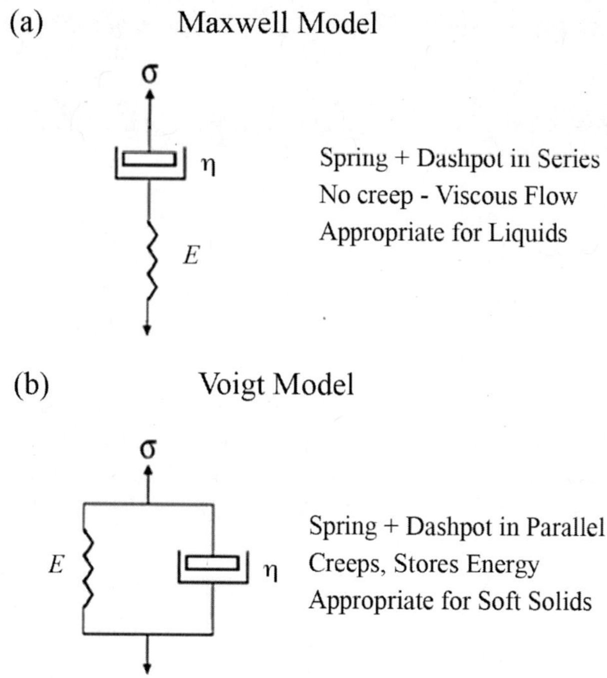

Figure 3. Schematics of common viscoelastic models. (a) Maxwell model, (b) Voigt model.

Voigt Model

While the Maxwell model is the simplest viscoelastic fluid model, the Voigt model (sometimes called the Kelvin-Voigt model) is the corresponding model for a solid. It consists of a dashpot and spring in series (Fig 3b), resulting in the stress-strain relationship

$$\sigma(t) = E\varepsilon(t) + \eta \frac{d\varepsilon(t)}{dt} \qquad [26]$$

which leads to

$$G^* = G' + iG'' = E + i\omega\eta \qquad [27]$$

for an applied sinusoidal strain.

Notice that, in contrast to the Maxwell model, the Voigt model will store elastic energy even while not under flow. Upon an applied constant stress, it will deform over a time dictated by η, with more viscous (higher η) materials deforming more slowly. The total deformation, however, is dependent only on E, the spring constant. Should the stress then be removed, the material will relax over the same period of time, until the strain returns to its initial value. The Voigt model, then, is appropriate for a solid material that is nearly elastic, but shows some viscous flow.

It is the standard model for thin films of polymers, supported vesicle layers, and other soft, thin films.

Equipped with these tools, we are now capable of using a QCM to make rheological measurements on either surface adsorbed material or bulk solutions. Alternatively, we can use rheological information to correct microgravimetry experiments, in order to obtain more accurate masses or kinetic information.

Applications of QCM-Based Rheology

While the theory surrounding acoustic rheology is well developed, as of the publication of this book, its application to scientific problems is uncommon. This is likely due to a lack of knowledge of the potential information discernable with rheological measurements, the scattered and inaccessible nature of the literature treating the theory of ultrasonic rheology, and the expense of QCM instrumentation. Despite these setbacks, labs already equipped with a QCM could begin implementation of rheological experiments with little effort – no additional data collection or analysis tools are necessary. With the equations above, the most basic QCM experiment can be translated into a rheological measurement.

As previously mentioned, rheological measurements provide valuable insight into the molecular and macromolecular structure of bulk fluids and films. The sensitivity, small sample size, and ease of use of the QCM make it an attractive tool to pursue these analytical problems with. The most powerful aspect of the QCM, however, is still its surface specificity. The QCM can perform rheological measurements on structures as thin as a lipid bilayer (*20*) – two molecular widths – and still extract significant information from a strong signal.

While the applications of QCM to acoustic rheology have been limited, they have proved its capabilities. The methods and results of a few groundbreaking experiments, along with a discussion of their significance, are presented in this section. We begin examining situations where rheology calls for and provides a correction for microgravimetry experiments, and then move to true rheological experiments on bulk and surface systems, respectively.

Microgravimetry Corrections

In many applications of the QCM, viscous losses in adsorbed films result in an underestimate of the mass when the Sauerbrey equation is applied. Because the Sauerbrey expression assumes a fully elastic film, when this condition is not met, application of Equation 1 will not result in an accurate mass. True Sauerbrey films are actually rather rare in modern QCM experimentation, which focuses on soft macromolecular films. Polymer, protein, or vesicle layers, if sufficiently thick, will all induce viscous losses that will create errors in measured masses. Well-hydrated layers tend to be the most viscous, and least suited for Sauerbrey treatment. Additionally, since these effects tend to scale with the quantity of material at the surface, kinetic measurements can also be distorted, as the more material that adsorbs, the greater the error in reported mass will be.

There are a number of ways to check for the breakdown of Sauerbrey conditions. If the dissipation value is significant, typically noted as a $\Delta D/\Delta f$ ratio greater than 10^{-7} Hz^{-1}, it is likely that the Sauerbrey equation is no longer valid. The best test, however, uses deviation in reported overtones. Since the Sauerbrey equation scales linearly with overtone number (see Eq. 1), if the values of $\Delta f_n/n$ are not equivalent, then the Sauerbrey model is invalid (*15*) (here, Δf_n represents the value of Δf at any harmonic number n). This can be easily represented by graphing $\Delta f_n/n$ vs. time and seeing if the different traces overlap. If they do, a Sauerbrey analysis is appropriate, if not, a viscoelastic model must be employed.

Claesson and coworkers have applied all of the discussed models to a system of polymer brushes (*37*). They adsorbed polymers consisting of a random blend of hydrophilic but uncharged $PEO_{45}MEMA$ and cationic METAC monomers to form polymer brush films with different water content (see original paper for structures and synthesis information). Changing the monomer ratio also affected the amount of material adsorbed. Brushes containing an even mix of both monomers tend to adsorb more to the surface, and contain more water than pure layers.

Table 1 shows the summary of results obtained by measuring the adsorption of these brush formulations using QCM. The notation $PEO_{45}MEMA$:METAC-*x* indicates the percentage *x* of monomers in the polymer that were METAC, as opposed to $PEO_{45}MEMA$. As can be seen, in almost all cases, the Sauerbrey analysis underestimates the mass, especially when the water content is high. At the greatest difference, the Sauerbrey mass differs from the more accurate Voinova mass by over 40%. Note that these values represent an extreme case, where the polymer brushes have a water content of about 80%, an exceptionally high value.

Interestingly, we can see that the Johannsmann analysis doesn't provide a better measure of the film mass than the Sauerbrey analysis. These masses were calculated by assuming $J^*(\omega)$ in Equation 18 was linear with respect to frequency. As can be seen in Figure 4, this is only approximately the case. This is one source of error in these measurements. Even attempting to minimize this error by choosing a minimal (largest negative) slope, such that the mass estimate is maximized, the Johannsmann mass reports a smaller value than the Voinova model. This is a direct result of the fact that the Johannsmann equation doesn't take into account the dissipative effect of the bulk fluid above the film. As previously mentioned, this results in an under-correction for the mass adsorbed.

In microgravimetry experiments conducted on well-hydrated or naturally viscous materials, a Sauerbrey analysis may fail. The case of a thin film of polymer brushes presented by Claesson and coworkers show this to be the case. It is facile to check for conditions under which the Sauerbrey equation is valid by plotting $\Delta f_n/n$ traces. When the Sauerbrey approximation becomes untenable, the Voigt model can easily be applied to acquire the parameters for film density ρ and thickness h (and therefore mass), along with both components of the complex shear modulus, G^*. For well hydrated polymer brushes, this allowed for an accurate analysis of the film mass, including coupled water.

Table 1. Comparison of Viscoelastic Models. Source: Iruthayaraj, J; Olanya, G; Claesson, P. M. *J. Phys. Chem. B.* 2008, *112*, 15028-15036.

Material	*Sauerbrey*	*m / ng cm⁻²* *Johannsmann*	*Voinova*	$m_{optical}$ */ ng cm⁻²*	*Mass% Water*	*%Diff S/V*[a]
poly(PEO_{45}MEMA)	380	380	432	70	83.8%	13.5%
PEO_{45}MEMA:METAC-2	1452	1504	1854	214	88.5%	27.7%
PEO_{45}MEMA:METAC-10	1648	1810	2085	202	90.3%	26.5%
PEO_{45}MEMA:METAC-25	1193	1214	1336	154	88.5%	12.0%
PEO_{45}MEMA:METAC-50	607	637	853	138	83.8%	40.7%
PEO_{45}MEMA:METAC-75	411	421	493	79	84.0%	20.0%
poly(METAC)	40	40	-	21	-	-

[a] Percentage difference in Sauerbrey and Voinova Masses

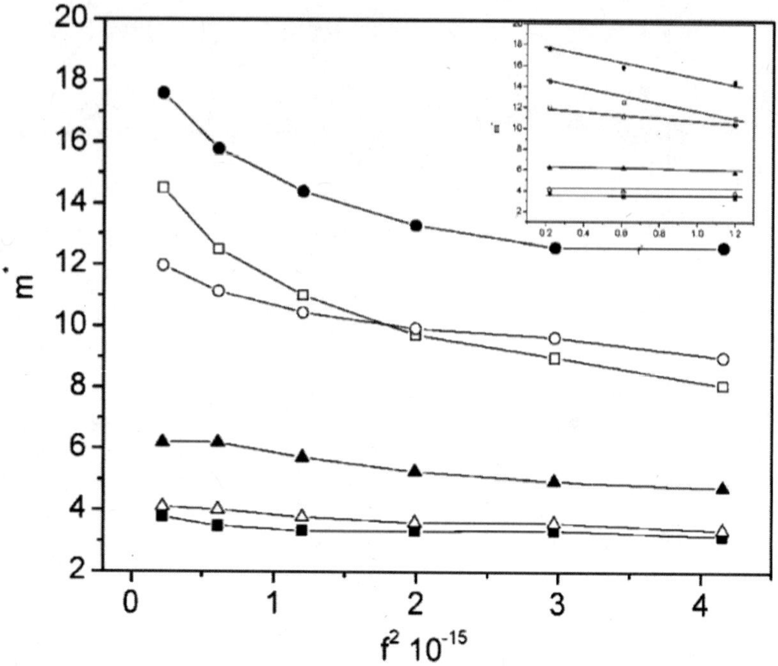

Figure 4. Plot of equivalent mass (m) vs. frequency squared (f^2) for the $PEO_{45}MEMA$:METAC-x series: x = 2% (□), x = 10% (●), x = 25% (○), x = 50% (▴), x = 75% (▵), and poly($PEO_{45}MEMA$) (■). (Inset) The frequency-independent sensed mass obtained by a linear fit to the mass calculated using the first three overtones. Reproduced from Ref. (37).

Bulk Rheology

The QCM functions surprisingly well as a bulk rheometer, and has a number of advantages over standard rheological instruments. Typical QCMs require a small sample size (usually ~1 mL minimum, though less is theoretically feasible), have the potential to perform automated experiments in series, and are highly precise, with standard deviations on the order of 0.01 to 0.1 mPa·s for viscosity measurements. Measurements tend to be more accurate and precise for lower viscosity substances, though this may be a failing in current theory to explicitly account for high-damping conditions in QCM measurements. The Maxwell model, in particular, seems to break down at high viscosities (above 3 mPa·s, errors typically exceed 10%, see Ref. (*38*)). Using the QCM as a rheometer does have two major drawbacks. First, the instrument is expensive to purchase and maintain. Second, the frequency range at which data can be taken is relatively small. This range is also high frequency, which while not an intrinsic negative, does limit the instrument to characterization of megahertz timescale features.

While the QCM can be used to simply measure the viscosity of a Newtonian or non-Newtonian (*e.g.* Maxwellian) fluid, and does so quite well, obtaining the G' and G'' parameters for a complex system is often much more illuminating. Below, the application of ultrasonic rheology on a QCM is demonstrated through two different studies, one on surfactant systems (*38*), and one on protein solutions (*39*). The first study, on surfactants, demonstrates the sensitivity and versatility of the procedure. The second shows how microscopic information can be extracted from bulk rheological parameters.

Surfactants and Microemulsions

Surfactants microstructures – micelles, emulsions, and reverse emulsions – are extremely interesting for their industrial applications to detergent formulations, their use in nanofabrication processes, and finally their ability to affect and mimic biological environments (*40*). These features arise from the ability for surfactant molecules to spontaneously self-assemble in solution, creating ordered microstructures. Studying the nature of these microstructures, both their physical origins and our ability to rationally modify them, is therefore quite interesting.

Molecular microstructures will change the rheological properties of the solution depending on their conformation, and are therefore probable by rheology. Different structures will lead to different molecular orientations. For instance, consider an aqueous surfactant solution. Assuming the monomer concentration is constant, if the average size of any micelles in solution grows, there will be more net interaction between the hydrophobic tails of the surfactant, and less contact with solvating water (Fig. 5) (*38*). The net relative positioning of molecules in the solution has changed, and so has the energy balance between the micelle structure and its surrounding solvent.

Rheological measurements will be sensitive to the tendency for adjacent molecules to either elastically store energy or dissipate it as heat upon physical perturbation. Differences in the positions of molecules in microstructures, then, will change the rheological parameters of the system. For instance, micelles will increase the ability of a solution to dissipate energy, since their structure provides a dissipative region for heat loss.

This is demonstrated by Figure 5, which shows rheological parameters as a function of SDS concentration, measured by the QCM. Notice that below the critical micelle concentration (CMC, above which surfactants form micelle aggregates), as the concentration of SDS increases, the system is becoming more elastic (G' is increasing). Above the CMC, however, the system returns quickly to Newtonian behavior. Notice how small the changes in the moduli and viscosity actually are – QCM provided the first rheological determination of the CMC of a surfactant, as previous techniques lacked the precision to perform such measurements. Indeed, this kind of measurement is right at the sensitivity limit for the QCM, as the error on this data is on the same order of magnitude as the separation between the data points themselves.

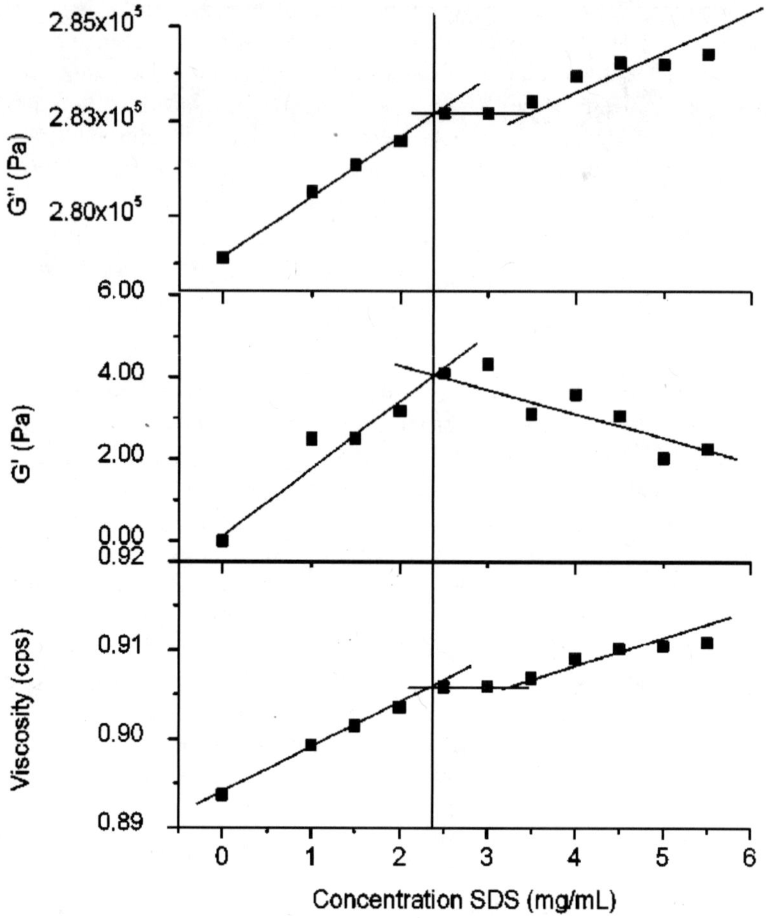

Figure 5. Viscoelastic properties of aqueous SDS solution near the CMC of SDS (2.4 mg/mL). Inflections near the CMC show the monomer to micelle transition. The large values the G" to G' ratio indicate a nearly Newtonian fluid. Lines are to guide the eye. Reproduced from Ref. (38).

The rheological parameters can also be affected by modifying the average size of micelles in solution. Figure 6 shows the change in viscosity, G', and G'' upon swelling of SDS micelles with toluene, forming a microemulsion. At each point along the abscissa, the concentration of SDS in solution is the same, but in the microemulsion trace, the average size of the micelles is larger. This leads to a larger viscous response in the same way that increasing the number of micelles does – the fraction of solution occupied by hydrophobic core is increased, resulting in a more viscous solution. Similar effects can be seen when swelling micelles with salt or a co-surfactant long chain alcohol (*38*).

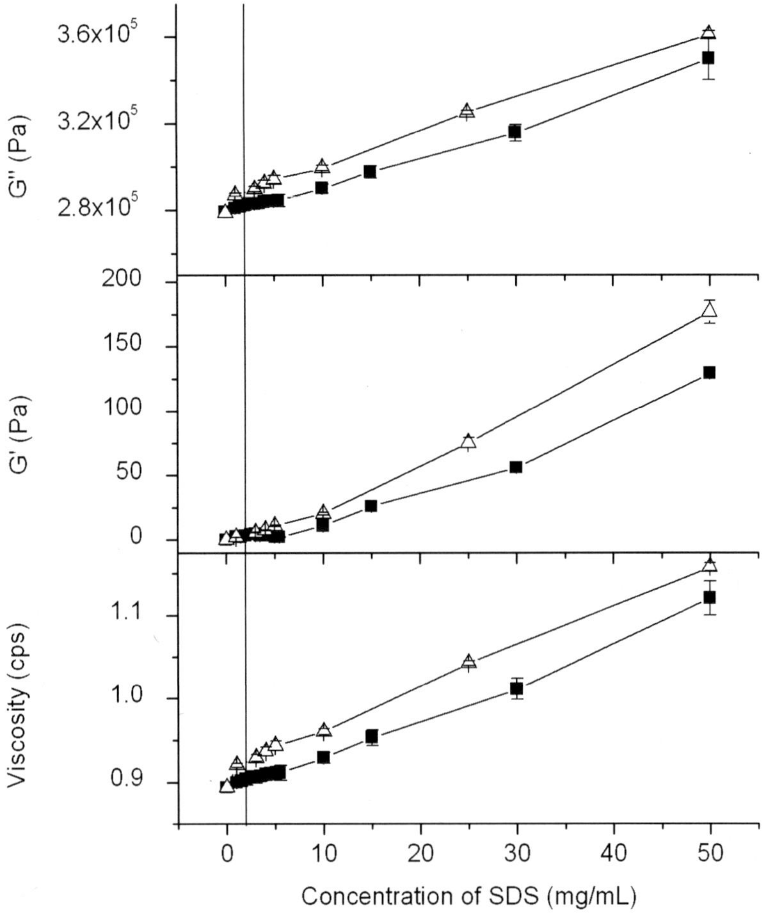

Figure 6. Viscoelastic properties of aqueous SDS (□) and an aqueous SDS saturated with toluene microemulsion (▵). Viscosity, G', and G'' values of the microemulsion are increased, especially above the CMC (2.4 mg/mL, vertical line). These increased values show the sensitivity of rheological measurements to the size and shape of solvated structures. Reproduced from Ref. (38).

Protein Solutions

There has also been a rising interest in the characterization of concentrated proteins (>100 mg/mL), since such formulations can form the basis for biologic drugs (*30*). Since most biologic drugs must be administered subcutaneously, it is beneficial to concentrate injected solutions as much as possible to minimize the volume of material added to the blood. However, this also increases the viscosity of the solution, as the proteins in the drug formula begin to aggregate. Since these drugs are typically administered via syringe, if the solution is too viscous, it can take a great deal of force and time to pump the solution into the body, which can be quite dangerous for the patient and equipment.

In order to ameliorate this problem, researchers have tested the viscosity and rheological parameters of protein solutions under various conditions. The QCM provides an excellent tool towards this end, mostly because it has such a small minimum sample volume, and highly concentrated biologic drug is extremely expensive. Pioneering studies by Amgen® have revealed that QCM measurements can be enlightening as to the conditions under which protein solution viscosity is minimized with respect to external conditions, for example pH, ionic strength, or temperature (*30*).

In addition to providing a quantifier for the ability of a protein solution to flow through a syringe, rheology of protein systems yields information about their physical behavior in solution. Specifically, an increase in viscosity, or the loss modulus, is indicative of protein aggregation. By monitoring the storage and loss moduli as a function of protein concentration, one can obtain quantitative information about the conditions under which protein aggregation occurs, and the extent of that aggregation.

Few studies have been done on the viscosity of protein solutions, and no clear framework for the quantification has arisen from those studies. This is likely due to the difficulty of generalizing protein behavior – different proteins will behave very differently in solution, aggregating at different concentrations to different extents. Additionally, the rheological data is difficult to trace back to simple first principles, such as protein structure. For instance, Figure 7 shows the rheological parameters for bovine serum albumin (BSA) (*38*), a model globular protein, as a function of concentration. The most striking feature in the data is the increase in both G' and G'' at a concentration around 400 mg/mL. Plotting the loss tangent shows a distinct change in a characteristic relaxation process at this concentration as well. Because of this rheological response, it is likely that a new mechanism of protein aggregation becomes dominant around that concentration, however we have no data about the nature of that structure. Such information would have to be collected from orthogonal methods, but might be informed by the rheology presented here.

Surface Rheology

Surface rheology provides similar information to bulk rheology, but limits that information to adsorbed material. These experiments are typically performed by fitting multiple harmonics of Δf and ΔD to obtain the unknown parameters. For a standard Voinova analysis, outlined in Equations 20 and 23 in order determine the film thickness h, the film density ρ, and the shear moduli G' and G''. Given at least two measured harmonics, and therefore four known values of Δf and ΔD, one can determine these four parameters based on an iterative error-reducing fitting process. Initial parameter guessing must be utilized, and accuracy of the fit can be increased by increasing confidence in these initial guesses, or by fixing one or more unknown parameters with externally determined data.

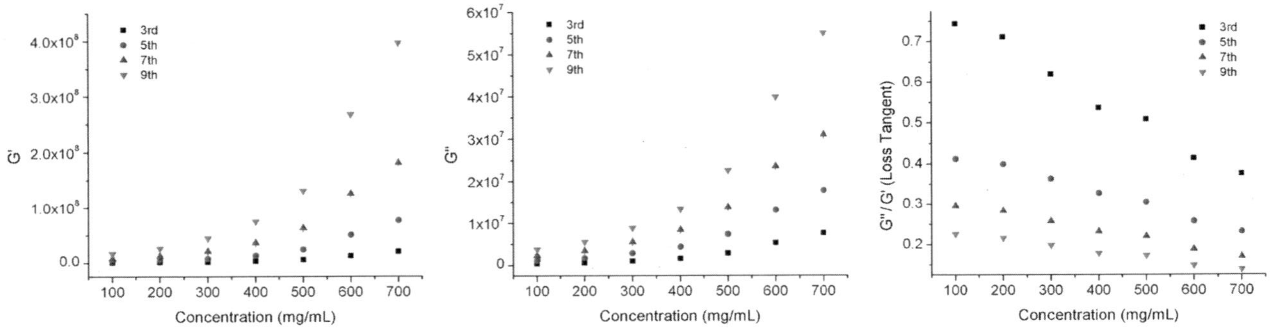

Figure 7. Rheological parameters of concentrated BSA solutions. Storage (G') and loss (G") moduli were measured at high concentrations of BSA for multiple harmonics of the QCM response. Solutions show increases in both parameters with concentration at all frequencies. Plotting the loss tangent (G"/G'), however, shows that there is a relaxation process at around 15MHz (3rd harmonic) that diminishes with increased concentration, probably due to protein aggregation (39).

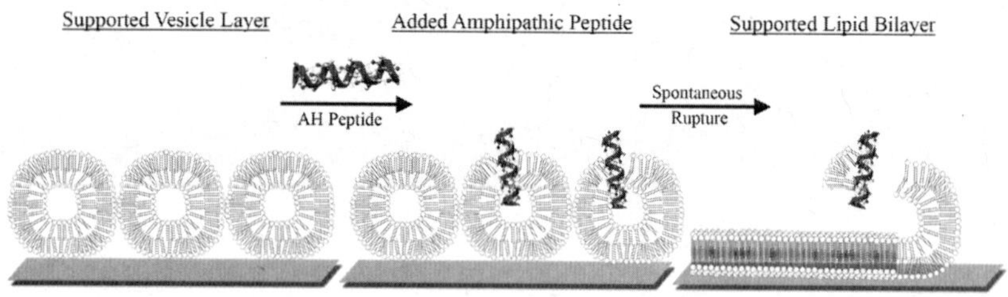

Figure 8. Schematic of the vesicle manipulation experiment conducted by Frank and coworkers, Ref. (41). First, a supported vesicular layer (SVL) was formed by adsorbing intact vesicles to the surface. Then this structure was transformed into a supported lipid bilayer (SLB) via the introduction of a specifically designed amphipathic (AH) peptide. The SVL layer will be very viscous, due to the size of the vesicles and the large amount of solvent they trap. In contrast, the SLB will be very rigid. See Fig. 9.

For instance, Frank and coworkers used this iterative process to model QCM data obtained during the deposition and rupture of adsorbed lipid vesicles (*41*). Vesicles, spherical aggregates of phospholipids, were adsorbed onto a gold surface. In this conformation, the vesicles form a very viscous surface layer, due to their large size and the quantity of water they carry with them. Then, the vesicles were ruptured with a specially designed amphipathic peptide, resulting in a loss of bound water, and forming a supported lipid bilayer (SLB). This entire process is outlined in Figure 8.

In contrast to the supported vesicle layer, the SLB is a non-dissipative structure. The differences in these structures were measured by using the Voinova model, with results shown in Figure 9. Instead of using the shear moduli G' and G'', the authors have used a Voigt model, resulting in the equivalent shear modulus μ and shear viscosity η_{sh}. Recall these terms are related via the frequency of oscillation (Eq. 27)

$$G = G' + iG'' = \mu + 2\pi i f \eta_{sh}$$

and are therefore easily inter-converted.

The Frank study demonstrates how easy it is to obtain viscoelastic information about an adsorbed substance using the QCM. Additionally, it is apparent that the QCM apparatus easily allows for modification of the system *in situ*, and can monitor such processes in real time. This is very powerful for dynamic or transient phenomena. As long as the process occurs on a timescale equal to or greater than the QCM's data acquisition speed (1 ms), the process is likely to be seen. Of course, in experiments near both the speed and sensitivity limits of the instrument, discerning subtle features of the data will be difficult if not impossible. Large changes, such as the demonstrated shift from a vesicle layer to SLB, are more amenable to experimental analysis.

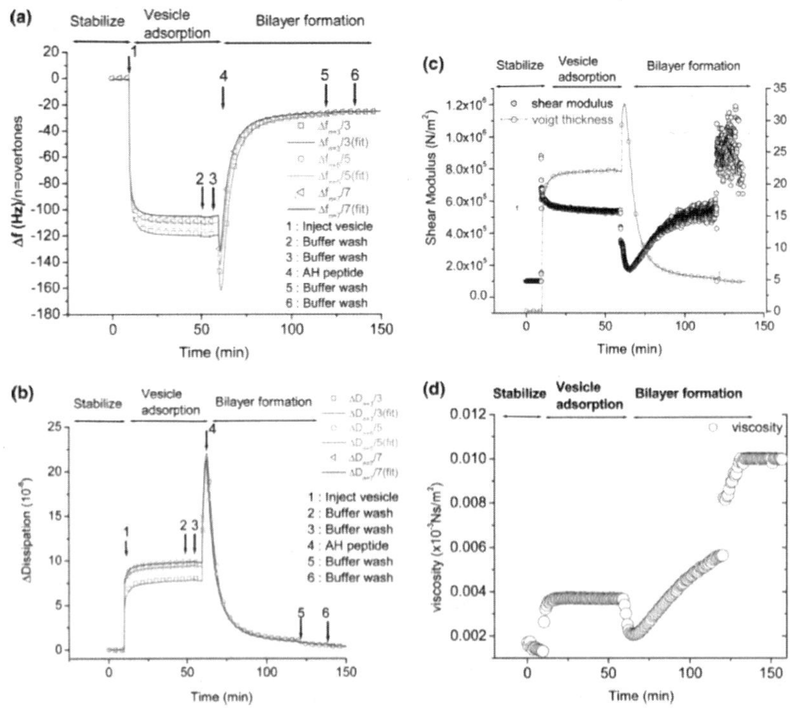

Figure 9. QCM-D responses for structural transformation from adsorbed vesicles to a bilayer induced by AH peptide. (a) QCM-D plot of $\Delta f_n/n$ vs. time for vesicle adsorption (1), followed by bilayer formation by the AH peptide (step 4, see Fig. 8). Lines show Voinova model fit. (b) Same as (a), but for ΔD. (c) Changes in the effective shear modulus and Voigt-based thickness as a function of time obtained from viscoelastic Voigt-Voinova model. (d) Shear viscosity as a function of time from the viscoelastic Voigt-Voinova model. Reproduced from Ref. (41).

Conclusions

Acoustic rheology, in the megahertz regime, provides facile access to molecular structure and dynamics. While the technique is experimentally simple, it is underutilized in current research, likely due to two factors – first, the theory surrounding the field has been presented in a technical manner by many authors who have not made attempts to reconcile actual or formal differences in their work. Secondly, as a purely rheological instrument, the purchase of a QCM would be expensive.

Despite these setbacks, there remains a large pool of current QCM users who would stand to benefit from performing a rheological analysis on their data. The most obvious use is the use of corrections for microgravimetry experiments – more accurate masses can be obtained for situations when the Sauerbrey condition breaks down. As the use of QCMs expand, especially to the biological sciences, it remains the imperative of physicist and physical chemists to rigorously review the use of published models, and ensure that these models are being applied appropriately.

Additionally, many labs with QCMs don't realize that their instrument can serve double duty as a rheometer. Even for simple viscosity measurements, a QCM will perform well. For labs with limited time or resources, the QCM can substitute for another instrument.

Finally, and most importantly, is the use of QCM as an actual rheometer, obtaining G' and G'', especially in surface situations. As demonstrated, these G^* parameters are highly sensitive to a number of physical properties of a system, and can be used to study a broad range of phenomena. Tapping the full power of the QCM requires a more advanced analysis than just plugging frequency values into the Sauerbrey equation. The field of surface chemistry would be greatly served if more scientists were aware of and utilized the powerful features of the QCM to push their research forward.

Acknowledgments

I would like to thank Mal Johal, my undergraduate research advisor, for his continued support of my research and education. Additionally, Matt Dixon provided invaluable feedback and helped analyze the BSA data. Connie Cheng and Noah Rosenberg contributed experimentally, collecting the BSA and *Bacillus anthrasis* PA data, respectively. Finally, I'd like to thank Q-Sense for their continued support of Pomona College's basic research on QCM theory and applications. The experimental data in this work was made possible through scholarships from the Arnold and Mabel Beckman Foundation and Barry M. Goldwater Foundation.

References

1. McLeish, T. C. B.; Milner, S. T. *Entangled Dynamics and Melt Flow of Branched Polymers in Branched Polymers II*; Advances in Polymer Science; Springer: Berlin, 1999; Vol. 143, pp 195−256.
2. Harris, J. *Rheology and Non-Newtonian Flow*; Longman Group Ltd.: London, 1977; Chapter 1.
3. Evans, D. F.; Wennerstrom, H. *The Colloidal Domain*, 2nd ed.; Wiley: Hoboken, NJ, 1999.
4. Maenosono, S.; Okubo, T.; Yamaguhci, Y. *J. Nanopart. Res.* **2003**, *5*, 5–15.
5. Squires, T. M.; Mason, T. G. *Annu. Rev. Fluid Mech.* **2010**, *42*, 413–438.
6. Wiertz, D. *Annu. Rev. Biophys.* **2009**, *38*, 301–326.
7. Hoffman, B. D.; Crocker, J. C. *Annu. Rev. Biomed. Eng.* **2009**, *11*, 259–288.
8. Waigh, T. A. *Rep. Prog. Phys.* **2005**, *68*, 685–742.
9. Gardel, M. L.; Weitz, D. A. In *Microscale Diagnostic Techniques*; Breuer, K. S., Ed.; Berlin: Springer, 2005; pp 1−50.
10. Mason, T. G. *Rheol. Acta* **2000**, *39*, 371–378.
11. Solomon, M. J.; Lu, Q. *Curr. Opin. Colloid Interface Sci.* **2001**, *6*, 430–437.
12. Rodahl, M.; Höök, F.; Krozer, A.; Brzezinski, P.; Kasemo, B. *Rev. Sci. Instrum.* **1995**, *66*, 3924–3930.

13. Mason, W. P.; Baker, W. O.; McSkimin, H. J.; Heiss, J. H. *Phys. Rev.* **1949**, *75*, 936.
14. Kanazawa, K. K. *Faraday Discuss.* **1997**, *107*, 77–90.
15. Sauerbrey, G. *Z. Phys.* **1959**, *155*, 206–222.
16. Nomura, T.; Okuhara, M. *Anal. Chim. Acta* **1982**, *142*, 281–284.
17. Marx, K. A. *Biomacromolecules* **2003**, *4*, 1099–1120.
18. Lane, T. J.; Fletcher, W. R.; Gormally, M. V.; Johal, M. S. *Langmuir* **2008**, *24*, 10633–10636.
19. Keller, C. A.; Kasemo, B. *Biophys. J.* **1998**, *75*, 1397–1402.
20. Voinova, M. V.; Jonson, M.; Kasemo, B. *Biosens. Bioelectron.* **2002**, *17*, 835–841.
21. Rodahl, M.; Hook, F.; Fredriksson, C.; Keller, C. A.; Krozer, A.; Brzezinski, P.; Voinova, M.; Kasemo, B. *Faraday Discuss.* **1997**, *107*, 229–246.
22. Mason, W. P. *Piezoelectric Crystals and Their Applications to Ultrasonics*; Van Nostrand: Princeton, 1948.
23. Tiersten, H. F. *J. Acoust. Soc. Am.* **1963**, *35*, 234.
24. Johannsmann, D.; Heim, L.-H. *J. Appl. Phys.* **2006**, *100*, 094505.
25. Unpublished data, collected by Rosenberg, N. M.
26. Johannsmann, D.; Mathauer, K.; Knoll, W. *Phys. Rev. B* **1992**, *46*, 7808–7815.
27. Kanazawa, K. K.; Gordon, J. G. *Anal. Chem.* **1985**, *57*, 1771–1772.
28. Rodahl, M.; Kasemo, B. *Sens. Actuators* **1996**, *54*, 448–456.
29. Stockbridge, C. D. *Vacuum Microbalance Techniques*; New York, NY, 1966; Vol. 5, pp 147–178.
30. Patel, A. R.; Kerwin, B. A.; Kanapuram, S. R. *J. Pharm. Sci.* **2008**, *10*, 1002.
31. Saluja, A.; Badkar, A. V.; Zeng, D. L.; Kalonia, D. S. *J. Pharm. Sci.* **2007**, *10*, 1002.
32. Kudryashov, E. D.; Hunt, N. T.; Arikainen, E. O.; Buckin, V. A. *J. Dairy Sci.* **2001**, *84*, 375–388.
33. Goodwin, J. W.; Goodwin, J. W.; Hughes, R. W. *Rheology for Chemists: an introduction*; Royal Society of Chemistry: Caimbridge, UK, 2008; Chapter 2.
34. See, e.g., Zwang, T. J.; Fletcher, W. R.; Lane, T. J.; Johal, M. S. *Langmuir* **2010**, *26*, 4598–-4601.
35. See, for more history: Buttry, D. A.; Ward, M. D. *Chem. Rev.* **1992**, *92*, 1355.
36. Cross, G. H.; Reeves, A.; Brand, S.; Swann, M. J.; Peel, L. L.; Freeman, N. J.; Lu, J. R. *J. Phys. D: Appl. Phys.* **2004**, *37*, 74–80.
37. Iruthayaraj, J.; Olanya, G.; Claesson, P. M. *J. Phys. Chem. C* **2008**, *112*, 15028–15036.
38. Lane, T. J.; Cheng, C. Y. H.; Dixon, M. C.; Oom, A.; Johal, M. S. *Anal. Chem.* **2008**, *80*, 7840–7845.
39. Unpublished data, acquired by Cheng, C. H. Y., with Q-Sense, Inc.
40. *Microemulsions: Properties and Applications*; Fanun, M., Ed.; Surfactant Science Series; Taylor & Francis, Inc.: Boca Raton, FL, 2008; Vol. 144.

41. Cho, N.-J.; Kanazawa, K. K.; Glenn, J. S.; Frank, C. W. *Anal. Chem.* **2007**, *79*, 7027–7035.

Chapter 11

Effect of Hydrophobization of Gold QCM-D Crystals on Surfactant Adsorption at the Solid-Liquid Interface

Mona Marie Knock[1,*] and Laurie S. Sanii[1,2]

[1]Clorox Services Company, Clorox Technical Center, 7200 Johnson Drive, Pleasanton, California 94588
[2]Current address: Kemira Chemicals, Inc., Atlanta R&D Center, 387 Technology Circle, Suite 300, Atlanta, Georgia 30313
*E-mail: mona.knock@clorox.com

The effect of hydrophobization of gold quartz crystal microbalance with dissipation (QCM-D) crystals is investigated in terms of its effect on surfactant adsorption at the solid-liquid interface. Hydrophilic gold crystals were modified with octadecanethiol to produce a hydrophobic surface. The gold surface was characterized prior to and after coating. Water contact angles were obtained to confirm hydrophobicity and atomic force microscopy (AFM) images were acquired to confirm surface modification and alkylthiol monolayer morphology. Modified crystals were tested for QCM-D response uniformity and then exposed to surfactant solutions. The effect of hydrophobization on the adsorption of surfactants is investigated for sodium dodecyl sulfate (SDS), 'dioctyl' sulfosuccinate sodium salt (AOT), hexadecyltrimethylammonium bromide (CTAB), and dihexadecyldimethylammonium bromide (DHDAB). Surfactant solutions were studied above their respective critical micelle concentration (CMC) in the absence of added electrolyte. Adsorption for SDS and AOT on hydrophilic gold was minimal and was below the detection limit for QCM-D at concentrations just above the CMC. For CTAB and DHDAB solutions at concentrations well above the CMC, adsorption on hydrophilic gold was substantial, forming more than one

© 2011 American Chemical Society

bilayer of coverage. On hydrophobed gold, SDS, AOT, CTAB, and DHDAB all adsorbed to form a surfactant monolayer.

Introduction

Surfactant adsorption at the solid-liquid interface is frequently studied by a selection of techniques, including classical adsorption isotherms on powders (*1*), neutron reflection (NR) (*2–5*), reflectometry (*6*), ellipsometry (*7–9*), surface plasmon resonance (SPR) (*10–12*), FTIR (*13*), atomic absorption spectroscopy (AAS) and X-ray photoelectron spectroscopy (XPS) (*14*), TIR Raman and sum-frequency spectroscopy (SFS) (*15*). Despite the widespread interest in surfactant adsorption at this interface, few results are published which rely upon quantitative surfactant adsorption data acquired by a quartz crystal microbalance with dissipation (QCM-D) (*6, 16, 17*). Instead, QCM-D is more frequently applied to quantifying the adsorption of larger molecules, such as proteins and polymers, as well as bulky biological samples such as lipid layers, cells, and bacteria. Nevertheless, QCM-D offers an ideal experimental setup for fast, flow-cell or static measurements of adsorption at the solid-liquid interface. Here, we explore the applicability of QCM-D to quantifying surfactant adsorption at the solid-solution interface. The effect of whether the sensor surface is hydrophilic or hydrophobic is compared with the effect of changing the surfactant tail from single to twin, and the surfactant charge from anionic to cationic. The surfactants are limited to those classic examples for which there is a large amount of comparative adsorption data: sodium dodecyl sulfate (SDS); 'dioctyl' sulfosuccinate sodium salt, Aerosol-OT (AOT); hexadecyltrimethylammonium bromide (CTAB); and dihexadecyldimethylammonium bromide (DHDAB). The hydrophilic surface is freshly cleaned gold, and the hydrophobic surface is octadecanethiolate self-assembled monolayers (SAMs) on gold. Great care is taken to ensure reproducibility of the sensor surface condition as well as the QCM-D experiment. Adsorption from aqueous anionic surfactant solutions was studied at 25°C, and adsorption from aqueous solutions of cationic surfactants was studied at 32°C, a temperature well above the 28°C Krafft point of the most water-insoluble surfactant, DHDAB (*18*). Surfactant solutions were studied above their respective critical micelle concentration (CMC) in the absence of added electrolyte. Surfactant adsorption results on hydrophilic and hydrophobed gold are reported out in three forms: measured ng/cm^{-2}, area per molecule based on the expected adsorption model, and adsorption model based on absolute adsorbed weight and published areas per molecule.

Materials and Methods

Surfactants

Surfactant solutions were made with ultrapure 18 MΩ resistivity water from a Barnstead EasyPure II UV/UF system. Different surfactants compared at same relative concentrations above their CMCs. SDS was obtained from Aldrich as SigmaUltra, ≥99.0%, and solutions were prepared at 13.67 mM, 1.67 times the critical micelle concentration (CMC) of 8.2 mM. AOT was obtained from Aldrich as SigmaUltra, ≥99%, and solutions were prepared at 4 mM, 1.67 times the CMC of 2.45 mM. CTAB was obtained as SigmaUltra, ≥99% pure. CTAB solutions were prepared at 20 mM, which is 20 times the CMC of 1 mM, and above the Krafft temperature of 24°C. DHDAB was from Fluka, ≥ 97% pure. DHDAB solutions were prepared at 0.8 mM, which is 20 times above the critical aggregation concentration of $4 \pm 2 \times 10^{-5}$ M, according to the protocol described by Tucker, et al. (*19*), and held at 32°C, well above the Krafft temperature of 28°C. All surfactants were used as received, without further purification, and without added electrolyte.

QCM-D Sensor Substrates and SAM Formation

Gold QSX 301 QCM-D sensors were obtained from Q-Sense. Prior to use or SAM formation, sensors were cleaned in plasma cleaner (Harrick PDC-32G) at medium RF, for 20-60 minutes. After plasma cleaning, sensors were soaked in 200 proof ethanol (200 proof, 99.5+% ACS reagent, Acros) for at least 30 minutes to reduce adsorbed oxygen (*16, 17*). After soaking in ethanol, sensors were dried with nitrogen prior to hydrophobic surface modification or use as the hydrophilic gold surface.

For hydrophobic SAM formation, clean gold sensors were dip-coated in a 5 mM or greater solution of 1-octadecanethiol (Sigma-Aldrich, purity 98%) dissolved completely in extra dry hexanes (Acros Organics, 99.9%, Extra Dry, AcroSeal) (*20*). Sensors were left in thiol solution overnight. Rinse solutions were hexanes, followed by absolute ethanol.

Substrate Uniformity: Contact Angles and AFM

QCM-D sensors were checked for uniformity through measurement of their contact angles with water. Sessile 5 μL water droplet contact angles were measured on Krüss DSA 10L drop shape analysis instrument. Drop shape was 'circle-fitted' for small angles, 'Tangent-2' fitted for large angles using the DSA Version 1.80.0.10 software. Hydrophobicity spot checks were done on all hydrophobic crystals prior to all experiments. All inadequate crystals were discarded.

QCM-D sensor uniformity was also examined by AFM, using the non-contact mode of a Veeco CP-II AFM. AFM tips were Veeco RTESPA-M5 (MPP11123-50) mounted tips, 1-10 Ohm-cm Phosphorus(n) doped Si, with a resonant frequency between 280-319 kHz, and spring constant between 20-80 N/m. The resonant frequency for each tip was determined experimentally and applied prior to image acquisition. Topography and phase data were acquired in both scan directions and compared for similarity. The scan rate was set between 0.25 and 1 Hz. Topography images were acquired using Proscan 1.8 software, and image analysis was performed using Gwyddion 2.20.

QCM-D Theory

QCM-D is based upon a thin quartz disc sandwiched between a pair of electrodes. When AC voltage is applied, the piezoelectric quartz crystal oscillates. The resonance frequency (*f*) of the crystal depends on the total oscillating mass, including water coupled to the oscillation. *f* decreases when a thin film is attached to the sensor crystal. This decrease is proportional to mass of film, if the film is thin and rigid. The mass (*m*) of the adhering layer can be calculated by using the Sauerbrey (*21*) relation, $\Delta m = -\Delta f\, C / n$, where C = 17.7 ng Hz^{-1} cm^{-2} for a ~5 MHz quartz crystal and n = 1, 3, 5, 7, 9, 11 is the overtone number. Soft, viscoelastic films do not fully couple to crystal oscillation, and for such films, the Sauerbrey relation underestimates mass on surface. Soft films dampen the oscillation of the crystal. This damping or dissipation (*D*) of the crystal's oscillation reveals the film's viscoelasticity. *D* is defined as $D = E_{dissipated} / (2\pi E_{stored})$, where $E_{dissipated}$ is dissipated or lost energy during one oscillation cycle, and E_{stored} is the total amount of energy stored in the oscillator. Soft, dissipating films, or films from viscoelastic fluids must be modeled, and cannot be analyzed by the Sauerbrey relation (*22*). The guidance from Q-Sense (*23*) is to use the Sauerbrey relation if $\Delta D \leq 2 \times 10^{-6}$. The results described here exclusively use the Sauerbrey relation, as the surfactant films we observe are not viscoelastic.

QCM-D Experimental Procedure

QCM-D was performed using a Q-Sense E4, which is designed to run four simultaneous experiments. In this work, four identical experiments were performed simultaneously. Clean, dry sensors were mounted in the four flow-through modules, and the experimental setup was equilibrated at 25.0 or 32.0 ± 0.02°C with ultrapure water flow at 150 uL/min until the frequency and dissipation baselines were level. Frequencies were monitored such that drift of 3rd harmonic was ≤ 1.5 Hz/hr. Each experiment began with flow-through of ultrapure water at 150 uL/min for 10 minutes to establish a reference zero baseline. After 10 minutes, the pump was stopped briefly in order to switch from ultrapure water to surfactant solution. Surfactant solution was pumped through the modules for 85 minutes. Thereafter, the pump was stopped and surfactant solution was allowed to sit on sensors overnight. Data collection stopped after 17 hours.

Data reproducibility was checked by four simultaneous experiments, using same solution container for all four inlet tubing inputs. The temperature of the measuring chamber was kept at 25.0 or 32.0 ± 0.02°C with a Peltier unit. The average relative humidity was 45%, and the average lab temperature was 21°C. The data from overtone frequencies 3, 5, 7, 9, and 11 was averaged over all self-consistent sensor outputs.

After each quadrupled experiment, 50/50 ethanol/water was pumped over sensors to remove surfactant from sensor surfaces. Sensors were then removed from their chambers and examined by AFM and water contact angle measurement. The four chambers were disassembled, and ethanol was rinsed through their inlet/outlet ports. Ethanol-inert Kalrez o-rings were soaked in ethanol for a few minutes.

O-rings and chambers were dried by nitrogen and then re-assembled. New inlet tubing was connected to avoid possible surfactant carry-over into next experiment.

Calculation of Surfactant Area per Molecule from Sauerbrey Mass Data

Adsorbed areal mass data obtained through the Sauerbrey relation allow for direct calculation of the area per adsorbed surfactant molecule, $A_{\text{surfactant}}$. This is only possible when it is known which molecules are adsorbed, their respective molecular weights, and their expected adsorption model. The calculation converts the mathematically determined Sauerbrey thin film areal mass result, Δm, in ng/cm^2, into Å^2/molecule as follows using the molecular weight, M.W., of each surfactant:

$$\left(\frac{1}{\left(\Delta m \frac{\text{ng}}{\text{cm}^2}\right)\left(10^{-9}\frac{\text{g}}{\text{ng}}\right)\left(\frac{1}{\text{M.W.}}\frac{\text{mol}}{\text{g}}\right)\left(N_{Avogadro}\frac{\text{molecule}}{\text{mol}}\right)}\right)\left(\frac{10^{16}\,\text{Å}^2}{\text{cm}^2}\right)$$

The resulting Å^2/molecule result is, by default, for one monolayer coverage of that surfactant molecule. If the expected adsorption model is multi-layered, one multiplies the result by the number of monolayers expected, e.g. × 2 for a bilayer, or × 4 for a double bilayer. The calculated area per molecule is therefore model dependent. If the assumption of a monolayer or bilayer is incorrect, the area per molecule will be unreasonable and similarly incorrect.

Results

Water Contact Angles

The contact angles of water with cleaned hydrophilic gold, freshly hydrophobed gold, and used hydrophobed gold are shown in Table I.

Table I. Water Contact Angles on Gold Sensors

	Gold	*Octadecanethiolate*	*Exposed Octadecanethiolate*
Surface History	plasma-clean, ethanol-soak	fresh SAM on gold	post-experiment, 50/50 ethanol/water rinse
∠ ± *std. dev.*	9.3 ± 1.69°	106.1 ± 5.24°	101.1 ± 5.41°

The contact angle of water on cleaned gold indicates our gold surface is hydrophilic. There is some debate as to whether clean gold – if it even exists under atmospheric conditions – is hydrophilic, or whether the hydrophilicity of clean gold is due to molecules adsorbed from the cleaning process itself (*17*, *24*, *25*). Whether our cleaned gold QCM-D sensors are hydrophilic due to adsorbed oxygen or water, or are bare gold, is not actually of importance to analysis of our data. We are primarily concerned with the hydrophilicity of the gold substrate, rather than its chemical composition. New freshly cleaned gold sensors were used for each hydrophilic gold experiment, and were not reused.

The contact angle of the octadecanethiolate SAM on gold indicates a uniformly hydrophobic surface which is resistant to surfactant exposure and rinse-cleaning post-experiment. Although hydrophobed crystals were not reused, they retained their hydrophobicity for months when stored under absolute ethanol (*26*).

AFM

AFM images acquired pre- and post-QCM-D experiment show a uniform gold grain structure for clean hydrophilic gold as well as for hydrophobed gold. The phase images showed that the surface compositional texture was uniform for cleaned hydrophilic gold, as well as for freshly hydrophobed gold. No significant difference was observed between images of freshly cleaned gold with and without the post-plasma ethanol soak, although images of the ethanol-soaked hydrophilic gold sensor were slightly sharper, possibly due to a less hydrated surface. Figure 1 shows the AFM topography image for ethanol-soaked hydrophilic gold. The average grain feature height for cleaned hydrophilic gold is 6 nm, with an average roughness (Ra) of 1.1 nm and root mean square roughness (RMS) of 1.4 nm. Figure 2 shows the topography image for freshly hydrophobed gold. The average grain feature height for the alkanethiolate SAM on gold is 8 nm, and the Ra is 1.6 nm, and the RMS is 2.0 nm. The increased feature height for the hydrophobed gold suggests that the octadecanethiolate SAM is 2 nm thick, which agrees with literature values obtained by other methods (*20*, *27*).

AFM images for hydrophobed gold that had been exposed to surfactant solution and then rinse-cleaned with 50/50 ethanol/water indicate that the surfactant is predominantly removed by the rinse-cleaning. An example of a sensor exposed to surfactant solution and then rinse-cleaned is shown in Figure 3. Hydrophobed sensors retained their hydrophobicity after the QCM-D experiment and rinse-cleaning. Nevertheless, consistent with the hydrophilic gold sensors, which were used only once, hydrophobed sensors exposed to surfactants were not reused.

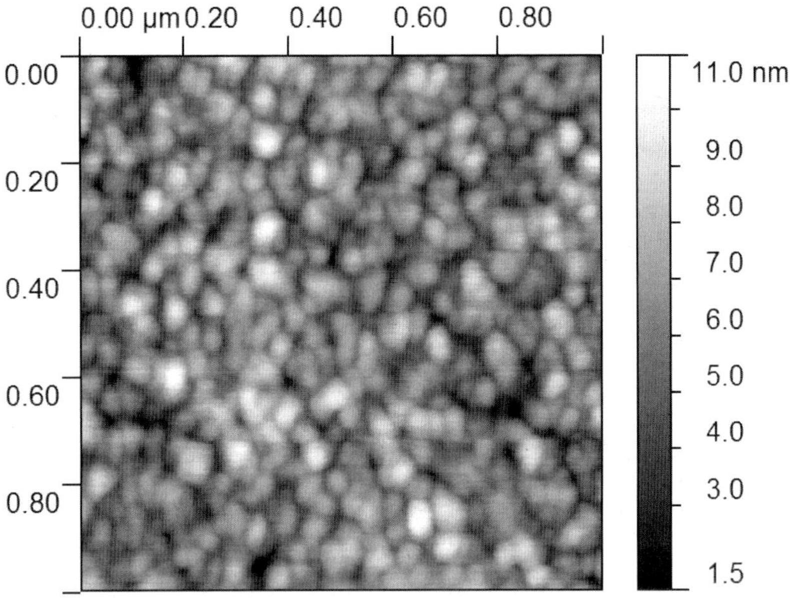

Figure 1. Non-contact AFM topography image of cleaned gold hydrophilic surface.

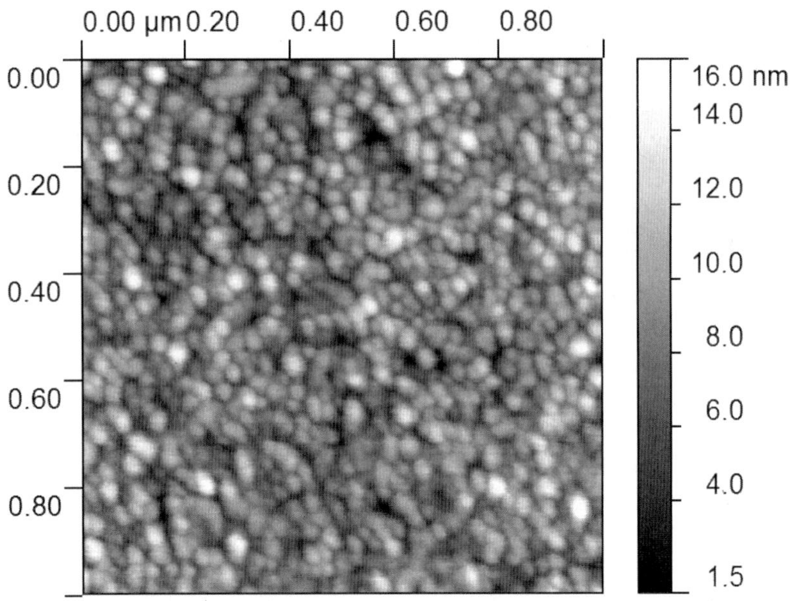

Figure 2. Non-contact AFM topography image of a gold surface freshly hydrophobed with an octadecanthiolate SAM.

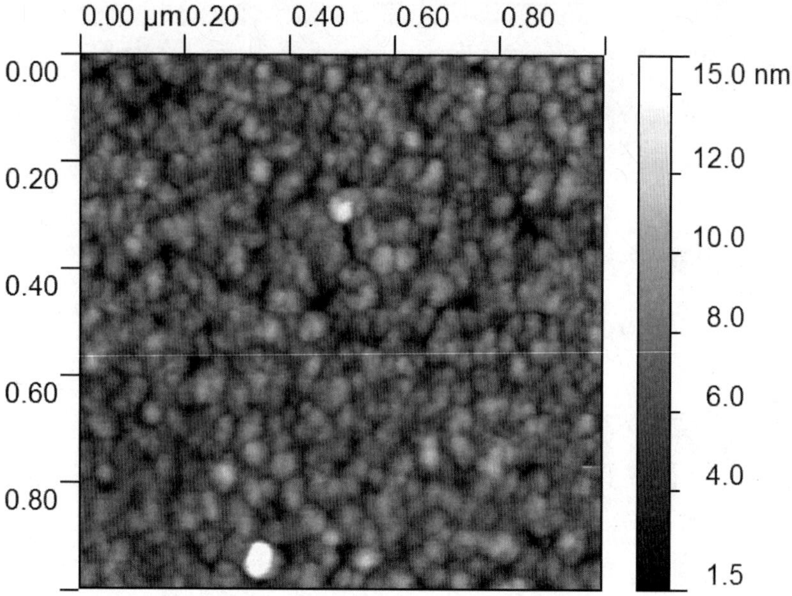

Figure 3. Non-contact AFM topography image of hydrophobed sensor after surfactant exposure and rinse-cleaning.

Effect of Gold Surface Roughness

Instead of directly calculating area per molecule from the QCM-D adsorbed areal mass data with respect to an expected adsorption model, some researchers attempt to calculate an absolute adsorbed mass per QCM-D crystal. This poses several difficulties, as the area probed by each overtone is different, with the fundamental frequency probing the entire sensor area, and the third, fifth, seventh, ninth, and eleventh overtones probing smaller concentric circles within that area (*23*). The sensitivity area decreases as the overtones increase. Nevertheless, most researchers do not use data from the fundamental frequency, as it is often erratic. Exclusively using data from the overtones complicates determination of the relevant sensor area. Although the diameter of the quartz crystal itself is 14 mm, the diameter of the gold surface is smaller, at 10 mm. It is expected that the o-rings compressed against the crystal do not significantly infringe upon the 10 mm diameter gold surface, and remain confined within 2 mm of the crystal's edge. The gold nominal surface area is therefore 0.7854 cm^2. The roughness-corrected actual area of the gold sensor has been determined through through AFM roughness profiling, as prescribed in the work of Gutig, Grady, and Striolo (*17*). The referenced authors found an increase in surface area to 1.39 cm^2 due to roughness caused by pre-experimental cleaning methods. We have attempted to replicate this result on our similarly cleaned gold sensors, but instead find our roughness-corrected area is only marginally different from the nominal area. Our calculation of roughness-corrected surface area follows for completeness, given that our results differ from the published value of 1.39 cm^2.

To calculate roughness-corrected surface area, two x-axis and two y-axis line profiles were taken for representative one micron AFM images of uncleaned, cleaned hydrophilic, and cleaned octadecanethiolate-coated gold surfaces. Roughness-corrected factors were determined by calculating the integrated length of each line profile arithmetically, and dividing that length by the nominal length of the line profile. The x- and y- roughness-corrected factors were multiplied, and the average of these two products was taken for each surface type. The roughness-corrected factors for each surface type are: 1.009855 for uncleaned gold, 1.015964 for cleaned hydrophilic gold, 1.066190 for hydrophobic octadecanethiolate-coated gold. Figure 4 shows the x- and y- line profiles used for the roughness calculations.

The roughness-corrected area results calculated from the line profiles in Figure 4 are the product of the roughness factor and the nominal area. The results are summarized as follows: gold crystals straight out of the box, uncleaned, $1.009855 \times 0.7854\ cm^2 = 0.79\ cm^2$; gold crystals after 60 minutes plasma cleaning, $1.015964 \times 0.7854\ cm^2 = 0.80\ cm^2$; gold crystals after plasma cleaning and octadecanethiol coating, $1.066190 \times 0.7854\ cm^2 = 0.84\ cm^2$. These results indicate that the roughness of gold QCM-D sensors should not be a significant factor in surfactant adsorption. Incidentally, the calculated roughness factors correspond well with the projected area statistics computed by the AFM analysis program, Gwyddion 2.20, for the full one micron scan areas: 1.009519 μm^2 for the uncleaned crystals; 1.017085 μm^2 for the gold crystals after 60 minutes plasma cleaning; 1.044906 μm^2 for gold crystals after plasma cleaning and octadecanethiol coating. We noted that small single data point importing errors of AFM line profile data can lead to large errors in area roughness calculations, resulting in increased roughness, which is a possible cause of the difference between our calculation and that in the literature.

These gold sensor surface areas can be used to determine the extent of surface coverage of a surfactant, given the Sauerbrey areal mass and an assumed area per molecule. The calculation below provides the number of molecules on the sensor:

$$\left(\Delta m \frac{\text{ng}}{\text{cm}^2}\right)\left(10^{-9}\frac{\text{g}}{\text{ng}}\right)\left(\frac{1}{\text{M.W.}}\frac{\text{mol}}{\text{g}}\right)\left(N_{Avogadro}\frac{\text{molecule}}{\text{mol}}\right)\left(A_{sensor}\text{cm}^2\right)$$

One can then calculate the extent of planar surfactant coverage by multiplying the number of molecules by the relevant $A_{\text{surfactant}}$ from the literature. Dividing that surfactant coverage area by the roughness corrected surface area provides the number of adsorbed surfactant monolayers (or equivalent, as no structural information is articulated). A result of one suggests a monolayer, two suggests a bilayer, etc. Quantifying non-planar adsorption by this method is more challenging.

QCM-D of SDS and AOT

For the anionic surfactants SDS and AOT, all adsorption experiments were performed at 25°C and at $1.67 \times$ their respective CMCs. Because the adsorption of SDS and AOT on hydrophilic gold was minimal, and approached the detection limit of QCM-D for surfactant adsorption, data from these experiments were analyzed at 20 minutes. Extending the experiment overnight with the pump off led to apparent desorption of these surfactants from hydrophilic gold, and the experimental drift over that time period exceeded the observed adsorption values. Figure 5 shows the mass adsorption behavior for SDS and AOT on hydrophilic gold and on hydrophobed gold after 20 minutes of surfactant flow, as calculated by the Sauerbrey relation. Unfortunately, the error for SDS and AOT adsorption on hydrophilic gold is too high for quantitative analysis. The standard deviations in ng/cm^{-2} are ±12.52 out of 39.84 for SDS and ±12.79 out of 34.61 for AOT. Anionic surfactants on (negative-to-neutral charge) hydrophilic surfaces adsorbed from micellar solutions are at the detection limit for QCM (*28*, *29*). This is surprising because the specifications for the Q-Sense E4 instrument state normal mass sensitivity in liquid of ~1.8 ng/cm^2, and mass sensitivity in liquid of ~0.5 ng/cm^2. These specifications do not appear to apply to the equilibrium adsorption-desorption behavior characteristic of lightweight surfactant molecules without electrostatic attraction to the substrate. Nevertheless, we can safely conclude that far less than a monolayer of SDS or AOT adsorbs to our hydrophilic gold surface.

In contrast to hydrophilic gold, SDS and AOT adsorb strongly to the octadecanethiolate hydrophobic surface. So while QCM-D does not detect an anionic surfactant monolayer (or bilayer) on hydrophilic gold, it does detect an anionic surfactant monolayer on hydrophobed gold. Using the areal mass adsorption data in Figure 5, we can calculate the area per molecule for SDS or AOT adsorbed on hydrophobed gold. We would expect that micellar solutions of AOT or SDS each form a surfactant monolayer at a hydrophobic surface. Based on this adsorption model, the A_{AOT} is 38 ± 2.5 Å^2/molecule, and the A_{SDS} is 32 ± 9.5 Å^2/molecule. A_{SDS} is somewhat smaller than the 42-59 Å^2/molecule range of molecular areas observed by other methods for micellar solutions of SDS adsorbed on hydrophobic surfaces (*10*, *30–32*). A_{AOT} is also smaller than the 60-71 Å^2/molecule range of published molecular areas observed by other methods (*33–36*). Although this could point towards lamellar ordering of AOT (*37*, *38*) at the hydrophobic interface, no lamellar phase should exist at a 4 mM (or 0.1778%) concentration at 25°C. Since our surfactants were used as received, and not purified, impurities carried over from the manufacturing process or through hydrolysis products of SDS (dodecanol) and AOT (dicarboxylic acids, monoesters, 2-ethylhexanol, sodium salts) could contribute significantly to our increased adsorbed mass. Cosolutes and impurities are known to have significant effects on solution phase (*39*) and adsorbed phase behavior (*16*), and alkanols such as dodecanol are known to concentrate at interfaces, notably increasing the packing density of the adsorbed layer (*40*). We attribute the compression of the measured (average) area per molecule for SDS and AOT to surface active impurities which concentrate at the interface and reduce the effective molecular area per adsorbed molecule. For SDS or AOT adsorbed on hydrophobed gold, we detect a more tightly packed, mixed monolayer than would be seen for pure SDS or AOT.

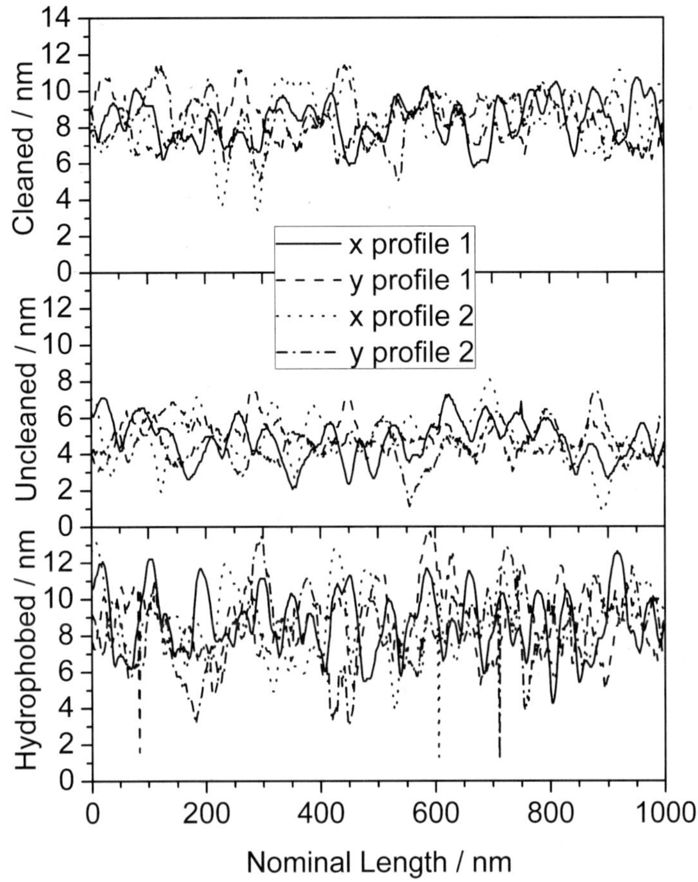

Figure 4. X- and y- line profiles from one micron 512 × 512 AFM images used to calculate surface roughness of 60-minute plasma cleaned hydrophilic gold, gold prior to cleaning, and cleaned gold which was made hydrophobic by octadecanethiolate.

We can recalculate the adsorption model using the roughness-corrected surface area adsorption data of 2.19×10^{14} molecules of AOT, and 2.67×10^{14} molecules of SDS. Assigning area per molecule values of $A_{AOT} = 60$ and $A_{SDS} = 42$ Å^2/molecule, our surface area coverage for AOT is 1.32 cm^2, and SDS is 1.12 cm^2. These surface coverage areas somewhat exceed that for a single-component monolayer over the 0.84 cm^2 hydrophobed gold area, especially for AOT. We attribute our increased adsorption mass and reduced area per surfactant molecule primarily to surface active impurities which are not accurately accounted for in either calculation approach.

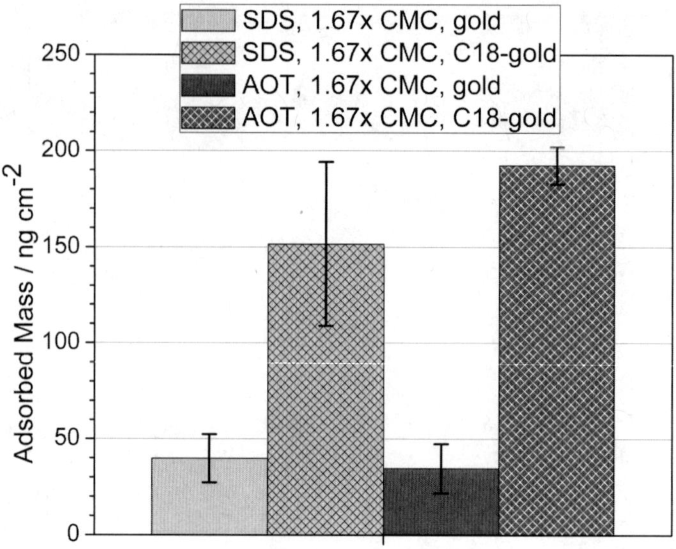

Figure 5. Adsorbed mass on hydrophilic gold and hydrophobed gold (C18-gold) after 20 minutes of exposure to 1.67 × CMC solutions of SDS or AOT.

QCM-D of CTAB and DHDAB

The surprising inability to quantitatively detect anionic surfactant adsorption on hydrophilic gold by QCM-D inspired us to extend this study to include cationic surfactants. This was quickly rewarded, as the adsorption of CTAB and DHDAB on hydrophilic gold is significantly greater than for SDS and AOT. Figure 6 presents the adsorption data for CTAB and DHDAB on hydrophilic gold and hydrophobed gold at 20 times their respective CMCs. Figure 6 presents three different adsorption timescales: pump-off after 85 minutes of flow-through, 17 hours of adsorption, and a variably timed maximum adsorption. Note that all three time points demonstrate similar amounts of adsorbed surfactant, and we can easily conclude that we are observing equilibrium adsorption. Using the maximum adsorption values for CTAB and DHDAB on bare hydrophilic gold and hydrophobed gold, areas per molecule are calculated for the anticipated adsorption model of a cationic monolayer on hydrophobed gold, and a cationic bilayer on hydrophilic gold. For CTAB on hydrophobed gold, a monolayer would result in 42 ± 6 Å^2/molecule. For DHDAB on hydrophobed gold, a monolayer would result in 59 ± 13 Å^2/molecule. For CTAB on hydrophilic gold, a bilayer would result in 24 ± 2.5 Å^2/molecule, which is atypically small. For DHDAB on hydrophilic gold, a bilayer would result in 36 ± 2 Å^2/molecule, also unusually small. It seems unreasonable for bilayers of a molecule to be 1.7 times more tightly packed than monolayers of the same molecule. Although our calculated areas per molecule for the CTAB and DHDAB monolayers on hydrophobed gold are in agreement with the literature, the areas per molecule we calculate on the assumption of a bilayer on hydrophilic gold are unrealistic (*41–45*). The data suggest that we have more than a bilayer of DHDAB and CTAB on hydrophilic gold.

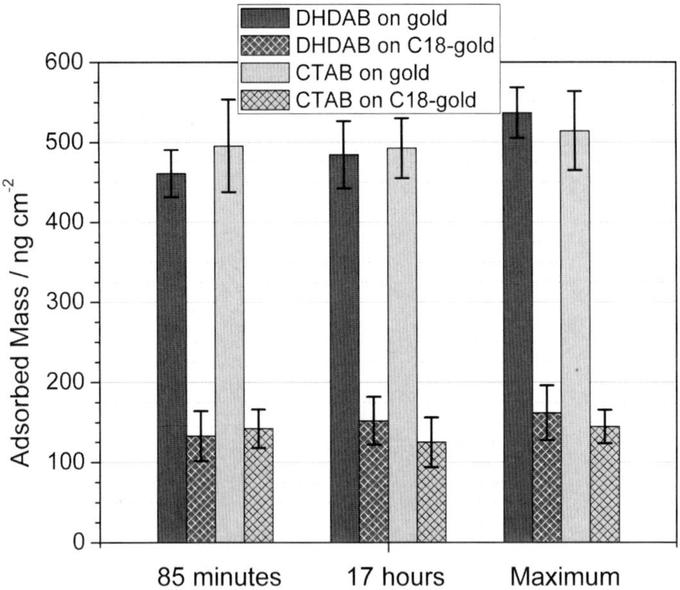

Figure 6. Adsorbed mass on hydrophilic gold and hydrophobed gold (C18-gold) after exposure to 20 × CMC solutions of CTAB or DHDAB.

At 20 times the CMC, 85 minutes of adsorption, and at 32°C, we get 495 ± 46 ng/cm^2 for CTAB on hydrophilic gold. With more than twice as much adsorbed as reported in the literature (*46*, *47*), we expect at least a bilayer for CTAB. In comparison to published results, our calculated bilayer model area per molecule appears unrealistically small at 24 ± 2.5 Å^2/molecule, and could indicate a large mass of entrapped hydration. If we attribute 30% of our mass to water, we obtain an area per molecule for CTAB of 35 Å^2/molecule, which is still somewhat small. If we attempt to attribute all of our additional mass in the bilayer model to entrapped hydration, our CTAB and DHDAB bilayers would contain 50-60% entrapped water. Attributing this much mass to entrapped hydration is unreasonable according to the work of Bingen, *et al* (*48*), who determined that the effect of hydration on QCM-D response decreases proportionally to the increase in surface adsorption coverage of the surface-active species. Given the strong adsorption demonstrated by CTAB and DHDAB on hydrophilic gold, the equilibrium experimental conditions, and the proportional decrease in QCM-D-sensed hydration as adsorption reaches its maximum value, we propose that our results on hydrophilic gold represent a double bilayer, bringing our A_{CTAB} to 47 ± 4.5 Å^2/molecule, which is much more reasonable.

Using CTAB and DHDAB maximum adsorption values on hydrophilic gold and hydrophobed gold, the following areas per molecule are obtained as shown in Table II. These areas per molecule are based upon the standard adsorption model of a surfactant monolayer on hydrophobed gold, and the proposed model of a cationic surfactant double bilayer on hydrophilic gold. Due to the correspondence of these calculated values with the literature, the proposed double bilayer model seems acceptable.

Table II. Calculated A_{CTAB} and A_{DHDAB} on Hydrophilic and Hydrophobed Gold

Adsorption Model	A_{CTAB} / Å^2	A_{DHDAB} / Å^2
Surfactant Double Bilayer on Hydrophilic Gold	47 ± 4.5	71 ± 4.5
Surfactant Monolayer On Octadecanethiolate-SAM	42 ± 6	59 ± 13

We can recalculate the hydrophilic gold adsorption model using the roughness-corrected surface area adsorption data of 4.49 × 10^{14} molecules of DHDAB, and 6.78 × 10^{14} molecules of CTAB. Assigning average values of A_{DHDAB} = 65 and A_{CTAB} = 46 Å^2/molecule, our surface area coverage for DHDAB would be 2.92 cm^2, and for CTAB would be 3.12 cm^2. Given the clean hydrophilic gold sensor roughness-corrected area of 0.80 cm^2, this comes to 3.66 layers for DHDAB, nearly a double bilayer, and 3.91 layers for CTAB, a double bilayer. Accounting for some uncertainty in the assigned molecular areas, we appear to be observing a double bilayer for CTAB and DHDAB on hydrophilic gold by the roughness-corrected surface area adsorption calculation as well. For hydrophobed gold, the corresponding recalculation uses the roughness corrected surface area adsorption data of 1.43 × 10^{14} molecules of DHDAB, and 2.00 × 10^{14} molecules of CTAB. Assigning the same values of A_{DHDAB} = 65 and A_{CTAB} = 46 Å^2/molecule, our surface area coverage on hydrophobed gold for DHDAB would be 0.93 cm^2, and for CTAB would be 0.92 cm^2. Since the roughness-corrected area for the hydrophobed gold sensor is 0.84 cm^2, these areas suggest just over one monolayer of coverage for DHDAB and CTAB on hydrophobed gold. Figure 7 illustrates the observed equilibrium adsorption of SDS, AOT, CTAB, and DHDAB as monolayers on hydrophobed gold, and CTAB and DHDAB as double bilayers on hydrophilic gold.

QCM-D of SDS at the Hydrophilic Solid-Liquid Interface

The data discussed thus far suggest that we have only trace, incomplete SDS adsorption on hydrophilic gold, certainly less than a monolayer of coverage. Is this because we are at a relatively low concentration of 1.67 × CMC? This concentration just above the CMC is convenient because the dissipation changes are ≤ 2 × 10^{-6}, enabling Sauerbrey analysis (*23*). Data collected for SDS at 4 × CMC gave mass adsorption of 48.94 ± 7.984 ng/cm^2, which is higher than observed for 1.67 × CMC. This concentration is borderline for Sauerbrey analysis since ΔD = 3-5 × 10^{-6} > 2 × 10^{-6}. If we compute the molecular area for this concentration anyway, we get 98 ± 32 Å^2/molecule for SDS at 4 × CMC, which still indicates an incomplete monolayer, but compares favorably with SPR results from Tulpar and Ducker at a hydrophilic SAM surface of near neutral charge (*29*). Looking to even higher concentrations, SDS at 20 × CMC has high ΔD and analyses by Sauerbrey or by Voigt modeling are not meaningful.

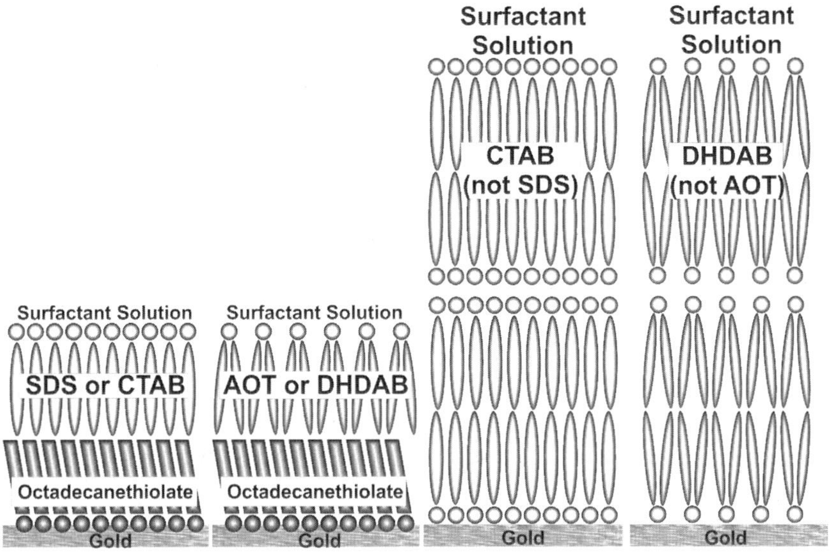

Figure 7. Cartoon of proposed model for equilibrium adsorption of SDS, AOT, CTAB, and DHDAB on hydrophobed gold and CTAB and DHDAB on hydrophilic gold.

Conclusion

QCM-D was used to quantify surfactant adsorption at the solid-liquid interface on hydrophilic gold and on an octadecanethiolate hydrophobic SAM on gold. For the anionic surfactants, SDS and AOT, QCM-D cannot detect a monolayer of coverage on hydrophilic gold at concentrations just above the CMC of each surfactant at 25°C. Further investigation at higher concentrations suggests that SDS and AOT barely adsorb onto hydrophilic surfaces of negative to neutral charge. SDS and AOT strongly adsorb to the hydrophobed gold surface, and at a monolayer of coverage, A_{SDS} and A_{AOT} are both smaller than typically reported. We attribute this enhanced adsorbed mass and compressed molecular area to the coadsorption of surface active impurities which are not accurately accounted for in the calculations. The cationic surfactants CTAB and DHDAB adsorb strongly to both hydrophilic gold and hydrophobed gold at 20 times their CMCs and at 32°C. On hydrophobed gold, their areas per molecule agree with published literature values and correspond with one monolayer of coverage. On hydrophilic gold, CTAB and DHDAB adsorb more strongly than literature analogues, and the results suggest that CTAB and DHDAB form double bilayers at the hydrophilic gold surface under equilibrium conditions.

Acknowledgments

The authors are grateful to The Clorox Company for supporting this research. We especially thank Keith Mainquist, Ph.D. of Clorox, for his helpful calculations of surface roughness. We also thank the following people for useful discussion:

David R. Scheuing and Marc W. Brutschy, of Clorox; Prof. Colin D. Bain, of the University of Durham; Prof. Brian Grady, and group, of the University of Oklahoma; and Matt Dixon, Ph.D., of Q-Sense.

References

1. Saleeb, F. Z. *Kolloid Z. Z. Polym.* **1970**, *239*, 602–605.
2. Fragneto, G.; Thomas, R. K.; Rennie, A. R.; Penfold, J. *Langmuir* **1996**, *12*, 6036–6043.
3. Li, Z. X.; Lu, J. R.; Fragneto, G.; Thomas, R. K.; Binks, B. P.; Fletcher, P. D. I.; Penfold, J. *Colloids Surf., A* **1998**, *135*, 277–281.
4. Fragneto, G.; Li, Z. X.; Thomas, R. K.; Rennie, A. R.; Penfold, J. *J. Colloid Interface Sci.* **1996**, *178*, 531–537.
5. Turner, S. F.; Clarke, S. M.; Rennie, A. R.; Thirtle, P. N.; Cooke, D. J.; Li, Z. X.; Thomas, R. K. *Langmuir* **1999**, *15*, 1017–1023.
6. Macakova, L.; Blomberg, E.; Claesson, P. M. *Langmuir* **2007**, *23*, 12436–12444.
7. Lundin, M.; Macakova, L.; Dedinaite, A.; Claesson, P. M. *Langmuir* **2008**, *24*, 3814–3827.
8. Besio, G. J.; Prud'homme, R. K.; Benziger, J. B. *Langmuir* **1988**, *4*, 140–144.
9. Stålgren, J. J. R.; Eriksson, J.; Boschkova, K. *J. Colloid Interface Sci.* **2002**, *253*, 190–195.
10. Sigal, G. B.; Mrksich, M.; Whitesides, G. M. *Langmuir* **1997**, *13*, 2749–2755.
11. Martinez, J.; Talroze, R.; Watkins, E.; Majewski, J. P.; Stroeve, P. *J. Phys. Chem. C* **2007**, *111*, 9211–9220.
12. Levchenko, A. A.; Argo, B. P.; Vidu, R.; Talroze, R. V.; Stroeve, P. *Langmuir* **2002**, *18*, 8464–8471.
13. Montgomery, M. E.; Wirth, M. J. *Langmuir* **1994**, *10*, 861–869.
14. Karlsson, P. M.; Palmqvist, A. E. C.; Holmberg, K. *Langmuir* **2008**, *24*, 13414–13419.
15. Tyrode, E.; Rutland, M. W.; Bain, C. D. *J. Am. Chem. Soc.* **2008**, *130*, 17434.
16. Shi, L.; Ghezzi, M.; Caminati, G.; Lo Nostro, P.; Grady, B. P.; Striolo, A. *Langmuir* **2009**, *25*, 5536–5544.
17. Gutig, C.; Grady, B. P.; Striolo, A. *Langmuir* **2008**, *24*, 4806–4816.
18. Grillo, I.; Penfold, J.; Tucker, I.; Cousin, F. *Langmuir* **2008**, *24*, 6509–6520.
19. Tucker, I.; Penfold, J.; Thomas, R. K.; Grillo, I.; Barker, J. G.; Mildner, D. F. R. *Langmuir* **2008**, *24*, 6509–6520.
20. Sur, U. K.; Lakshminarayanan, V. *J. Electroanal. Chem.* **2004**, *565*, 343–350.
21. Sauerbrey, G. *Z. Phys.* **1959**, *15*, 206–222.
22. Höök, F.; Rodahl, M.; Brzezinski, P.; Kasemo, B. *Langmuir* **1998**, *14*, 729–734.
23. Discussion with Matt Dixon, Ph.D., Q-Sense.
24. Bain, C. D.; Troughton, E. B.; Tao, Y. T.; Evall, J.; Whitesides, G. M.; Nuzzo, R. G. *J. Am. Chem. Soc.* **1989**, *111*, 321–335.

25. Geddes, N. J.; Paschinger, E. M.; Furlong, D. N.; Caruso, F.; Hoffmann, C. L.; Rabolt, J. F. *Thin Solid Films* **1995**, *260*, 192–199.
26. Willey, T. M.; Vance, A. L.; van Buuren, T.; Bostedt, C.; Terminello, L. J.; Fadley, C. S. *Surf. Sci.* **2005**, *576*, 188–196.
27. Nuzzo, R. G.; Dubois, L. H.; Allara, D. L. *J. Am. Chem. Soc.* **1990**, *112*, 558–569.
28. NR also found no AOT adsorption onto a hydrophilic surface, even in presence of 0.1 M NaBr. Fragneto, G.; Li, Z. X.; Thomas, R. K.; Rennie, A. R.; Penfold, J. *J. Colloid Interface Sci.* **1996**, *178*, 531–537.
29. SPR found A_{SDS} adsorbed on a hydrophilic SAM surface of near neutral charge to be ~ 100 Å^2/molecule, which compares well with our results. Tulpar, A.; Ducker, W. A. *J. Phys. Chem. B* **2004**, *108*, 1667–1676.
30. SPR of SDS on undecanethiol found 47 Å^2/molecule. Martinez, R.; Talroze, R.; Watkins, E.; Majewski, J. R.; Stroeve, P. *J. Phys. Chem. C* **2007**, *111*, 9211–9220.
31. NR of SDS on polystyrene found 42-47 Å^2/molecule. Turner, S. F.; Clarke, S. M.; Rennie, A. R.; Thirtle, P. N.; Cooke, D. J.; Li, Z. X.; Thomas, R. K. *Langmuir* **1999**, *15*, 1017–1023.
32. SPR of SDS on hexadecanethiol found 59 Å^2/molecule. Sigal, G. B.; Mrksich, M.; Whitesides, G. M. *Langmuir* **1997**, *13*, 2749–2755.
33. AOT adsorption isotherm on carbon black found 71 Å^2/molecule. Reference (*1*).
34. X-ray diffraction of the AOT lamellar phase found 61.5 Å^2/molecule. Fontell, K. *J. Colloid Interface Sci.* **1973**, *44* (2), 318–329.
35. NR of AOT at C10/water interface found 66 ± 4 Å^2/molecule. Reference (*28*).
36. NR of AOT at OTS/water interface found 78-82 ± 5 Å^2/molecule. Li, Z. X.; Lu, J. R.; Fragneto, G.; Thomas, R. K.; Binks, B. P.; Fletcher, P. D. I.; Penfold, J. *Colloids Surf., A* **1998**, *135*, 277–281.
37. Li, Z. X.; Lu, J. R.; Thomas, R. K.; Weller, A. *Langmuir* **2001**, *17*, 5858–5864.
38. Li, Z. X.; Weller, A.; Thomas, R. K. *J. Phys. Chem. B* **1999**, *103*, 10800–10806.
39. Sager, W. F. C. *Langmuir* **1998**, *14*, 6385–6395.
40. Casson, B. D.; Bain, C. D. *J. Phys. Chem. B* **1998**, *102*, 7434–7441.
41. NR found CTAB at 60 ± 5 Å^2/molecule on smooth silicon and 66 ± 4 on rough silicon. Reference (*2*).
42. DHDAB at 60 Å^2/molecule at 30°C at the air-water interface. Tucker, I.; Penfold, J.; Thomas, R. K.; Grillo, I.; Barker, J. G.; Mildner, D. F. R. *Langmuir* **2008**, *24*, 6509–6520.
43. DHDAB at 61 Å^2/molecule at 35°C at the air-water interface. Penfold, J.; Sivia, D. S.; Staples, E.; Tucker, I.; Thomas, R. K. *Langmuir* **2004**, *20*, 2265–2269.
44. DHDAB at 60 Å^2/molecule at 20°C on a Langmuir Blodgett trough. Hąc-Wydro, K.; Wydro, P.; Dynarowicz-Łątka, P. *J. Colloid Interface Sci.* **2005**, *286*, 504–510.

45. DHDAB at 74 Å^2/molecule at 30°C on a Langmuir Blodgett trough. Dynarowicz, P.; Romeu, N. V.; Trillo, J. S. *Colloids Surf., A* **1998**, *131*, 249–256.
46. CTAB at 180 ± 30 ng/cm^2 at 1.5 × CMC, 25°C, on gold by QCM-D. Reference (*16*).
47. CTAB at 270 ± 12 ng/cm^2 on silica at 1.2 × CMC, 25°C, by QCM-D and at 190 ± 10 ng/cm^2 by reflectometry. Reference (*6*).
48. Bingen, P.; Wang, G.; Steinmetz, N. F.; Rodahl, M.; Richter, R. P. *Anal. Chem.* **2008**, *80*, 8880–8890.

Chapter 12

Thermally Treated Octadecylphosphonic Acid Thin Film Grown on SS316L and Its Stability in Aqueous Environment

Min Soo Lim,[1,*] Katelyn J. Smiley,[1] and Ellen S. Gawalt[2]

[1]Department of Chemistry, Slippery Rock University, Slippery Rock, PA 16057, USA
[2]Department of Chemistry and Biochemistry, Duquesne University, Pittsburgh, PA 15282, USA
[*]min.lim@sru.edu

An octadecylphosphonic acid (ODPA) film was formed on the surface of stainless steel 316L and thermally treated at 100 – 120 °C for 30 – 60 minutes. AFM images, and water contact angles revealed that the film was stabilized enough after the thermal treatment to survive in water flush. DRIFT spectra showed that thermal treatment enhanced P-O stretches of the film. A model is suggested to explain the stability of ODPA film enhanced by thermal treatment. An ODPA molecule is physisorbed to the surface weakly by hydrogen bonding. Heating drives away water molecules leading to the formation of strong monodentate or mixed mono/bi-dentate bonds of ODPA molecules to the surface.

Introduction

Biomaterials are defined as synthetic materials that are used to replace or restore function to body tissues and are continuously or intermittently in contact with body fluids (*1*, *2*). They can be metals or metal alloys (titanium, stainless steel 316 L, and Co-Cr alloys), ceramics (alumina, zirconia, and carbon), or polymers (polyethylene and polyurethane). The range of their applications is large and wide including orthopedics, cardiovascular applications, ophthalmics, and dental applications (*2*).

© 2011 American Chemical Society

Stainless steel 316L (SS316L) has been extensively used for surgical implant materials such as cranial plates, orthopedic fracture plates, dental implants, spinal rods, and joint replacement prostheses due to its lower cost, ease of fabrication, ductility, and viable biocompatibility (*3*, *4*). However, concern about the continued usage of it has been brought up because the corrosion of stainless steel in the body can cause mechanical failure of the implants due to the cracks and cavities initiated by the corrosion process (*3–5*). Furthermore, electrochemical corrosion of the stainless steel implants in the physiological environment releases iron, nickel, and chromium ions which can act as allergens and carcinogens (*3*, *4*). Therefore, the surface properties of SS316L determine its success and failure as a biomaterial and thus the surface needs to be treated properly for better corrosion resistance and enhanced biocompatibility without losing its physical and mechanical properties. Several attempts have been made to make a thin protective coating on the surface with the self-assembly method (*6–8*).

Molecular self-assembly is a method of thin film formation that drives amphiphilic organic molecules to be spontaneously adsorbed and organized on a solid inorganic surface. A few good examples are alkanethiols on noble metals such as gold (*9–13*) and silver (*14*, *15*), alkyl carboxylic acids on metal oxides (*16–20*), alkyl phosphonic acids on mica (*21*, *22*), and alkyl silanes on SiO_2 surfaces (*23*, *24*). Typically, an adsorbing molecule consists of a head group such as thiol, carboxylic acid, phosphonic acid, or silane that initiates the molecule to anchor to a solid surface, a methylene backbone chain that influences the packing density of a film by its length (*16*, *18*, *25*), and a terminal group that determines the surface properties of the thin film (*9–11*). It is generally accepted that a minimum backbone chainlength of 11 – 12 carbons is required for the close packed monolayer (*16*, *25*). The subtle balance between substrate-head group interactions (*14*) and chain length-dependent intermolecular interactions determines the growth kinetics of the film (*16–18*, *26*, *27*). Formation of monolayers of alkyl thiols, for example, indicates that molecule-substrate interaction is the dominating force for short chain thiols (C_n, $n < 10$) while chain-chain interactions play a significant role for long chain thiols (C_n, $n > 12$) (*25*). The terminating functionality of monolayer has been widely used for changing frictional properties (*9–11*) and hydrophobicity (*28*) of surfaces, grafting overlayer in surface molecular engineering (*29*), serving as molecular recognition layer (*30*), and immobilizing bioactive molecules to surface (*31*).

Functionalized (*31*, *32*) and *normal* alkyl phosphonic acids (*7*, *8*, *33*, *34*) have been utilized for the formation of thin films on the metal alloy and metal oxide surfaces due to strong interactions between adsorbing molecules and the substrate surfaces. The thin films formed on the surfaces of biomaterials need to exhibit stability in an aqueous environment since the implants must interact with biological systems for long time periods with minimal failure. However, little work has been done to understand the stability of these films in the aqueous environment. The stability of self-assembeld monolayers (SAMs) of functionalized alkyl phosphonic acids formed on metals such as titanium and gold (*35*) and metal alloy such as stainless steel (*36*) was studied under physiological conditions. It was discovered that the SAMs was stable upto 1 to 7 days but suffered substantially from desorption afterwards. Several research groups have noted that thermally treating the films (100 – 120 °C) before using them for further studies (*29*, *31*, *33*, *34*) enhances the stability of films but these studies lack visual evidence. The present study is the first systematic attempt to document the thermally driven stabilization of ocatadecylphosphonic acid (ODPA) films self-organized on the surface of stainless steel. In this study, ODPA films have been grown on stainless steel 316L surfaces and the enhanced stability of ODPA films after thermal treatment at 100-120 °C has been tested by flushing modified substrates extensively with deionized water before and after thermal treatment and characterized using atomic force microscopy (AFM), contact angle goniometry, and infrared spectroscopy.

Experimental Section

Materials

Octadecylphosphonic acid (ODPA, 99.0 % purity) was purchased from Alfa Aesar and used without further purification. THF and methanol were purchased from Aldrich Chemical Co. Stainless steel 316L foil (0.5 mm thickness, 99.99% purity, wt%: Fe 66%, Cr 19 %, Ni 10 %, Mn 3 %, and Mo 2 %) was purchased from Goodfellow, Inc.

Substrate Preparation

SS316L substrates were cut into 1 cm × 1 cm coupons, sanded with 150, 320, 400, and 600 grit sandpapers, and then polished with 400, 800, 1200 grit silicon carbide papers, and 1 μm diamond suspension using Buehler Ecomet 4 mechanical polisher. The polished substrates were flushed extensively with acetone and methanol, cleaned by ultrasonication in acetone and methanol, and immersed in boiling methanol for 10 minutes to remove traces of impurities. The cleaned substrates were stored in an oven at 100 °C for 30 – 60 minutes.

Film Formation

ODPA thin films were grown on the SS316L substrates as follows: clean substrates were taken from the oven and cooled to room temperature. They were then incubated in 1 mM ODPA in dry THF for 3 - 10 minutes. The first group of samples was thermally treated in an oven at 100-120 °C for 30 – 60 minutes and then cooled to room temperature. They were flushed in THF to remove any physisorbed ODPA. A second group of samples was dried in air without thermal treatment, lightly rinsed in THF. Each sample from both groups was then flushed extensively and independently with deionized water, and air dried. For clarification, "samples being flushed with water" means leaving the samples in a rapid stream of water and "samples being lightly rinsed with THF" means "dipping samples in THF and slowly removing them" to remove lightly bound material.

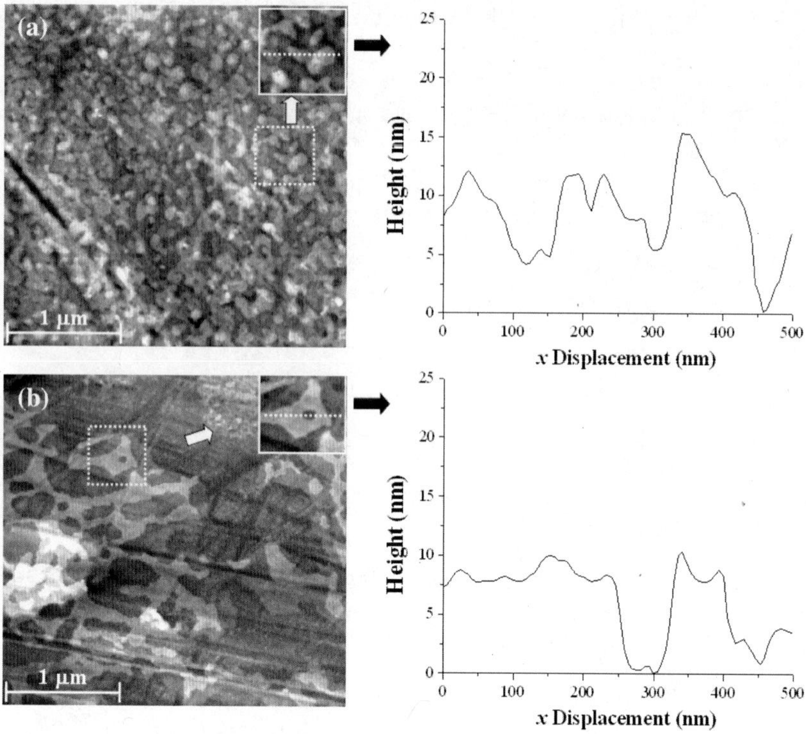

Figure 1. AFM images were obtained from ODPA films (a) with and (b) without thermal treatment after the films were flushed extensively with deionized water. An 500 nm × 500 nm region (a dotted box) in each image was expanded and shown in an inset. A cross-sectional profile was also taken from the dotted white cross-sectional line on each inset image.

Characterization of Monolayers

Atomic Force Microscopy (AFM; Pico SPM, Molecular Imaging). An AFM was used to analyze the change of surface coverage of the samples before and after they were flushed with deionized water. Topographic images of the surfaces were acquired in *non*-contact mode at ambient condition using silicon tips (Veeco Metrology, LLC) with a resonance frequency of 182 – 195 kHz and a force constant of 42 – 48 N/m.

Contact Angle Goniometry. Static contact angles were obtained using a FTA 100 instrument and software from First Ten Angstroms Inc. (Portsmouth, VA) to analyze the wettability of SAMs on the surface. Deionized water droplets of 2 – 5 µl (NANOpure Infinity System, Barnstead/Thermolyne Co.) were placed on the surface in the ambient environment. Measurements were performed within a day of sample preparation and at least 8 replicates were performed.

Diffuse Reflectance Infrared Fourier Transform Spectroscopy (DRIFT; Thermo Nicolet-NEXUS 470 FT-IR). DRIFT was used to analyze the alkyl chain ordering of the ODPA molecules on the SS316L substrates. The spectra of each sample were collected under a nitrogen environment to eliminate any background signal due to CO_2 and H_2O absorption bands. Cleaned, but unmodified, SS316L substrates were used for background spectra. One thousand twenty four scans were collected of each sample.

Results and Discussion

Following adsorption of ODPA to SS316L one group of samples were thermally treated in an oven at 100 - 120 °C for 30 – 60 minutes. They were then flushed with THF. A second group of samples were rinsed lightly with THF to remove multilayers and dried in air without thermal treatment. Samples of both groups were then flushed independently with deionized water.

This study employed atomic force microscopy and contact angle goniometry for the first time to investigate the influence of thermal treatment on ODPA film self-organization and stability on the SS316L surface. Figure 1(a) is an AFM image of a thermally treated ODPA film that was taken after it was flushed extensively with water. It shows that ODPA molecules adsorb on the SS316L surface and form islands of irregular shapes and sizes. It is posited that the shapes and sizes are dependent on those of irregular topographic features of underlying substrate. Several polish marks that were left in the course of polishing process are seen in the image.

Another AFM image in Figure 1(b) was taken from an ODPA film with no thermal treatment after the film was also flushed extensively with water. A significant portion of the film was lost during the water flush, revealing the underlying substrate surface. It is clear from these two images that thermal treatment enhances the stability of the ODPA molecules on the SS316L substrate and that attached molecules are resistant to water flushing.

A 500 nm × 500 nm region (a dotted box) was taken arbitrarily in each image and expanded as shown in an inset. A cross-sectional profile that was taken from the dotted white cross-sectional line from each inset image reveals that each film is likely to be a multilayer. While these thin films are not monolayers as in other reports, the AFM images still show that self-assembled films are much more stable when processed thermally.

Surface wettability estimated from water contact angles can be employed as a measure of surface coverage of hydrophobic and hydrophlic films (*37*). The influence of thermal treatment on the stability of ODPA films in an aqueous environment was also investigated with contact angle goniometry. Control and modified substrates were immersed in a dynamic flow of deionized water. Each sample was taken from the water at 0, 3, 30, 60, 120, and 180 minutes for contact angle measurements. The contact angle of water on the control substrates (cleaned and diamond polished SS316L which was kept in the oven at 100°C) was 0°. As shown in Figure 2, contact angles were measured from film surfaces and plotted as a function of immersion time in water. Before immersion in water, both samples exhibit comparable magnitude of contact angles of approximately 97 ° due the hydrophobic nature of both films. As immersion time increases, the contact angle of the thermally treated film was initially reduced by 5 – 7 ° and remained in the range of 90 – 92 ° after 60 minutes. The contact angles of

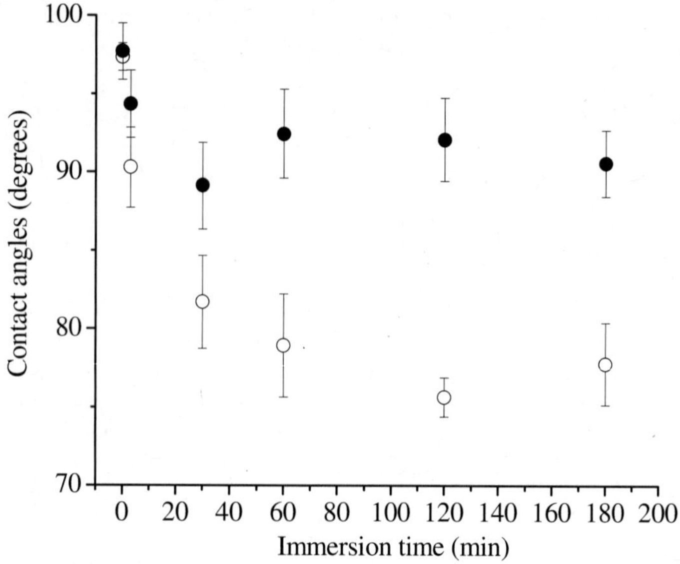

Figure 2. Contact angles were measured from the ODPA films with (filled circles) and without thermal treatment (open circles) and plotted as a function of immersion time in water.

the non-thermally treated samples were reduced by 20 – 22 ° and maintained in the range of 75 – 78 ° after a 60 minute immersion. The thermally treated film preserved hydrophobicity while the untreated one lost a substantial degree of hydrophobicity. The results show that the thermally treated films do not expose the surface to H_2O while the untreated films do. These results confirm the AFM data that showed the enhancement of stability of ODPA film by thermal treatment.

Infrared spectra (Figure 3) of the C–H stretching region of the ODPA film on SS316L (a) after thermal treatment followed by a THF flush to remove physical layer and (b) after a water flush without thermal treatment, respectively are significantly different. Peak positions for the methylene symmetric and asymmetric stretching modes are indicative of the order of methylene backbone chains in the substrate surface (*7*, *8*, *33*, *34*). Aliphatic monolayers characterized by $\nu_{CH2asym} < 2918$ cm^{-1}, and $\nu_{CH2sym} < 2850$ cm^{-1} are considered to be ordered with an all-trans chain configuration. Therefore, peaks in the spectra corresponding $\nu_{CH2asym} = 2915$ cm^{-1}, and $\nu_{CH2sym} = 2849$ cm^{-1} indicate that the films before and after water flushes are ordered on the surface.

Infrared spectra (Figure 4) of the P–O stretching region of (a) ODPA films on SS316L without thermal treatment, (b) with thermal treatment and (c) solid ODPA, respectively indicate surface bonding. All three spectra exhibit bands at 1230, 1232, and 1255 cm^{-1} which are indicative of the presence of P=O bond. Peaks at 900 – 950 cm^{-1} are assigned to P–O–H stretches. Spectra 4a contains a peak at 934 cm^{-1} which is smaller in position to the peaks in the solid spectrum at 947 cm^{-1} and 935 cm^{-1}. Therefore, this peak is assigned to P-O-H and indicates that little has changed in the head group of the physisorped molecule. Therefore, most likely the molecules are attached through hydrogen bonding when the films

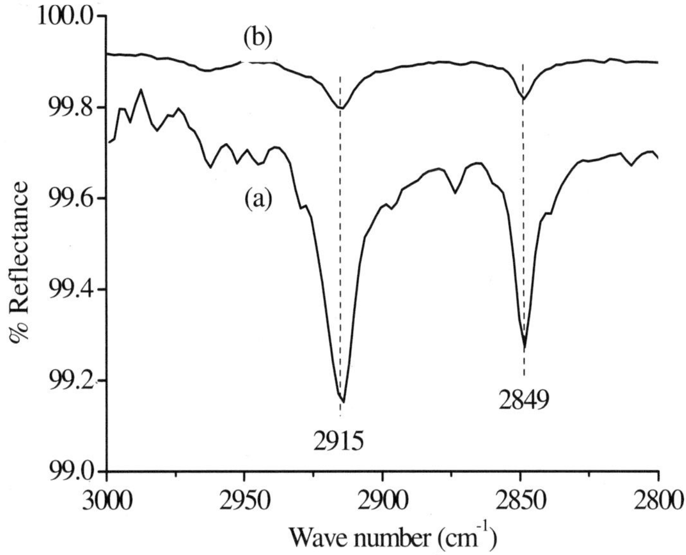

Figure 3. DRIFT spectra of the C–H stretching region of ODPA film (a) with (b) without thermal treatment after a water flush

are not thermally treated. The Spectrum of the thermally treated film contains a significantly shifted peak at 904 cm^{-1}.Since multilayers are present on these surfaces, even after thermal treatment with flushing, it is not surprising to see the P-O-H peak persist in some form. Peaks in the 1000-1150 cm^{-1} range can be assigned to P-O stretches. Therefore, in 4a and 4c, it can be seen that there are at least two peaks corresponding to group P-O stretches.

This data indicates that there is little difference in the head group between the solid state and when the molecule is deposited onto the surface without thermal treatment. However, in spectrum 4b(thermally treated), only one, broad peak centered at 1078 cm^{-1} is seen. This dramatic change is indicative of a chemical change in the phosphonic acid head group indicating chemisorption to the surface.This emphasizes that ODPA molecule-surface interactions are substantially enhanced after the thermal treatment.

It has been reported previously that bonding alkanephosphonic acid to a metal oxide surface is facilitated simply by heating the films (*29*, *34*). Using DRIFT it has been demonstrated that the ODPA film strongly bound to the surface of native oxide of titanium resists solvent rinsing and mechanical peel testing after thermal treatment at 120 °C for 18 hours (*29*, *34*). In contrast, they found that the same film without thermal treatment was easily removed by solvent rinsing. Their observations agree with what is reported in this study, in that the heating process drives the stability of alkanephosphonic acid on metal and metal oxide surfaces. The data in this study in conjunction with the results of Gawalt (*34*), Danahy (*29*), and Raman (*7*) provides a basis for speculation on the mechanism of stabilizing the ODPA film by a thermal treatment. An ODPA molecule is weakly bound to SS316L surface through physisorption onto a surface-hydroxyl group forming a hydrogen bonding pair (Figure 5). This hydrogen bonding interaction between the ODPA molecules and the surface is weak so that the untreated film can be easily removed by water flushes. However, heating the samples drives off water molecules from the surface facilitating the chemisorption of ODPA molecules to the native oxide on the surface. The presence of the peaks for P=O, P–O, and P–O–H as shown in Figure 4(c) leads to the conclusion that an ODPA molecule is chemisorbed on the SS316L surface in a monodentate or mixed mono/bi-dentate manner after thermal treatment, which is strong enough to resist aggressive water flushes.

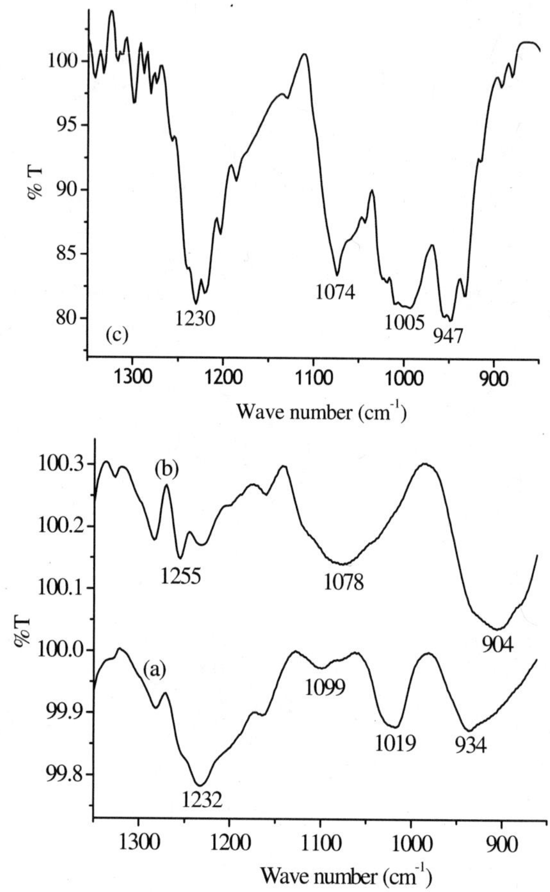

Figure 4. DRIFT spectra of the P-O stretching region of (a) ODPA film without thermal treatment, (b) ODPA film with thermal treatment, and (c) solid ODPA.

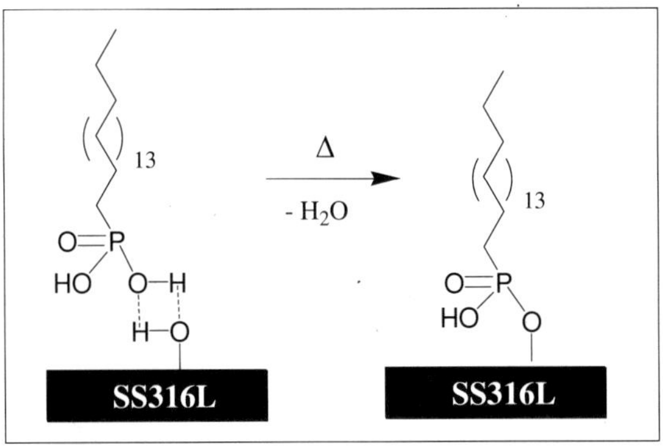

Figure 5. A suggested model for the thermally driven stability of ODPA film on the SS316L surface. An ODPA molecule is physisorbed weakly on the surface through hydrogen bonding interaction. Thermal treatment drives away a water molecule, leading to monodentate bonding of the ODPA molecule to the surface.

Conclusion

The influence of the thermal treatment of ODPA films on SS316L on its stability was investigated in this study. Thermally treated films exhibited substantial stability as measured by resistance to water flushes as characterized by AFM. However, non-thermally-treated films lose a significant amount of ODPA monomers after flushes with water exposing the underlying substrate. The stability of the film in aqueous environment was investigated with contact angle goniometry. The thermally treated films retained their high hydrophobic character while the untreated samples lost their initial hydrophobicity following immersion in a dynamic flow of deionized water for 0 – 180 minutes. The peak positions corresponding to methylene stretching in DRIFT spectra revealed that the remains of thermally untreated films after water flushes were alkyl-chain ordered like the thermally treated films after the flushes. The presence of the peaks for P=O, P–O, and P–O–H in a DRIFT spectra taken from thermally treated film indicates that ODPA molecules binds to the SS316L surface in a monodentate or mixed mono/bi-dentate manner. A mechanism for thermally stabilizing ODPA film on SS316L has been proposed, that is, ODPA molecules binds covalently to the surface by thermal removal of water molecules from the surface.

Acknowledgments

The authors acknowledge the ACS-Petroleum Research Fund and its donors and National Institutes of Health (NIAMS 1 R15 AR056864-01) for funding of the project. KS thanks NSF-REU/DOD-ASSURE program for a summer fellowship. MSL is supported in part by CHES Research Award (Award number: 200071540) at Slippery Rock University. Professor Aaron Timperman at Department of Chemistry, West Virginia University is acknowledged for the generosity of allowing M. S. Lim to use a contact angle goniometer.

References

1. Agrawal, C. M. Reconstructing the Human Body Using Biomaterials. *Journal of the Minerals, Metals and Materials Society* **1998**, *50* (1), 31–35.
2. Davis, J. R., Ed. *Handbook of Materials for Medical Devices*; ASM International: Materials Park, OH, USA, 2003; pp 1−11.
3. Mudali, U. K.; Sridhar, T. M.; Raj, B. Corrosion of bio implants. *Sadhana* **2003**, *28* (3−4), 601–637.
4. Hansen, D. C. Metal Corrosion in the Human Body: The Ultimate Bio-Corrosion Scenario. *The Electrochemical Society Interface* **2008**, *17* (2), 31–34.
5. Taira, M.; Lautenschlager, E. P. In vitro corrosion fatigue of 316L cold worked stainless steel. *J. Biomed. Mater. Res.* **1992**, *26* (9), 1131–1139.
6. Mahapatro, A.; Johnson, D. M.; Patel, D. N.; Feldman, M. D.; Ayon, A. A.; Agrawal, C. M. Surface Modification of Functional Self-Assembled Monolayers on 316L Stainless Steel via Lipase Catalys. *Langmuir* **2006**, *22* (3), 901–905.
7. Raman, A.; Dubey, M.; Gouzman, I.; Gawalt, E. S. Formation of Self-Assembled Monolayers of Alkylphosphonic Acid on the Native Oxide Surface of SS316L. *Langmuir* **2006**, *22* (15), 6469–6472.
8. Raman, A.; Gawalt, E. S. Self-Assembled Monolayers of Alkanoic Acids on the Native Oxide Surface of SS316L by Solution Deposition. *Langmuir* **2007**, *23* (5), 2284–2288.
9. Kim, H. I.; Koini, T.; Lee, T. R.; Perry, S. S. Systematic Studies of the Frictional Properties of Fluorinated Monolayers with Atomic Force Microscopy: Comparison of CF3- and CH3-Terminated Films. *Langmuir* **1997**, *13*, 7192–7196.
10. Kim, H. I.; Koini, T.; Lee, T. R.; Perry, S. S. Molecular contribution to the frictional properties of fluorinated self-assembled monolayers. *Tribol. Lett.* **1998**, *4*, 137–140.
11. Kim, H. I.; Graupe, M.; Oloba, O.; Koini, T.; Imaduddin, S.; Lee, T. R.; Perry, S. S. Molecularly Specific Studies of the Frictional Properties of Monolayer Films: A Systematic Comparison of CF3-, (CH3)2CH-, and CH3-Terminated Films. *Langmuir* **1999**, *15*, 3179–3185.
12. Lee, S.; Shon, Y.-S.; Colorado, J. R.; Guenard, R. L.; Lee, T. R.; Perry, S. S. The Influence of Packing Densities and Surface Order on the Frictional Properties of Alkanethiol Self-Assembled Monolayers (SAMs) on Gold: A Comparison of SAMs Derived from Normal and Spiroalkanedithiols. *Langmuir* **2000**, *16*, 2220–2224.
13. Shon, Y.-S.; Lee, T. R. Desorption and Exchange of Self-Assembled Monolayers (SAMs) on Gold Generated from Chelating Alkanedithiols. *J. Phys. Chem. B* **2000**, *104*, 8192–8200.
14. Rieley, H.; Kendall, G. K. X-ray Studies of Self-Assembled Monolayers on Coinage Metals. 3. Angularly Resolved Near Edge X-ray Absorption Fine Structure Determination of the Orientation in 1-Octanethiol SAMs on Ag(111) and Cu(111). *Langmuir* **1999**, *15*, 8867–8875.

15. Rieley, H.; Kendall, G. K.; Jones, R. G.; Woodruff, D. P. X-ray Studies of Self-Assembled Monolayers on Coinage Metals. 2. Surface Adsorption Structures in 1-Octanethiol on Cu(111) and Ag(111) and Their Determination by the Normal Incidence X-ray Standing Wave Technique. *Langmuir* **1999**, *15*, 8856–8866.
16. Allara, D. L.; Nuzzo, R. G. Spontaneously Organized Molecular Assemblies. 1. Formation, Dynamics, and Physical Properties of n -Alkanoic Acids Adsorbed from Solution on an Oxidized Aluminum Surface. *Langmuir* **1985**, *1*, 45–52.
17. Allara, D. L.; Nuzzo, R. G. Spontaneously Organized Molecular Assemblies. 2. Quantitative Infrared Spectroscopic Determination of Equilibrium Structures of Solution-Adsorbed n Alkanoic Acids on an Oxidized Aluminum Surface. *Langmuir* **1985**, *1*, 52–66.
18. Tao, Y.-T. Structural Comparison of Self-Assembled Monolayers of n-Alkanoic Acids on the Surfaces of Silver, Copper, and Aluminum. *J. Am. Chem. Soc.* **1993**, *115*, 4350–4358.
19. Taylor, C. E.; Schwartz, D. K. Octadecanoic Acid Self-Assembled Monolayer Growth at Sapphire Surfaces. *Langmuir* **2003**, *19*, 2665–2672.
20. Lim, M. S.; Feng, K.; Chen, X.; Wu, N.; Raman, A.; Nightingale, J.; Gawalt, E. S.; Korakakis, D.; Hornak, L. A.; Timperman, A. T. Adsorption and Desorption of Stearic Acid Self-Assembled Monolayers on Aluminum Oxide. *Langmuir* **2007**, *23* (5), 2444–2452.
21. Doudevski, I.; Schwartz, D. K. Mechanisms of Self-Assembled Monolayer Desorption Determined Using in Situ Atomic Force Microscopy. *Langmuir* **2000**, *16*, 9381–9384.
22. Nie, H.-Y.; Miller, D. J.; Francis, J. T.; Walzak, M. J.; McIntyre, N. S. Robust Self-Assembled Octadecylphosphonic Acid Monolayers on a Mica Substrate. *Langmuir* **2005**, *21*, 2773–2778.
23. Grange, J. D. L.; Markham, J. L. Effects of Surface Hydration on the Deposition of Silane Monolayers on Silica. *Langmuir* **1993**, *9*, 1749–1763.
24. Rühe, J.; Novotny, V. J.; Kanazawa, K. K.; Clarke, T.; Street, G. B. Structure and Tribological Properties of Ultrathin Alkylsilane Films Chemisorbed to Solid Surfaces. *Langmuir* **1993**, *9*, 2383–2388.
25. Carpick, R. W.; Salmeron, M. Scratching the Surface: Fundamental Investigations of Tribology with Atomic Force Microscopy. *Chem. Rev.* **1997**, *97* (4), 1163–1194.
26. McDermott, M. T.; Green, J.-B. D.; Porter, M. D. Scanning Force Microscopic Exploration of the Lubrication Capabilities of n-Alkanethiolate Monolayers Chemisorbed at Gold: Structural Basis of Microscopic Friction and Wear. *Langmuir* **1997**, *13*, 2504–2510.
27. Xiao, X.; Hu, J.; Charych, D. H.; Salmeron, M. Chain Length Dependence of the Frictional Properties of Alkylsilane Molecules Self-Assembled on Mica Studied by Atomic Force Microscopy. *Langmuir* **1996**, *12*, 235–237.
28. Wasserman, S. R.; Tao, Y.-T.; Whitesides, G. M. Structure and Reactivity of Alkylsiloxane Monolayers Formed by Reaction of Alkyltrichlorosilanes on Silicon Substrates. *Langmuir* **1989**, *5* (4), 1074–1087.

29. Danahy, M. P.; Avaltroni, M. J.; Midwood, K. S.; Schwarzbauer, J. E.; Schwartz, J. Self-assembled Monolayers of α,ϖ-Diphosphonic Acids on Ti Enable Complete or Spatially Controlled Surface Derivatization. *Langmuir* **2004**, *20* (13), 5333–5337.
30. Huisman, B.-H.; Kooyman, R. P. H.; Veggel, F. C. J. M. v.; Reinhoudt, D. N. Molecular recognition by self-assembled monolayers detected with surface plasmon resonance. *Adv. Mater.* **1996**, *8* (7), 561–564.
31. Adden, N.; Gamble, L. J.; Castner, D. G.; Hoffmann, A.; Gross, G.; Menzel, H. Phosphonic Acid Monolayers for Binding of Bioactive Molecules to Titanium Surfaces. *Langmuir* **2006**, *22* (19), 8197–8204.
32. Quin˜ones, R.; Gawalt, E. S. Polystyrene Formation on Monolayer-Modified Nitinol Effectively Controls Corrosion. *Langmuir* **2008**, *24* (19), 10858–10864.
33. Gao, W.; Dickinson, L.; Grozinger, C.; Morin, F. G.; Reven, L. Self-Assembled Monolayers of Alkylphosphonic Acids on Metal Oxides. *Langmuir* **1996**, *12* (26), 6429–6435.
34. Gawalt, E. S.; Avaltroni, M. J.; Koch, N.; Schwartz, J. Self-Assembly and Bonding of Alkanephosphonic Acids on the Native Oxide Surface of Titanium. *Langmuir* **2001**, *17* (19), 5736–5738.
35. Mani, G.; Johnson, D. M.; Marton, D.; Dougherty, V. L.; Feldman, M. D.; Patel, D.; Ayon, A. A.; Agrawal, C. M. Stability of Self-Assembled Monolayers on Titanium and Gold. *Langmuir* **2008**, *24*, 6774–6784.
36. Kaufmann, C. R.; Mani, G.; Marton, D.; Johnson, D. M.; Agrawal, C. M. Long-term stability of self-assembled monolayers on 316 L stainless steel. *Biomed. Mater.* **2010**, *5*, 025008/1–025008/10.
37. Demirel, G.; Çağlayan, M. O.; Garipcan, B.; Duman, M.; Pişkin, E. Formation and Organization of Amino Terminated Self-assembled Layers on Si(001) Surface. *Nanoscale Res. Lett.* **2007**, *2* (7), 350–354.

Chapter 13

Amphiphilic Invertible Polymers (AIPs)

Micellization and Self-Assembly in Aqueous Solutions

A. Kohut,[1] L. Sieburg,[1,2] S. Vasylyev,[3] O. Kudina,[4] I. Hevus,[1] S. Stafslien,[5] J. Daniels,[5] V. Kislenko,[6] and A. Voronov[1,*]

[1]Department of Coatings and Polymeric Materials, North Dakota State University, Fargo, ND
[2]Department of Chemistry, University of Nebraska-Lincoln, Lincoln, NE
[3]Institute of Particle Technology, University of Erlangen-Nuremberg, Erlangen, Germany
[4]Department of Organic Chemistry, Lviv Polytechnic National University, Lviv, Ukraine
[5]Center of Nanoscale Science and Engineering, Fargo, ND
[6]Department of General Chemistry, Lviv Polytechnic National University, Lviv, Ukraine
[*]andriy.voronov@ndsu.edu

Amphiphilic invertible polymers (AIPs) are composed of the hydrophilic, poly(ethylene glycol), and hydrophobic, aliphatic dicarboxylic acid moieties, alternately distributed along the main backbone. They build micelles, self-assemble by increasing concentration and are capable of accommodating otherwise insoluble substances in aqueous medium. The resulting complex nanostructures can undergo inverse conformational changes and are functional in both polar and non-polar environments, giving potentially broad use in many applications, especially in the pharmaceutical industry, agriculture, and cosmetics.

Introduction

Interest in nanomaterials derived from amphiphilic self-assembling systems, including polymers, has dramatically increased over the past few years. The main challenge for scientists is to find new systems and materials that have enhanced

© 2011 American Chemical Society

properties and the potential to replace existing classic systems. These new materials have to face new and increasingly sophisticated demands of industry, where nanomaterials are finding wider niches.

Responsive polymers adapt to surrounding environments, can regulate transport of ions and molecules, change wettability and adhesion of different species on external stimuli etc (*1*). Amphiphilic molecules, which are molecules with polar and non-polar moieties, arrange themselves at interfaces and tend to build aggregates in solution (*2*). Polymeric surfactants, for example, PEG-based copolymers (e.g., Pluronics) have been studied in a broad range of applications in aqueous environments that include crystallization (*3*, *4*), surface modification for biocompatibility (*5*, *6*), control of particle aggregation in solution (*7*, *8*), and drug delivery (*9*). At low polymer concentration, Pluronics exist in solution as isolated polymer coils or unimers even in a very dilute solution (*10*, *11*). As polymer concentration or temperature increases, micelle formation is initiated (*12–14*). Further increase in temperature or polymer concentration results in the formation of micellar gels or crystals due to the hydrophobic effect (*15*). The behavior of amphiphilic copolymers in organic solvents is, in many respects, similar to that of surfactants in water (*16–18*). For example, they form micelles and can lower the surface tension at the interface between organic solvents and water (*19*).

Recently, the focus of research in the field of polymeric materials has moved toward the design of polymers with smart responsive behavior. For this reason, our research group has recently carried out a broad study on synthesis, characterization and properties of amphiphilic **invertible** polymers (AIPs) composed of the hydrophilic and hydrophobic constituents alternately distributed along the main backbone. The preparation of self-assembled functional micellar nanostructures from AIPs is a one novel and promising approach. Made from hydrophobic and hydrophilic constituents that are alternately distributed along the polymer backbone, AIPs form invertible polymeric micelles or self-arranged aggregates of micelles in response to tuning of the polymer concentration and environmental polarity (*20–31*). The resulting complex nanostructures have a controlled size and morphology and are functional in both polar and non-polar environments, giving them potentially broad use in many chemical and biochemical applications. In particular, they can be used as nanoreactors for the synthesis of nanoparticles that have a protective shell made from both hydrophilic and hydrophobic polymeric fragments, and, thus, will be highly stable in both polar and non-polar environments (*24*, *25*, *28*). The ability to encapsulate insoluble substances makes their use as polymeric nanocontainers especially promising for the pharmaceutical industry (controlled release, targeted devices), agriculture (micronutrient delivery, development of fertilizers), and cosmetics (encapsulation of functional ingredients and their delivery).

Concept of Amphiphilic Invertible Polymers

The most prominent advantage of amphiphilic invertible polymers is a rapid switching of the polymeric molecular conformation in solvents differing

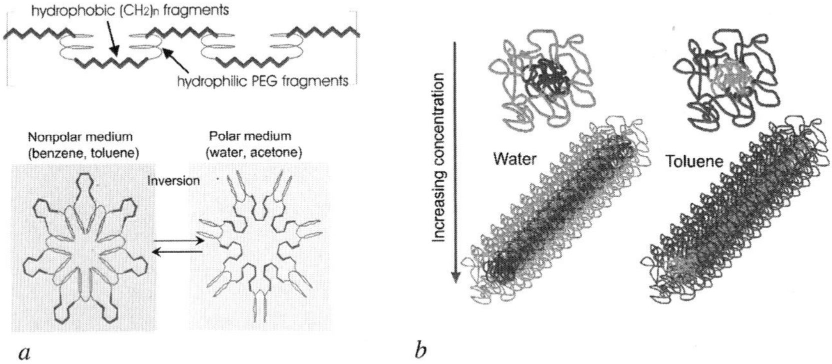

Figure 1. AIPs structure and scheme of AIPs micelle in polar and non-polar media (a). AIPs self-assembly by increasing concentration in aqueous and organic solutions (b). (see color insert)

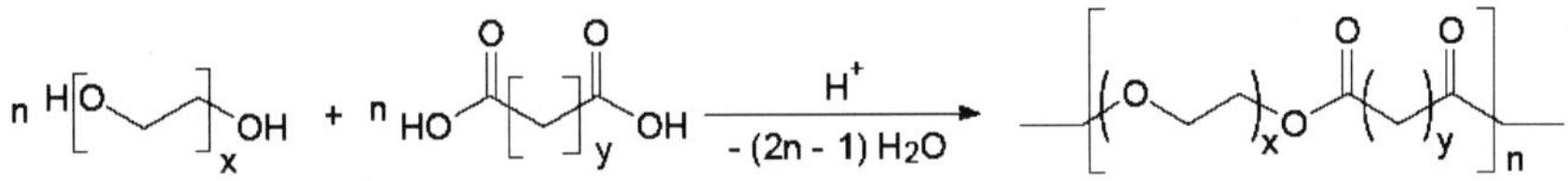

Figure 2. Synthesis of AIPs.

by polarity (Fig.1a), and the ability of self-assembly with increasing polymer concentration (Fig.1b).

The AIPs are recently designed from poly(ethylene glycol, PEG) (hydrophilic constituent) and aliphatic dicarboxylic acids (hydrophobic constituent). Synthesis of different compositions has been performed through the polycondensation of decanedioic or sebacic acid and PEG-300 (average M_N 300 g/mol), PEG-600 (molecular weight 600 g/mol), PEG-1000 (molecular weight 1000 g/mol) (Fig.2). The reaction results in alternating invertible amphiphilic polyester structures with various ratios of hydrophilic and hydrophobic constituents.

In a polar medium, poly(ethylene glycol) fragments (*blue* in Fig.1a) form the outer layer of the polymer architecture, and the dicarboxylic aliphatic acid fragments (*red* in Fig.1a) are collapsed and screened inside the architecture. By changing the polarity of an environment (from polar to non-polar), a change in the AIPs conformation occurs. The PEG constituents now collapse and form the interior of the micelle, and the aliphatic dicarboxylic acid fragments switch their location to the outer side of the micelle. It is assumed that conformational changes in AIP macromolecules, by changing polarity, are achieved by the presence of oxygen atoms along the polymer main backbone.

AIPs Self-Assembly and Application in Aqueous Solutions

Special attention has been devoted to tuning the properties of self-assembled AIPs micellar nanostructures in aqueous environment. A predictable and variable size and morphology towards their ability to encapsulate insoluble compounds

Table I. AIPs Composition and Characteristics

Amphiphilic Invertible Polymer	*M_w (g/mol), (PDI)*	*HLB*[a]
S10 poly(ethylene glycol)-1000–sebacate	11,700 (1.95)	24
D10 poly(ethylene glycol)-1000–dodecanedioate	7,600 (1,53)	23
S6 poly(ethylene glycol)-600–sebacate	16,400 (1,41)	21
D6 poly(ethylene glycol)-600–dodecanedioate	11,800 (1,46)	19
S3 poly(ethylene glycol)-300–sebacate	10,900 (1,58)	18
D3 poly(ethylene glycol)-300–dodecanedioate	10,100 (1,55)	17

[a] NOTE: Hydrophilic-lipophilic balance (HLB) is calculated according to Ref. (*32*).

in the micellar interior can facilitate the careful design of AIPs nanostructures. An ability to manipulate the AIP macromolecules by changing concentration, temperature, polymer composition etc. is very important in understanding the functional properties, processing the AIPs nanostructures and their use.

Existing AIP Library

The polycondensation of dicarboxylic acids with diols is known to be an equilibrium process; therefore, the water evolved during the reaction must be continuously removed from the reactive mixture to shift the equilibrium toward the desired polymer formation. The acylation of PEG by dicarboxylic acids has been carried out in toluene as an azeotrope former; a Dean-Stark trap has been used to remove water. The reaction has been performed in the presence of catalytic amounts of sulfuric acid (1.25 mol %). The progress of the reaction has been monitored by the amount of water released during polycondensation. The characteristics and composition of the synthesized AIPs are shown in a Table I. The chemical structure of the synthesized amphiphilic polymer is confirmed by FTIR and ^{1}H NMR spectroscopy (Fig.3).

AIPs Micellization in Aqueous Solution

Surface Activity Study

AIPs surface activity and theirs ability to form micelles in aqueous solution have been studied by measurements of surface tension to evaluate critical micelle concentration (cmc) for each AIP aqueous solution. Typically, the surface tension of an aqueous surfactant solution decreases with increasing surfactant concentration reflecting the increase in surface activity of the surfactant. The

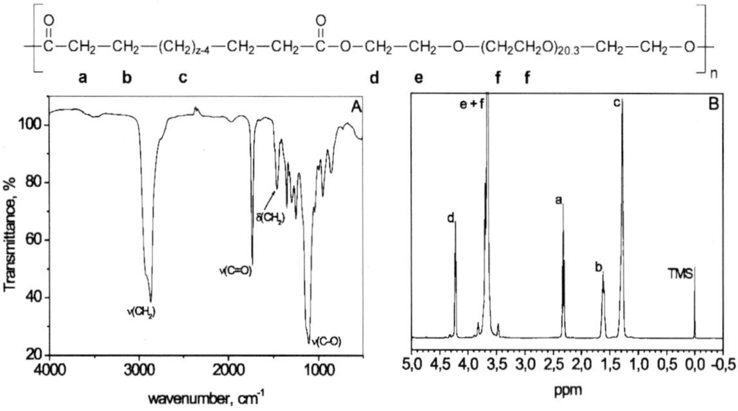

Figure 3. FTIR (A) and ¹H NMR (B) spectra of the D10. The letters on ¹H NMR spectra correspond to letters on the AIP structural scheme. TMS – tetramethylsilane (internal standard).

data of surface tension changing with increasing AIP concentration in aqueous solution are plotted semi-logarithmically (Fig.4) for AIPs based on sebacic acid (A) and fragments of dodecanedioic acid (B). It is clearly seen that both polymer ranges are surface active and decrease the surface tension of water up to 45 mN/m. Nevertheless, a small change in the AIP chemical composition (length of hydrophobic fragment in macromolecule), results in a significant difference in surface tension isotherm recorded for AIPs based on dodecanedioic acid (D-polymers) in comparison to AIPs based on sebacic acid (S-polymers).

The data on surface tension isotherm indicate that surface activity of amphiphilic invertible polymers containing dodecanedioic acid as a hydrophobic constituent decreases when length of PEG (hydrophilic) fragment increases D3 > D6 > D10 (Fig.4A). There is a slight decrease of surface tension at very small polymer concentration on isotherms recorded for these three polymers, and a rapid drop of surface tension ending in a cmc point. We assume that at very small concentration (usually from 10^{-7} to 10^{-4} % w/v depending on AIP composition) the macromolecules being introduced in water form molecular arrangements having hydrophobic pockets and external hydrophilic shell due to the interaction of hydrophobic fragments and flexibility of the alternated macromolecule. Formation of thermodynamically stable assemblies from the copolymers has been shown theoretically and experimentally in the literature both for the polar and nonpolar solvents at low concentration (*33*, *34*). This effect is expected to be more pronounced for the macromolecules of a higher molecular weight and longer hydrophobic fragments. In our case, the hydrophobic constituents of the alternated macromolecules undergo intensive interaction leading to the formation of assemblies with a hydrophobic inner part and hydrophilic outer part. Similar behavior was observed for Pluronics that have demonstrated two clear break points appearing when the surface tension of aqueous solution of Pluronics is plotted as a function of the concentration (*35*, *36*).

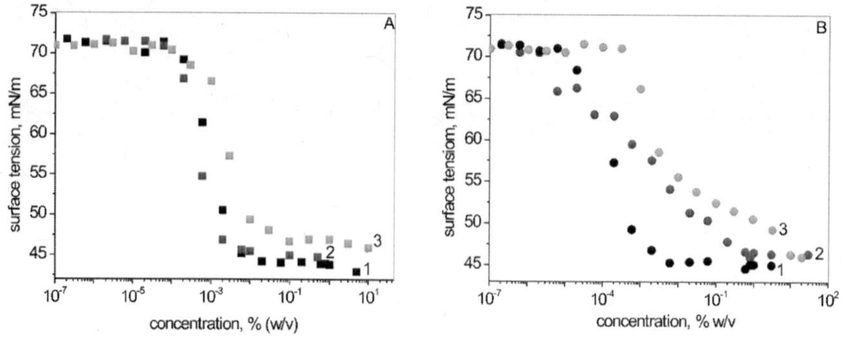

Figure 4. Critical micelle concentration for D-polymers (A) (D3 – 1, D6 – 2, D10 – 3) and S-polymers (B) (S3 – 1, S6 – 2, S10 - 3).

Similar to D3, D6 and D10, the polymers based on sebacic acid form micelles more readily when length of hydrophilic PEG residues in the macromolecule is smaller (Fig.4B). For S6 and S10 no slope at the very small concentration has been observed indicating that these polymers do not form molecular arrangements at low concentration as other amphiphilic invertible polymers do.

The slope at low concentrations of most AIPs may indicate that after the formation of molecular arrangements, the polymer assemblies migrate to the interface, adsorb at the air/water interface and form monomolecular adsorption layer (*28*). Following fast decrease of surface tension ends in a cmc point for each AIP and indicates formation of polymolecular micelles.

While the slope on the surface tension vs. concentration plot was well-expressed for most polymers, and cmc can be easily determined, the experimental data were not convincing enough for the most hydrophilic S10. To ensure an ability of S10 to form micelles in aqueous solution, an alternative approach to cmc determination has been used for all AIP compositions, when critical micelle concentration measurements have been carried out by solubilization of pyrene, a well-known fluorescent probe for studying the association behavior of amphiphilic polymers (*37, 38*). Depending on the environment of the pyrene, a red shift of the absorption band with enhanced excitation intensity is observed due to the migration of the probe from the hydrophilic to the hydrophobic region of the polymer micelles. In our experiments, it was possible to determine cmc of AIP compositions by recording excitation spectra of pyrene in the polymer aqueous solutions at different polymer concentrations (Table II).

^{1}H NMR-Spectroscopy Study

The formation of micelles by two AIPs, more hydrophilic S6 and more hydrophobic D3 in water has been extensively studied with ^{1}H NMR spectroscopy. Samples for ^{1}H NMR spectroscopy were prepared by dissolving an appropriate amount of polyester in deuterium oxide under gentle agitation. For comparison,

Table II. Critical Micelle Concentration of AIPs

AIP	*by surface tension, % w/v*	*by pyrene solubilization, % w/v*
S10	n/a	0.5
D10	0.01	0.008
S6	0.4	0.2
D6	0.002	0.002
S3	0.01	0.01
D3	0.006	0.006

^{1}H NMR spectra were also taken using $CDCl_3$. The solutions were left for at least 16 h to equilibrate at 25 °C before measuring. All NMR spectra were recorded at 400 MHz on a Varian VXR-400 NMR spectrometer.

Fig. 5 shows the local expanded ^{1}H NMR spectra of each peak region of 1% w/v S6 and D3 solutions. In $CDCl_3$, all groups show a distinct hyperfine structure indicating that amphiphilic polyesters S6 and D3 are dissolved as unimers, the macromolecules are expanded, and the segments of the chains can move freely (*39*). However, when D_2O has been used, the signal width corresponding to methylene groups localized in the internal part of the hydrophobic fragment (peak c in Fig.5) increases as compared with those in $CDCl_3$. The broadening of the signal indicates that the methylene groups avoiding contact with an aqueous medium form a micellar core, thus reducing the mobility of the protons in a hydrophobic polyester fragment.

Each $-CH_2-$ group in the α and β positions in relation to the carbonyl groups in the dicarboxylic acid moieties (Fig.5, protons a and b, respectively) shows two different signals in D_2O. The significant upfield shifts of a part of protons a (signals at 2.15 ppm) and b (signals at 1.53) imply that these protons are apparently in a nonpolar micellar core. The chemical shift is known to be sensitive to the chemical nature of the related protons, and transferring part of protons a and b to the nonpolar microenvironment induces the shift toward lower ppm values as a result of the change in magnetic susceptibility of the protons (*40*). The sudden ^{1}H downfield shifts experienced by retaining protons a and b indicate that they are transferred to a polar aqueous medium. Because the interaction with water enhances the deshielding effect of the C-H protons, it results in the appearance of the peaks at 2.41 and 1.61 ppm for S6 (a and b, respectively) and at 2.34 and 1.60 ppm for D3 (a, and b, respectively). Thus, protons a and b are located partially in a nonpolar micellar core and partially in a polar aqueous medium. The reason for this is a strong -I-effect of the carbonyl groups resulting in enhanced polarizability of the C-H bonds in the α and β positions.

The peaks that are due to the poly(ethylene glycol) fragments of polyester S6 (protons d-f) move considerably downfield, indicating that the PEG units are at the micellar outer surface and hence in contact with water. The peaks remain sharp, showing that PEG fragments are long enough to move freely. The PEG fragment length in D3 is much lower in comparison with those in S6 (on average, 6.4 ethylene oxide links per PEG fragment in D3 and 13.2 links in S6). The moving

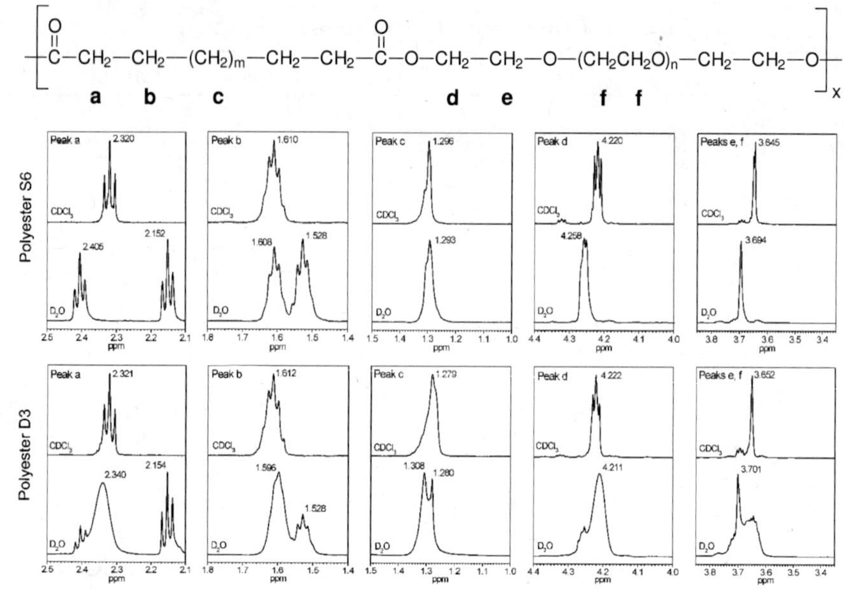

Figure 5. ¹H NMR spectra of 1% w/v solutions of amphiphilic polyesters S6 (top) and D3 (bottom) in D_2O and $CDCl_3$.

downfield signals attributed to the PEG protons of D3 broaden considerably. The PEG fragments in D3 are rather short, and their motions are limited.

AIP Micellar Size Study

To evaluate the AIP micellar size in aqueous solutions, the viscosity study has been carried out. The specific viscosity of aqueous AIP solutions with a concentration of 1 g/100 mL has been measured with a Ubbelohde viscometer, and the hydrodynamic radius of the micelles has been calculated according to the following equation:

$$R_m = (3V_m / 4\pi)^{1/3}$$

where V_m is the volume of micelles.

$$V_m = \omega \cdot M / (c \cdot N_A)$$

where ω is the volume fraction of macromolecules in solution, M is polymer molecular weight, c is polymer concentration, and N_A is Avogadro's number.

According to Newton's equation for spherical particles:

$$\omega = (\eta/\eta_o - 1)/\alpha$$

where η and η_o are the viscosities of the polymer solution and water respectively, and α is the coefficient equal to 2.5 for spherical particles.

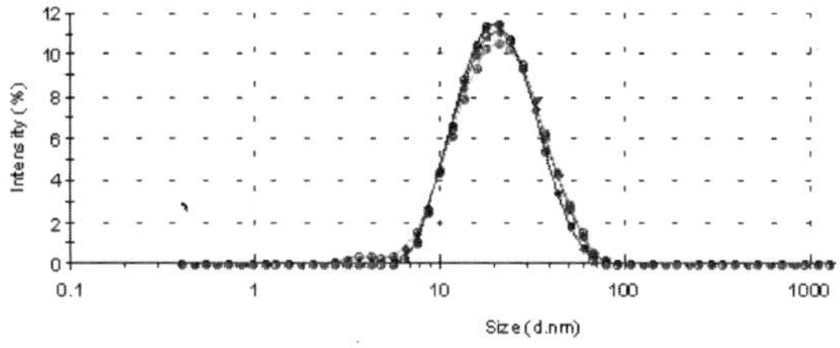

Figure 6. DLS measurement for micelles formed in 1% w/v D3 aqueous solution.

The calculated hydrodynamic radii of the AIPs in water are 10.3 nm for D3, 19.5 nm for S3, 13.9 nm for S6, 21.7 nm for D10 and 28.7 nm for S10. The hydrodynamic radii of the AIPs micelles is seen to increase as the length of the PEG fragment increases. This implies that the macromolecules of more hydrophilic AIPs have more extended conformation in water. The hydrodynamic radii of the micelles measured by dynamic light scattering have shown a good coincidence with the viscosity data. The selected measurement of hydrodynamic radius of D3 micelles is shown in Fig.6. The DLS data (9.4 nm) is very close to 10.3 nm (calculated from the viscosity).

AIP Self-Assembly in Aqueous Solution

^{1}H NMR-Spectroscopy Study

To study the self-assembly of the AIPs, the ^{1}H NMR spectra of S6 and D3 have been taken in D_2O over a wide range of polyester concentration. At low concentration in aqueous solution, S6 and D3 form micelles with a hydrophobic inner part built by dicarboxylic acid moieties and with an outer part made up of the hydrated PEG fragments. The highest investigated concentrations of S6 in D_2O are 10 and 5% for D3 (Fig.7) because the latter is more hydrophobic and it does not form a 10% solution in water. It can be concluded from the data presented in Figure 7 that the polyester micelles aggregate with increasing concentration to form structures containing hydrophilic and lipophilic domains.

Increasing polyester concentration in water leads to a broadening of the signals attributed to the PEG protons d-f (drastically for the polyester D3 with the shorter PEG fragments) indicating that the motions of the PEG fragments became limited due to their close packing in the hydrophilic domain. Disappearing of the hyperfine structures of the methylene groups a and b located in the area of the PEG fragments supports the idea that the mobility of the polar units decreases with increasing polyester concentration. A slight shift of the signal in d-f toward lower ppm values implies that the polarity within the hydrophilic domain is lower as compared with those in the outer part of polyester micelles.

The signals of hydrophobic protons c experience a slight narrowing that shows inessential enhancement in the mobility of the $-(CH_2)_n-$ groups when they are forming a hydrophobic domain in water. Interestingly, the c signals practically do not shift with increasing polyester concentration in D_2O, indicating that in terms

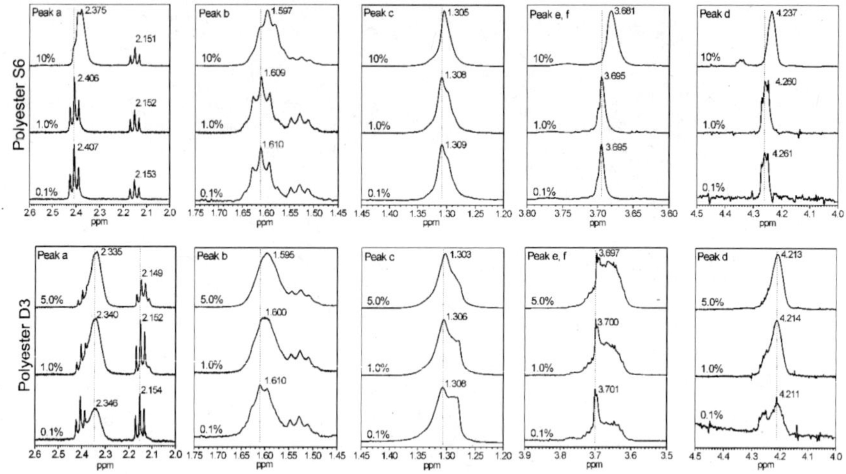

Figure 7. ^{1}H NMR spectra of S6 (top) and D3 (bottom) solutions in D_2O at different concentrations.

of polarity there is no essential difference between the micelle inner part and the hydrophobic domain of the hierarchical structures.

AIPs Functional Properties in Aqueous Solution

Adsolubilization Study

Adsolubilization is the incorporation of sparingly water-soluble organic compounds into surfactant layers adsorbed at the solid/solution interface (*41*). This phenomenon has been widely used in the development of various industrial products such as paints, pharmaceuticals, cosmetics and health care products, preparation of polymer thin films, wastewater treatment etc (*42–46*). Polymers S10 and D10 have been used in an adsolubilization of 2-naphthol to show ability of invertible polymers to adsolubilize poorly-water soluble compounds on nanoparticulate substrate differing by polarity. Potentially, it can be used also in biomedical applications, in particular, fabrication of based on AIP-modified nanoparticles amphiphilic adsorbents (enterosorbents) and selective removal of toxins (for example, bilirubin) can be of great interest (*47*, *48*).

Fig.8 shows the adsorption isotherms of S10 and D10 on hydrophilic fumed silica (AEROSIL® OX50) and hydrophobized silica nanoparticles (AEROSIL® NAX50, treated with HMDS, Evonik Degussa Corporation). Silica particles were shaken with polymer solutions of various concentrations, than separated by centrifugation. The polymer concentration in the supernatant was measured by referring to the tannic acid method (*49*), and the adsorbed amount was calculated from the difference in the concentration before and after the adsorption.

AIP macromolecules readily adsorb onto the both substrates. Hydrophobized silica nanoparticles themselves cannot be dispersed in water and remains floating at the air/water interface. When it was added to the AIP aqueous solution and then agitated, dispersion was formed within 24 h.

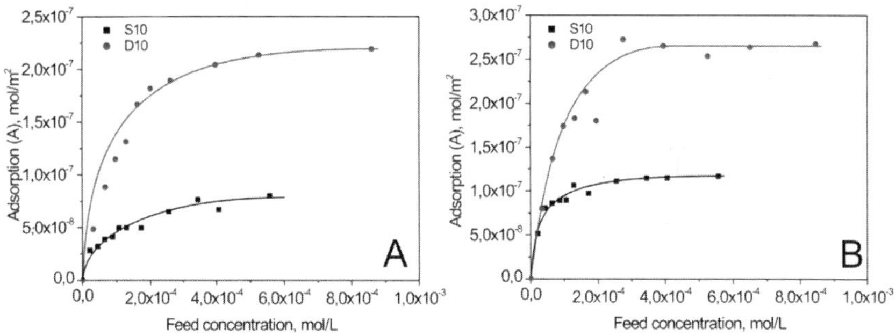

Figure 8. Adsorption isotherm of S10 (■) and D10 (●) on hydrophilic (A) and hydrophobized (B) silica nanoparticles.

The higher adsorption amount on hydrophilic silica for the composition with longer hydrophobic sequences (D10) indicates that, due to energetically less favorable contact with water, the polymer more readily adsorbs onto the polar silica surface compared to the more hydrophilic S10 (Fig.8A). A much faster increase in the adsorption amount is recorded for polymers adsorbed on hydrophobized nanoparticles (Fig.8B). It is generally observed when polymers possess high affinity to substrate. We assume that, opposite of the adsorption on polar silica, when PEG fragments are responsible for the formation of physical bonding between AIP and substrate, the dominant driving force for the adsorption on the hydrophobized silica is the hydrophobic interaction between the sebacic/dodecanedioic acid moieties and the substrate. The higher adsorption amount for more hydrophobic composition can be explained by the higher surface activity of D10 resulting in the adsorption of more compact micellar architectures than in the case of bulkier S10 micellar adsorption.

Sparingly water-soluble 2-naphthol was chosen to show the ability of AIPs to adsolubilize hydrophobic substances on substrates differing by polarity.

After equilibration, the supernatant was replaced by aqueous solution of 2-naphthol, and the resulting suspension was shaken for 24 h. The silica particles were separated by centrifugation and the concentration of 2-naphthol in the supernatant was determined by measuring the absorbance at 328 nm using a UV-Vis spectroscopy. The adsolubilized amount of 2-naphthol was calculated from the difference in the concentration before and after adsolubilization (Fig.9). In the same experiment, the desorbed amount of the polymer was calculated from the concentration of free polymer in the supernatant after the adsolubilization stage (e.g., after the supernatant replacement) (Fig.10).

Hydrophobic 2-naphthol does not adsorb onto hydrophilic silica nanoparticles (*50*). In the presence of adsorbed AIP, adsolubilization has been observed for both substrates. The amount of adsolubilized 2-naphthol increases as the S10 and D10 polyester concentration increases and reaches plateau values (Fig.9). Similar behavior has been reported by other previously reported adsolubilization studies (*50, 51*). Nevertheless, there is a significant difference in adsolubilization behavior between the hydrophilic and hydrophobized silica nanoparticles. On a polar substrate, there is a slow increase in adsolubilized 2-naphthol once the D10 and S10 are adsorbed. If AIPs are adsorbed on hydophobized nanoparticles, then

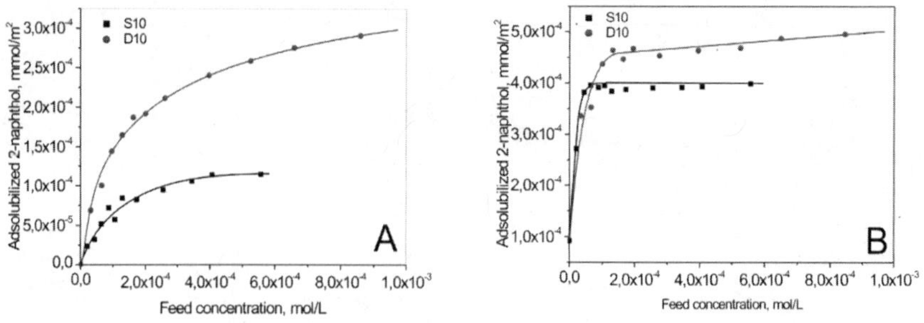

Figure 9. Amount of 2-naphthol adsolubilized into the adsorbed S10 (■) and D10 (●) layer on hydrophilic (A) and hydrophobized (B) silica.

the adsolubilized amount increases very rapidly and reaches plateau values at significantly smaller polymer concentration.

Fig.10 shows data on desorption of D10 and S10 from silica during the adsolubilization. For hydrophilic silica, the desorption was higher for S10, as it is expected for more hydrophilic polymer. On hydrophobized silica nanoparticles, the desorption of more hydrophobic D10 practically did not change, while almost three times less S10 has been desorbed. These results confirm that different fragments in AIP macromolecule are responsible for the interaction between the macromolecules and substrates during the adsorption. Significantly weaker physical bonding is formed when PEG fragments interact with a substrate compared to the bonding formed between the polymers' alkyl fragments and silica.

Considering the equilibrium adsorption, one can apply the Langmuir equation to describe the experimental data recorded in the adsorption and adsolubilization stages, respectively:

$$c_e/A = k_d/k_a + 1/A_m\, c_e$$

where A is the AIP adsorption at the first stage (A_1) and at the second stage (adsolubilization of 2-naphtol) (A_2), k_a and k_d are the rate constants of polymer adsorption/desorption, and A_m is the maximum polymer adsorption on the silica surface, c_e is the equilibrium polymer concentration.

The c_e at the first stage (AIP adsorption) can be calculated according to the following equation:

$$c_e = c_f - Ac_{s1},$$

where c_f is a feed polymer concentration, A is the AIP adsorption at the first stage, and c_{s1} is the silica concentration in the mixture at the first stage.

The c_e at the second stage (adsolubilization of 2-naphthol) can be calculated according to the following equation:

$$c_e = (A_1 - A_2)c_{s2},$$

where c_{s2} is the silica concentration in the mixture at the second stage.

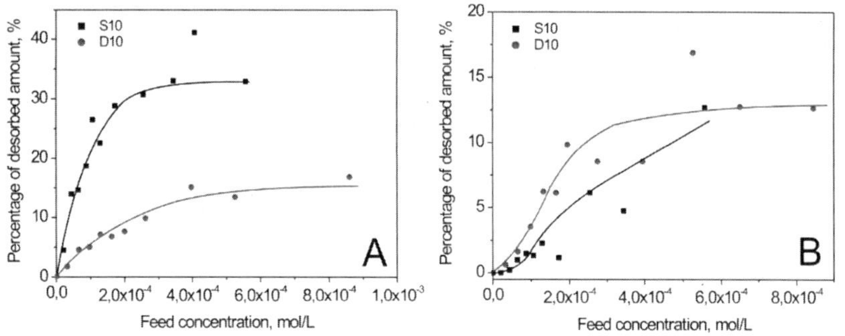

Figure 10. Desorption of S10 (■) and D10 (●) from hydrophilic (A) and hydrophobized (B) silica nanoparticles vs. feed polymer concentration.

Table III. Correlation coefficients, equilibrium constants and A_m for AIPs adsorption on hydrophilic and hydrophobized nanoparticles

AIP	*Substrate*	R^2	k_d/k_a, m^2/l	A_m^{-1}, m^2/mol	*Area per molecule, nm^2*
			AIP adsorption (1st stage)		
S-10	Hydrophilic	0.973	678	$1.19 \cdot 10^7$	19.8
S-10	Hydrophobized	0.998	112	$8.41 \cdot 10^6$	14.0
D-10	Hydrophilic	0.999	150	$4.35 \cdot 10^6$	7.2
D-10	Hydrophobized	0.995	73.9	$3.66 \cdot 10^6$	6.1
			2-naphthol adsolubilization (2nd stage)		
S-10	Hydrophilic	0.919	34.9	$1.93 \cdot 10^7$	32.1
S-10	Hydrophobized	0.999	0.3	$9.64 \cdot 10^6$	16.0
D-10	Hydrophilic	0.989	10.8	$4.86 \cdot 10^6$	8.1
D-10	Hydrophobized	0.963	3.7	$4.24 \cdot 10^6$	7.0

The calculated data (Table III) show that in case of D10 the area per molecule values are close for the hydrophilic and hydrophobic substrates, both after adsorption and adsolubilization. No changes in D10 conformation occur upon adsolubilization of 2-naphthol. In turn, the surface area of S10 is changed when 2-naphthol is incorporated into the adsorbed polymer layer. This indicates conformational changes in S10 and, in fact, weaker bonding between the S10 fragments and silica. This effect is less pronounced for the S10 adsorbed on a hydrophobized substrate where stronger bonding between the alkyl fragments and the substrate is formed in comparison to PEG fragments that form the bond with polar silica.

Finally, higher k_d/k_a values for S10 adsorbed on hydrophilic silica confirm that S10 macromolecules are more mobile in the adsorption layer in comparison to D10. In addition, the adsolubilization of 2-naphthol into the AIP fragments changes the macromolecular conformation in the AIP adsorption layer and, obviously, the nature of bonding between the macromolecule and the substrate, thus decreasing the k_d/k_a as well as the rate of polymer desorption.

AIPs-Promoted Accommodation of Lipophilic Nile Red

Knowledge of the composition of self-assembled AIP interior and exterior is an important characteristic as it may help to evaluate the architecture formed by the invertible macromolecules by various concentrations. In order to verify the composition of the AIP assembly, we carried out solubilization experiment and determined polarity of the solubilization sites and its changes by the increase of AIP concentration in solution. For this purpose, the lipophilic stain, Nile Red, has been chosen as guest molecules for accommodation (solubilization) into the AIP micelles. Nile Red is a solvatochromic dye and absorbs light at about 591 nm in dilute aqueous solutions. However, the water-solubility of Nile Red is extremely poor. The structure of Nile Red consists of an electron donor (diethylamino group) and an electron acceptor (a carbonyl group). The strong solvatochromic behavior arises from the fact that it undergoes large dipole moment changes during transition between two electronic states corresponding to charge transfer between diethylamino group and the aromatic acceptor system (quinoid), producing a twisted intramolecular charge transfer state that is stabilized in more polar solvents (*52*). This solvatochromic behavior of the Nile Red is such that polar media cause a red shift of absorbance spectra.

Samples for the accommodation of Nile Red were prepared by dissolving an appropriate amount of polyester in a saline solution under gentle agitation. Then a calculated volume of a 0.1 mg/mL Nile Red solution in ethanol was added to afford a final solution with the Nile Red concentration of 1 ug/mL. All UV-Vis spectra were recorded on a Varian Cary 5000 spectrometer.

Fig.11 shows visible absorption spectra two different AIP concentrations (cmc (A) and 1 w/v % (B) that is above cmc for all polymers) in saline. As it can be seen, visible absorption spectra of Nile Red in AIP solutions at cmc are blue shifted as compared to the dye spectra in water (maximum at 591 nm). Blue shifts indicate host-guest interaction between polymer and dye and incorporation of the Nile Red molecules into the polymeric micelles. The strongest shift has been observed for the most hydrophobic D3 (maximum at 557 nm). Decrease an in absorption maximum (corresponding to decrease in polarity of micellar solubilization sites) correlates with a length of hydrophobic fragment and to hydrophilic lipophilic balance of the polymer.

Absorbance intensity of the recorded visible spectra increases significantly with an increasing AIP concentration in saline (Fig.11B). It indicates that more Nile Red guest molecules are incorporated into self-assembled host AIP nanostructures compared to solutions at cmc. There is a blue shift for each polymeric composition at 1% confirming that polarity of solubilization cites at this polymer concentration is lower than at cmc. Thus, the self-assembly of AIP

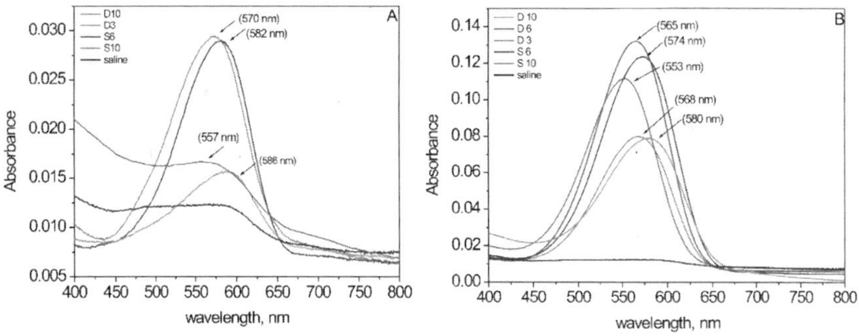

Figure 11. Visible absorption spectra of Nile Red in saline solution of different AIPs at two concentrations (A – at cmc, B – above cmc (1 w/v %)

macromolecules upon increasing concentration occurs by the interaction between AIP hydrophobic constituents. Similar to the experiment at cmc, the absorption maximum correlates with a length of hydrophobic fragment.

AIPs-Promoted Cellular Accumulation

The ability of the AIPs to effectively accommodate compounds that exhibit poor aqueous solubility can potentially be leveraged for a variety of pharmaceutical applications, including their use as vehicles for the delivery of chemotherapeutic agents to areas of pathology in the body (*53*). In particular, polymer micelles have been investigated for their ability to encapsulate antineoplastic compounds with limited solubility to treat various types of cancer (*54–56*). To investigate the potential of AIPs to serve as vehicles for drug delivery, cellular accumulation of Nile Red encapsulated AIPs (NR-AIPs) in L929 mouse fibroblast cells was examined. L929 cells were resuspended in Eagles minimal essential medium (EMEM), supplemented with 10% fetal bovine serum, penicillin (10,000 U/ml), streptomycin (10,000 U/ml) and amphotericin B (25μg/ml), and seeded in 24-well plates (50,000 cells/well). Upon reaching confluency (48 hrs of incubation at 37°C, 5% CO_2), the L929 cells were rinsed three times with Hanks balanced salt solution (HBSS) and 0.5 ml of NR-AIP solution in HBSS was added. The rate of encapsulated Nile Red was 0.2 μg / 1 g of AIP. After the desired duration of NR-AIP exposure (1hr – 96 hrs at 37°C, 5% CO_2), the NR-AIP solutions were removed and the L929 cells were examined using fluorescence microscopy (excitation: 535 nm; emission: 610 nm).

Fig.12 shows the results of the cellular accumulation studies in L929 mouse fibroblast cells. It is clear from the examination of the fluorescence images that S10 was much more effective at delivering the lipophilic dye, Nile Red, to the L929 cells than D6, D10 and S6. S10 was rapidly up taken by the L929 cells as evidenced by the relatively high cell-associated fluorescence observed after only 1 hr of exposure. In contrast, only a slight degree of cell-associated fluorescence was observed for S6 after 1 hr exposure, while no cell-associated fluorescence was observed for D6 and D10. Detailed examination of the images captured after longer durations of exposure (i.e., >24 hrs) indicate that all AIPs exhibit some

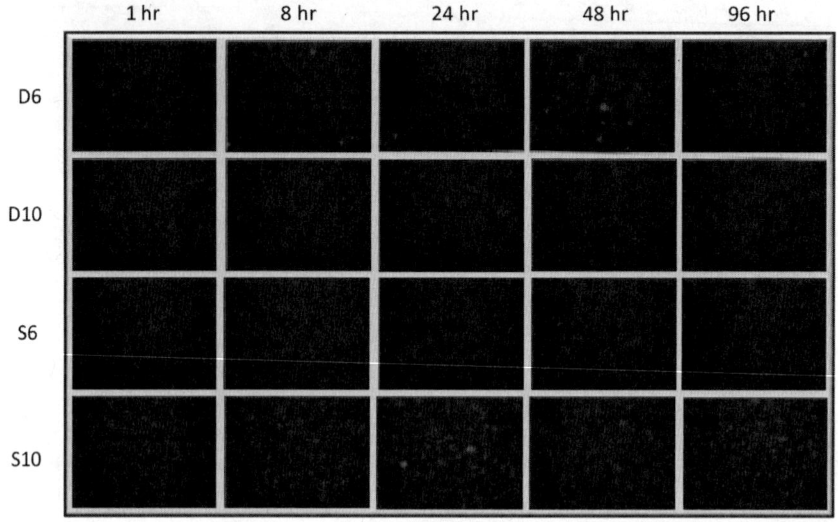

Figure 12. Uptake of Nile Red accommodated AIPs in L929 mouse fibroblast cells. A representative fluorescence image (20x magnification) of each NR-AIP is shown after exposure to L929 cells for 1, 8, 24, 48 and 96 hours.

degree of cell-associated fluorescence, with the AIPs based on sebacic acid (S6 and S10) showing the highest degree of cellular accumulation of Nile Red.

These results suggest that the AIPs were able to adequately adsorb from solution onto the surface of the L929 cells and subsequently deliver the encapsulated Nile Red molecules into their interior. In this regard, the AIPs may have undergone a conformational change due to the shift in environmental polarity upon adsorbing onto the cell surface, enhancing an interaction between the polymers and cells and enabling the release of the hydrophobic dye molecules for transport through the cell membrane. These results provide evidence that the AIPs possess the ability to switch molecular conformation and release hydrophobic substances at targeted areas located in aqueous environments.

In summary, amphiphilic invertible polymers (AIPs) composed of the poly(ethylene glycol) hydrophilic and dicarboxylic aliphatic acid constituents alternately distributed along the main backbone build micelles and self-assemble in aqueous medium by increasing concentration. The resulting complex nanostructures undergo inverse conformational changes and are functional in both polar and non-polar environments, giving them potentially broad use in many applications including pharmaceutical industry, agriculture and cosmetics.

Acknowledgments

The authors would like to thank the Office of Naval Research for financial support under grant N00014-07-1-1099 and North Dakota EPSCoR and National Science Foundation Grant EPS-0814442.

References

1. Cohen Stuart, M. A.; Huck, W. T. S.; Genzer, J.; Müller, M.; Ober, C.; Stamm, M.; Sukhorukov, G. B.; Szleifer, I.; Tsukruk, V. V.; Urban, M.; Winnik, F.; Zauscher, S.; Luzinov, I.; Minko, S. Emerging applications of stimuli-responsive polymer materials. *Nat. Mater.* **2010**, *9*, 101–113.
2. Hamley, I. W. *Introduction to Soft Matter: Polymers, Colloids, Amphiphilies and Liquid Crystals*; Wiley: Chichester, U.K., 2000.
3. McPherson, A. Crystallization of macromolecules: general principles. *Methods Enzymol.* **1985**, *114*, 112–120.
4. Cudney, R.; Patel, S.; Weisgraber, K.; Newhouse, Y.; McPherson, A. Screening and optimization strategies for macromolecular crystal growth. *Acta Crystallogr.* **1994**, *50*, 414–423.
5. Harris, J. M. In *Poly(Ethylene Glycol) Chemistry: Biotechnical and Biomedical Applications*; Harris, J. M., Ed.; Plenum Press: New York, 1992.
6. Andrade, J. D.; Hlady, V.; Jeon, S.-I. In *Hydrophilic Polymers: Performance with Environmental Acceptance*; Glass, J. E., Ed.; Advances in Chemistry Series 248; American Chemical Society: Washington, DC, 1996; pp 51−59.
7. Sakai, T.; Alexandridis, P. Size- and shape-controlled synthesis of colloidal gold through autoreduction of the auric cation by poly(ethylene oxide)–poly(propylene oxide) block copolymers in aqueous solutions at ambient conditions. *Nanotechnology* **2005**, *16*, S344–S353.
8. Alexandridis, P. Poly(ethylene oxide)/poly(propylene oxide) block copolymer surfactants. *Curr. Opin. Colloid Interface Sci.* **1997**, *2*, 478–489.
9. Adams, M. L.; Lavasanifar, A.; Kwon, G. S. Amphiphilic block copolymers for drug delivery. *J. Pharm. Sci.* **2003**, *92*, 1343–1355.
10. Mortensen, K. In *Amphiphilic Block Copolymers*; Alexandridis P., Lindman, B., Eds.; Elsevier: New York, 2000.
11. Rosch, H. In *Nonionic surfactants*; Schick, M. J., Ed.; Dekker: New York, 1967.
12. Alexandridis, P.; Holzwarth, J. F.; Hatton, T. A. Micellization of poly(ethylene oxide)-poly(propylene oxide)-poly(ethylene oxide) triblock copolymers in aqueous solutions: Thermodynamics of copolymer association. *Macromolecules* **1994**, *27*, 2414–2425.
13. Pedersen, J. S.; Gerstenberg, M. C. The structure of P85 Pluronic block copolymer micelles determined by small-angle neutron scattering. *Colloids Surf., A* **2003**, *213*, 175–187.
14. Wanka, G.; Hoffmann, H.; Ulbricht, W. Phase Diagrams and Aggregation Behavior of Poly(oxyethylene)-Poly(oxypropylene)-Poly(oxyethylene) Triblock Copolymers in Aqueous Solutions. *Macromolecules* **1994**, *27*, 4145–4159.
15. Tanford, C. *The Hydrophobic Effect: Formation of Micelles and Biological Membranes*, 2nd ed.; John Wiley & Sons: New York, 1980.
16. Liu, S. Y.; Armes, S. P. Polymeric surfactants for the new millennium: A pH-responsive, zwitterionic, schizophrenic diblock copolymer. *Angew. Chem., Int. Ed.* **2002**, *41*, 1413–1416.

17. Cogan, K. A.; Gast, A. P. Effect of water on diblock copolymers in oil: large aggregates, micelles, and microemulsions. *Macromolecules* **1990**, *23*, 745–753.
18. Halperin, A.; Tirrell, M.; Lodge, T. P. Tethered chains in polymer microstructures. *Adv. Polym. Sci.* **1991**, *100*, 31–71.
19. Winsor, P. A. In *Liquid Crystals and Plastic Crystals*; Gray, G. W., Winsor, P. A., Eds.; Ellis Harwood Ltd.: Chichester, U.K., 1974; Vol. 1, Chapter 5.
20. Voronov, A.; Kohut, A.; Peukert, W.; Voronov, S.; Gevus, O.; Tokarev, V. Invertible Architectures from Amphiphilic Polyesters. *Langmuir* **2006**, *22* (5), 1946–1948.
21. Basu, S.; Vutukuri, D. R.; Shyamroy, S.; Sandanaraj, B. S.; Thayumanavan, S. Invertible Amphiphilic Homopolymers. *J. Am. Chem. Soc.* **2004**, *126*, 9890–9891.
22. Kohut, A.; Ranjan, S.; Voronov, A.; Peukert, W.; Tokarev, V.; Bednarska, O.; Gevus, O.; Voronov, S. Design of a New Invertible Polymer Coating on a Solid Surface and Its Effect on Dispersion Colloidal Stability. *Langmuir* **2006**, *22* (15), 6498–6506.
23. Kohut, A.; Voronov, A.; Peukert, W. An Effective Way to Stabilize Colloidal Particles Dispersed in Polar and Nonpolar Media. *Langmuir* **2007**, *23* (2), 504–508.
24. Voronov, A.; Kohut, A.; Peukert, W. Synthesis of Amphiphilic Silver Nanoparticles in Nanoreactors from Invertible Polyester. *Langmuir* **2007**, *23* (2), 360–363.
25. Kohut, A.; Voronov, A.; Samaryk, V.; Peukert, W. Amphiphilic Invertible Polyesters as Reducing and Stabilizing Agents in the Formation of Metal Nanoparticles. *Macromol. Rapid Commun.* **2007**, *28*, 1410–1414.
26. Martinez Tomalino, L.; Voronov, A.; Kohut, A.; Peukert, W. Study of Amphiphilic Polyester Micelles by Hyper-Rayleigh Scattering: Invertibility and Phase Transfer. *J. Phys. Chem. B* **2008**, *112* (20), 6338–6343.
27. Voronov, A.; Vasylyev, S.; Kohut, A.; Peukert, W. Surface activity of new invertible amphiphilic polyesters based on poly(ethylene glycol) and aliphatic dicarboxylic acids. *J. Colloid Interface Sci.* **2008**, *323* (2), 379–385.
28. Voronov, A.; Kohut, A.; Vasylyev, S.; Peukert, W. Mechanism of silver ion reduction in concentrated solutions amphiphilic invertible polyesters in nonpolar solvent at room temperature. *Langmuir* **2008**, *24* (21), 12587–12594.
29. Kohut, A.; Voronov, A. Hierarchical Micellar Structures from Amphiphilic Invertible Polyesters: ^{1}H NMR Spectroscopic Study. *Langmuir* **2009**, *25* (8), 4356–4360.
30. Shafranska, O.; Voronov, A.; Kohut, A.; Wu, X.-F.; Akhatov, I. S. Polymer–metal complexes as a catalyst for the growth of carbon nanostructures. *Carbon* **2009**, *47* (13), 3137–3139.
31. Arumugam, S.; Vutukuri, D. R.; Thayumanavan, S.; Ramamurthy, V. Amphiphilic Homopolymer as a Reaction Medium in Water: Product Selectivity within Polymeric Nanopockets. *J. Am. Chem. Soc.* **2005**, *127*, 13200–13206.

32. Davies, J. T.; Rideal, E. K. *Interfacial Phenomena*; Academic Press: New York, 1961; p 371.
33. Sadron, Ch. *Angew. Chem., Int. Ed.* **1962**, *2*, 5.
34. Tuzar, Z.; Kratochvil, P. *Adv. Colloid Interface Sci.* **1976**, *6*, 201.
35. Saito, Y. J. *Jpn. Oil Chem. Soc. (Oleo Science)* **2000**, *49*, 1071.
36. Suzuki, K.; Saito, Y.; Horiuchi, T.; Tokuoka, Y.; Abe, M.; Sato, T. *J. Soc. Cosmet. Chem. Jpn.* **1996**, *30*, 330.
37. Schmitz, C.; Mourran, A.; Keul, H.; Möller, M. *Macromol. Chem. Phys.* **2008**, *209*, 1859–1871.
38. Wilhelm, M.; Zhao, C. L.; Wang, Y.; Xu, R.; Winnik, M. A.; Mura, J. L.; Riess, G.; Croucher, M. D. *Macromolecules* **1991**, *24*, 1033.
39. Wanka, G.; Hoffmann, H.; Ulbricht, W. *Macromolecules* **1994**, *27*, 4145–4159.
40. Kim, B. J.; Im, S. S.; Oh, S. G. *Langmuir* **2001**, *17*, 565–566.
41. Wu, J.; Harwell, J. H.; O'Rear, E. A. *Langmuir* **1987**, *3*, 531–537.
42. Zhu, B. Y.; Xhao, X.; Gu, T. *J. Chem. Soc., Faraday Trans. 1* **1988**, *84*, 3951–3960.
43. Esumi, K.; Yamada, T. *Langmuir* **1993**, *9*, 622–624.
44. Sakai, K.; Nakajima, E.; Takamatsu, Y.; Sharma, S. C.; Torigoe, K.; Yoshimura, T.; Esumi, K.; Sakai, H.; Abe, M. *J. Oleo Sci.* **2008**, *57*, 423–429.
45. Sakhalkar, S. S.; Hirt, D. E. *Langmuir* **1995**, *11*, 3369–3373.
46. Rosen, M. J.; Li, F. *J. Colloid Interface Sci.* **2001**, *234*, 418–424.
47. Zhao, W. R.; Gu, J. L.; Zhang, L. X.; Chen, H. R.; Shi, J. L. *J. Am. Chem. Soc.* **2005**, *127*, 8916–8917.
48. Zhu, Y. F.; Shi, J. L.; Shen, W. H.; Dong, X. P.; Feng, J. W.; Ruan, M. L.; Li, Y. S. Stimuli- responsive controlled drug release from a hollow mesoporous silica sphere/polyelectrolyte multilayer core-shell structure. *Angew. Chem., Int. Ed.* **2005**, *44*, 5083–5087.
49. Nuysink, J.; Koopal, L. K. The effect of polyethylene oxide molecular weight on determination of its concentration in aqueous solutions. *Talanta* **1982**, *29*, 495–501.
50. Tsurumi, D.; Yoshimura, T.; Esumi, K. Adsolubilization of 2-naphthol into adsorbed layer of PEO–PPO–PEO triblock copolymers on hydrophilic silica. *J. Colloid Interface Sci.* **2006**, *297*, 465–469.
51. Tsurumi, D.; Sakai, K.; Yoshimura, T.; Esumi, K. Adsolubilization of 2-naphthol into adsorbed layers of triblock PEO–PPO–PEO copolymers on hydrophobic silica particles. *J. Colloid Interface Sci.* **2006**, *302*, 82–86.
52. Ghoneim, N. Photophysics of Nile Red in solution- Steady state Spectroscopy. *Spectrochim. Acta, Part A* **2000**, *56*, 1003.
53. Torchilin, V. P. Targeted polymeric micelles for delivery of poorly soluble drugs. *Cell. Mol. Life Sci.* **2004**, *61* (19-20), 2549–2559.
54. Sinha, R.; Kim, G. J.; Nie, S.; Shin, D. M. Nanotechnology in cancer therapeutics: bioconjugated nanoparticles for drug delivery. *Mol. Cancer Ther.* **2006**, *5* (8), 1909–1917.
55. Das, M.; Mohanty, C.; Sahoo, S. K. Ligand-based targeted therapy for cancer tissue. *Expert Opin. Drug Delivery* **2009**, *6* (3), 285–304.

56. Sutton, D.; Nasongkla, N.; Blanco, E.; Gao, J. Functionalized Micellar Systems for Cancer Targeted Drug Delivery. *Pharm. Res.* **2007**, *24* (6), 1029–1046.

Chapter 14

Surface-Attached, Polymerized Vesicles Exposing Adhesive Peptide Functionalities

Markus Biesalski,[1,*] Kamlesh Shroff,[2] and Pieter Samyn[2]

[1]Ernst-Berl-Institute of Technical and Macromolecular Chemistry, Laboratory for Macromolecular Chemistry and Paper Chemistry, Darmstadt University of Technology, Petersenstrasse 22, D-64278 Darmstadt, Germany
[2]Department of Microsystems Engineering (IMTEK), Laboratory of Chemistry and Physics of Interfaces, University of Freiburg, D-79110 Freiburg, Germany
[*]biesalski@tu-darmstadt.de

Bio-inspired adhesive vesicles are designed by self-assembly of di-acetylenic fatty acids that were modified at the carboxylic acid head-group with either cell-adhesive or mussel-adhesive peptide sequences. The amphiphiles are stabilised by UV-light induced polymerisation of the hydrophobic tail into well-defined structures, and allow for controling the presentation of the adhesive groups at the vesicular surface. The vescicles can be immobilized on a surface-attached polymer hydrogel support in laterally confined structures (micro-arrays). The adhesive properties towards living cells and artificial oxidic substrates depend on the vesicle composition, concentration and presentation of the adhesive groups at the interface. This technique may serve as an interesting platform for creating functional adhesive interlayers.

Introduction

With ongoing trends in miniaturization of micro-engineered materials and systems during recent years, a great challenge includes the development of strategies for implementing new adhesive components and sensing mechanisms in functional micro-devices. Any new functionality should be firmly incorporated

© 2011 American Chemical Society

in miniaturized devices and presented in a precisely controlled fashion and in a stable manner. Application areas for such novel materials include, e.g., templates for engineering tissues (*1*) and targeted drug delivery vehicles (*2*) to model systems for the investigation of complex biological phenomena (*3*) and novel biosensor devices (*4*). With respect to the latter, researchers have started to simplify biological recognition phenomena to the interplay of small biological modules (*5*): e.g., the interaction of small peptide ligands found in extracellular matrix proteins illustrates the role of distinct biospecific activity. In particular, the engineering of synthetic materials with incorporated nature's lock-and-key mechanism may be a critical parameter to construct novel bioactive materials and devices. On the other hand, the adhesive properties of specific protein functions may be further exploited to control the technical challenges in glueing at the nanoscale. At present, one of the main challenges arising to adhesive technology includes the local dispense of very small amounts of adhesives into pads with small dimensions (*6*). Theoretically, the smallest local adhesive interaction happens at a single-molecular scale (*7*) or at diatomic scale (*8*), as being scientifically proven but technically difficult to control.

The following article is organized as follows: first we will give an extended introduction that progresses from diacetylenic amphiphiles as building blocks for functional materials, to peptide-functionalities to promote cell-adhesion, and mussel-protein inspired strategies to develop novel artificial adhesives. At the end of the introduction, we will shortly review on possible ways to immobilize diacetylenic-based materials on solid substrates. In the main section of the article, we will outline current and past efforts of our own group that were aimed for the design of peptide-functional polydiacetylenic materials, as well as two recent examples, where such materials have been shown to be promising candidates for the development of novel bioactive surface-coatings, as well as the development of novel adhesive, nanoscale glueing pads. In the first approach, polymerized vesicles from modified diacetylenic amphiphiles are used as for promoting the adhesion of living cells. Such controlled organic molecular assemblies have attracted great interest not only from the viewpoint of fundamental issues in physics, chemistry and biology but also from their future applications for molecular based devices, including their sensing capacities (*9*). The further immobilization of such self-assembled materials onto soft structured substrates will allow them to be incorporated in microsystem applications. There is need to develop strategies to immobilize the functional elements precisely and quantitatively into stable constructs that can be easily prepared and offer high sensitivity. Microcontact printing, for example, offers an advantage as a fast and reliable technique to produce arrays on a variety of substrates (*10*). However, the preparation of substrates and stamps is often difficult. A new strategy to form very stable constructs will therefore be implemented and is based on the immobilization of amphiphilic vesicles after printing to a soft polymer hydrogel support that is attached to a substrate. In the second approach, we review recent results on polymerized vesicles that expose peptides originating from mussel-proteins, known to promote adhesion to a variety of different surface-chemistries (metals, oxidic surfaces, and biological tissue). We investigated the possibility to utilize polymerized vesicles as a platform for locally confined, and chemically defined adhesion pads on solid substrates, and we show that these materials may be promising systems, if ever smaller glueing pads in engineered devices are required.

Diacetylenes as Building Blocks for Functional Materials

Polydiacetylenes form a unique class of polymeric materials that couple highly aligned and conjugated backbones with tailorable pendant side-groups and terminal functionalities. The carboxylic acid head group can be used to couple numerous chemical moieties, while the diacetylenic amphiphiles can be further structured in numerous forms, such as bulk single crystals (*11*), monolayer films by Langmuir-Blodgett deposition (*12*) or self-assembly (*13*), multilayer films (*14*), macromolecular sheets (*15*), or nanowires (*16*). They can also form nanocomposites after incorporation in inorganic host matrices (*17*).

Since the first studies by Wegner (*18*) on the polymerization mechanisms of diacetylenes, different amphiphiles carrying a diacetylenic polymerizable function received special interest due to their variability in chemistries, high stability of the polymerized assemblies, and ability to deliver sensing elements without sophisticated labeling. The diactylenic fatty acid moieties can be polymerized under the influence of ultraviolet light. In contrast to other polymeric systems that polymerize in solution, diacetylenic monomers only polymerize when the fatty acids are well-organized (crystalline state) with a typical inter-molecular distance of < 5 Å and at an angle of around 45°. This can be particularly achieved by preparing polydiacetylene films using the Langmuir-Blodgett technique. The two-dimensional structures result from polymerization into a conjugated polymer where the conjugation runs parallel to the film surface (*19–21*). The reaction proceeds along a 1-4 addition reaction of C_1 to the C_4 of the adjacent monomers and involves formation of a conjugated backbone with alternating double and triple bonds. The formation of an electron-rich conjugated π-electron system causes the polymer assembly to absorb light at specific wavelengths in the visible range, depending on the conjugation length and backbone conformation (*22*). In case of a well-ordered or aligned system with large conjugation lengths (i.e. high molar mass or close packing of the side chains), it shows a typical maximum of absorbance at wavelengths of around 640 nm (blue). The absorption shows a red-shift towards a maximum at 550 nm (red) when polymerized diacetylene backbone experiences reorganization due to internal or external stress situations, induced by changes in temperature or pH, mechanical solicitation and the binding of an analyte to functional sites exposed at the surface. This colorimetric transition is believed to result from conformational relaxation of the polymer backbone and shortening of the effective conjugation length (*23*, *24*). Polymerization of compressed and oriented mono- or multi-layer structures with diacetylene functionality has already produced a range of colored, robust, highly oriented, and environmentally responsive films (*25*).

Peptide Functionalities Controlling Adhesion of Living Cells

The adhesive interactions of cells to their extracellular matrix control and regulate among further important physiological properties of living cells, such as proliferation and differentiation. The particular requirement of cells to adhere to surfaces in order to undergo growth and division is generally referred to as 'anchorage dependence' (*26*). In order to further control cell adhesion by surface modifications that direct the specific attachment of living cells, the different interaction mechanisms between cells and extracellular matrixes should be considered: (i) non-specific interactions are due to the physico-chemical environment of the cell-membrane, and (ii) specific interactions are mediated by cell adhesion molecules situated within the cell membranes (*27*).

The non-specific interactions are controlled by a 20 to 50 nm thick repulsive structure surrounding the plasma membrane, which is called the glycocalyx (carbohydrate rich) cell coat. The negative charge at the cell surfaces arises from various glycoproteins, including immoglobulines and mucins that contain long carbohydrate chains terminated by negatively charged sialic acid, and other charged polysaccharides like glycosaminoglycans (*28*). The initial approaches and first binding between cells or cell-matrix may thus be hindered by the negatively charged surfaces, but prolonged contact causes glycocalyx reorganization and subsequently lower repulsion. Researchers have shown that the adhesion may be enhanced by reducing the initial surface charge, either by lowering the content of sialic acid (*29*) or coating with positively charged polylysine (*30*).

The specific interactions in cell-cell or cell-matrix contacts generally dominate adhesion mechanisms and are formed between dedicated cell adhesion molecules (CAM's) that are located within the plasma membrane. Among them, integrins are the most important class that bind to the extracellular matrix proteins via specialized cell junctions stabilizing cell interactions and promoting communication (*31*). The integrins are heterodimers of two non-covalently linked subunits (*32*), including fibronectin and other proteins like collagen, laminin, virtonectin etc that allows for binding to several matrix molecules. Fibronectin promotes the attachment of fibroplasts and other cells to the matrix in the connective tissue, while laminin promotes the attachment of epithelial cells to the basal lamina. Both processes are governed by the interaction of integrin-receptors that bind to small peptide-ligands in the respective proteins. The most prominent example for such a cell-adhesion promoting peptide-ligand is the minimal cell recognition sequence "RGD" (R = Arginine, G = Glycine, D = Aspartic acid) which is found in many extracellular matrix proteins (*33*). Since the discovery of peptide-ligands involved in cell adhesion mechanisms, scientists started to manipulate cellular behaviour by engineering those ligands into a variety of artificial materials, e.g. by directly attaching the peptide ligands onto grafted surfaces (*34*, *35*) or by linking them onto polymer substrates (*36*). The peptide modified polymers emerged as a relatively successful approach to control adhesion of various cells to a number of chemically different surfaces, especially by incorporating short RGD-peptide sequences on polyacrylamide (*37*), polyvinyl alcohol (*38*), polytetrafluoroethylene (*39*) and polyurethane (*40*). The cell adhesion on various RGD modified polymers was recently reviewed

(*41*). Apart from the immobilization of ligands, the design of RGD peptide as well as its density and arrangement on surfaces contribute to cell attachment. However, the RGD tri-peptide loses affinity and specificity when taken out of the context of a protein (*42*). The single RGD tri-peptide showed almost no effect in a cell detachment assay, but activity was retained by blocking the C-terminal carboxyl group of adding amino acids according to the natural sequence, wherein the activity of GRGDSP (G = Glycine, R = Arginine D = Aspartic acid, S = Serine, P = Proline) was the highest (*43*). When using short peptide ligands, the affinity is compromised but the selectivity is significantly enhanced. However, the GRGDSP-ligand was found to be thousand times less effective as compared to its parent protein fibronectin, but it was very selective in binding to the fibronectin receptor (*44*).

The above summary illustrates that peptide ligands can be used to target adhesion of cells by specifically engineering the ligand sequence. Also coupling of ligands to the underlaying substrates or to polymers is a complex task and several fabrication steps are essential to create surfaces for guiding or attaching living cells. Control of ligand spacing, clustering and presentation on artificial substrates is a key factor in controlling and designing cell-adhesive surfaces. A critical problem arises from the fact that the ligand cannot be easily presented in a well-defined way, and some ligands may get buried in a polymer film without being exposed at the surface.

Peptide Functionalities Controlling Adhesion of Artificial Substrates

A mechanism used by blue mussels (*M. Edulis*) for flexible adhesion to various substrates is based on the structure and composition of the byssal threads connected to a foot plaque. These threads have a stiffness with gradual variations and carry mussel foot proteins (*Mefp*) with decapeptide composition Ala-Lys-Pro-Ser-Tyr-Hyp*-diHyp*-Thr-DOPA*-Lys (*45–47*). The constituents and adhesive properties of the *Mefp* were investigated by many researchers (*48–50*) and mimicked in a variety of functional materials (*51–53*), reporting the key role of catechol groups in 3,4-dihydroxy-L-phenyl-alanine (DOPA) in controlling adhesive interactions. The DOPA is an enzymatic modification of tyrosine (*54*), which acts as an amino acid precursor. The direct extraction of the functional DOPA substances from mussels is rather inefficient (*55*), due to the relatively low quantities of DOPA present in *Mefp* (maximum 17 mol-%): cloning and expressing soluble recombinant *Mefp* in its full length required approximately 10,000 mussels to yield 1g of protein. Moreover, the degree of conversion from tyrosine to DOPA can be difficultly controlled and results in a very low yield (only 7 %). An improved method for converting tyrosine-containing polypeptide sequences into peptidyl DOPA was performed by hydroxylation (*56*), incorporating borate into the reaction mixture with mushroom tyrosinase at relatively high enzyme-to-substrate ratios to avoid the formation of trihydroxyphenylalanine side products. In practice, direct chemical routes are more often applied to incorporate functional DOPA moieties into synthetic polymers. Some specific conformations presenting the DOPA as either cohesive or adhesive functionalities are summarized below and illustrate the versatility of those materials. Different studies were made on cross-linking and curing of biopolymer hydrogels containing functional DOPA groups (*57–59*). The oxidation of DOPA plays an important role in cross-linking reactions to form a gel network, but the non-oxidized form of DOPA is believed to present stronger adhesive properties to metallic surfaces (*60*). Therefore, alternative procedures for gel formation other than oxidation were developed including procedures that are stimulated by temperature, pH, concentration or UV light: e.g., the DOPA can be incorporated into methacrylic block copolymers (PMMA-PMAA-PMMA) that spontaneously form a hydrogel during exposure to water vapour (*61*), or amphiphilic block copolymers with DOPA contents up to 10 wt.-% were prepared in aqueous solutions that form hydrogels by photopolymerization of methacrylate end groups attached to the hydrophobic segments (*62*). Otherwise, poly(dopamine methacryl-amide-co-methoxyethyl acrylate) adhesive films were directly synthesized by free-radical polymerization of the monomers (*63*), with the disadvantage that the copolymers are insoluble in water. There is a recent trend in incorporating the DOPA-residues into peptide sequences, where e.g. three DOPA-residues were coupled to mono-methoxy-terminated polyethylene glycol (PEG) conjugates that adsorb onto oxidized surfaces under different conditions (*64*). Also the end groups of a four-armed PEG polymer could be functionalized with single DOPA groups at each terminus, resulting in better mucoadsorption onto mucin coated TiO_2 substrates (*65*). In general, the synthesis of $(PEG)_n$-$(DOPA)_m$ linear molecules with $n = 113$ and $m = 1$ to 3 can be performed according to standard carbodiimide coupling chemistry (*66*). The thermal gelation of DOPA-functionalised PEG precursors may effectively contribute to form an adhesive hydrogel with better adhesion (*67*). Sometimes, the DOPA modified peptides were coupled to poly-N-substituted glycine oligomers peptoides and used as anchoring domains (*68*). Many of these materials have proven to be efficient in anti-fouling studies, where thin films were prepared by spin-coating or matrix-assisted pulsed-laser evaporation processes. The DOPA modified peptides also effectively served as anchoring points for immobilizating the polymer on TiO_2 surfaces. In general, a universal affinity fusion tag (e.g. poly-$(DOPA)_4$-tags) has been developed for immobilizing proteins on biomaterials (*69*).

The different mechanisms mediating the adhesive interactions of functional DOPA groups to metallic surfaces include hydrogen bonding within or between peptide molecules, chelate complexation with a metal counterface, oxidation and generation of radicals that coupling reactions or Michael-type addition reactions with lysine compounds. According to single-molecule experiments (*70*), the interfacial adhesion with metallic surfaces strongly depends on the pH-value. It appears that strong and reversible binding between DOPA and titania substrates occurred at pH < 7, with rupture at the interfacial bond between the catechol and metallic substrate. This observation suggested the formation of metal-oxygen coordination bonds in the adhesive interface. A regime of lower adhesive forces was measured at high pH due to the presence of quinone groups and its resonance structures. An equilibrium between dopa and dopa-quinone structures should thus be considered: with a value of pKa = 9.2 for the dopa hydroxyl group, the dopa structure is favored at low pH and the dopa-quinone structure predominantly appears at higher pH. It was therefore concluded that oxidation and cross-linking of dopa at high pH substantially reduces adhesion, while an optimum pH is between neutral and slightly alkaline conditions (pH = 8.3 in seawater). The strong adhesive interactions in contact with organic surfaces (e.g., NH_2 functionalised silicon wafer) were attributed to permanent covalent binding between the catechol and organic moieties.

The versatility of DOPA groups in interfering interfacial curing and adhesion reactions requires a good control over the content and the location of the adhesive groups, presented into the bulk (cross-linking) or at the surface (adhesion) of functional materials.

Surface Immobilization

Besides the local presentation of adhesive groups as discussed before, the deposition conditions relatively to the surface strongly determines the accessability of the functional moieties. The immobilization of adhesive mediating compounds onto surfaces or within miniaturized devices is a critical step in controlling their effectiveness and presentation to the environment in a precise and stable manner (*71*). Using polymerized diacetylenic vesicles as sensing elements, vesicles with thiol-groups could be immobilized at the surface of gold substrates (*72*). A limitation of this particular approach comprises the thermal instability of the generated gold-sulfur bond. The immobilization of amine-modified vesicles onto aldehyde functional surfaces could be achieved by the formation of a Schiff base (*73*). Also the immobilization of NHS-ester modified vesicles on amine-functional surfaces was reported (*74*), as well as the use of an automated micro-array spotter for preparation of laterally confined polymerized vesicles (*75*). The vesicle preparation procedure included the use of strong shear forces and heat, which may result in partial cleavage of the active ester groups, and thus limits a stable attachment of the vesicles to solid substrates. In addition, it was shown that immobilization of already polymerized diacetylenic vesicles can result in colorimetric response (i.e. blue-to-red shift) of the vesicles, which limits the use of such devices in further sensing applications (*76*). If the vesicles were first adsorbed onto the substrate, and polymerization was achieved in a second, subsequent step, the vesicles remained in their original blue-colored state and could be further used for sensing external stress. An open challenge still comprises the immobilization of already polymerized vesicles onto solid substrates, and retaining the capability of using them as stress sensors. For this purpose, the vesicles may be more favourably immobilized onto a soft (i.e. visco-elastic) support rather than on hard surfaces.

An interesting strategy for surface-immobilization of soft materials such as synthetic polymers, has been introduced by Rühe et al. (*77*). Polymer molecules can be immobilized through a photochemical reaction of benzophenone functions immobilized on planar surfaces. Upon UV irradiation, the benzophenone groups undergo a n-π^* transition with formation of a triplet-state. The latter is essentially a long-living bi-radical that can abstract an aliphatic proton in the close vicinity, e.g. from polymer molecules that are coated onto a benzophenone monolayer at a solid substrate. Upon recombination of the radicals within the aliphatic and benzophenone group, respectively, a chemical bond is formed between the polymer and the surface. If the benzophenone groups are being incorporated into the polymer itself, a UV irradiation of such films also results in the cross-linking of molecules within the layer. Hence, by this photo-chemical approach, surface-immobilized polymer monolayers (*77–79*), polymer networks (*80*), and bioactive peptide-polymer surfaces (*81*, *82*) have been reported. At present, the benzophenone photochemistry may be considered as an alternative pathway for immobilizing polymerized vesicles compared to direct linking of the vesicles to a solid substrate by specific chemical reactions. Micro-arrays are aimed at being prepared by micro-contact printing and micro-spotting, followed by UV irradiation.

Experimental: Materials and Characterization

The controlled presentation and access of ligand structures that govern interfacial adhesion needs a model system for defining the peptide functionality, morphology, distribution and conformation in the interface. The grafting of surfaces with peptide ligands or peptide-polymer hybrids meet some of these criteria, but they do not allow for precise control over the synthesis and presentation of ligands. The tunable architectures in the form of stable and bio-active "peptide-hybrid materials" can be built following a "bottom-up" approach of directed self-assembly of ligand modified monomeric units (*83*). The realization of functional supramolecular assemblies requires a careful design and choice of both, the adhesive portion and an element that directs self-assembly. The addition of an alkyl chain to the N-terminus of the functional peptide (lipidation) results in a molecular conjugate that is referred to as a peptide amphiphile (*84*, *85*). These peptide amphiphiles have gained much interest as molecular building blocks for a new generation of materials with a high degree of order at nanoscale level (*86*): they self-assemble into a variety of planar (e.g., monolayers, bilayers (*87–89*)) or three-dimensional (e.g. vesicles, fibers, rod-like micelles (*90*)) morphologies of which the stability can be tuned by changing the properties of the hydrophobic tail. When coupling the peptide moieties to a polymerizable hydrophobic tail, the latter does not only provide a driving force for self-assembly but also allows for stabilizing the structure through polymerization. As such, the polymerized peptide amphiphiles allow for a good definition of density and presentation (*84–89*) as opposed to other methods of ligand presentation, for e.g. attachment of peptide ligand(s) within the polymer networks (*41*), or ligands bound to inorganic nanoparticles (*91*).

Synthesis of Polymerizable Peptide Amphiphiles

The synthesis of polymerizable peptide amphiphiles has been outlined elsewhere (*93–95*). In the following we briefly summarize important steps in the synthesis of these materials. In brief, polymerizable peptide amphiphiles were synthesized by coupling an unsaturated fatty acid to either a methyl ester or a peptide-functional head-group. The commericially available 10,12-tricosadiynoic acid (TDA) was specifically chosen as a polymerizable group since it contains

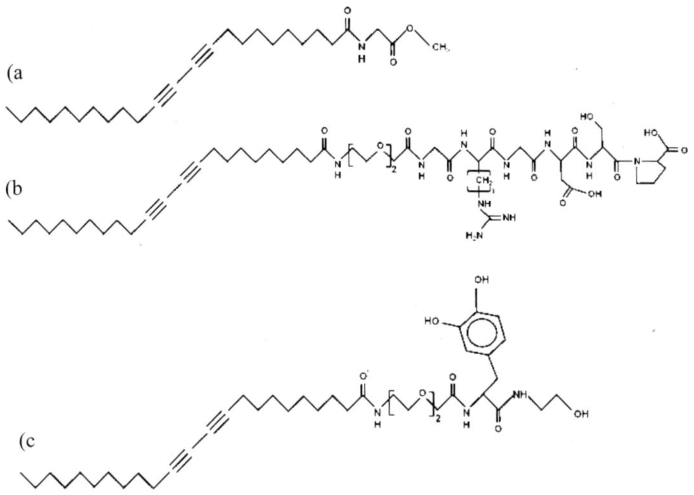

Figure 1. Peptide amphiphiles with polymerizable diacetylenic tails and bio-inspired adhesive head-group functionalities, (a) neutral matrix amphiphile PA1, (b) cell-adhesive amphiphile PA2, (c) substrate-adhesive amphiphile PA3.

an alkyl chain with two conjugated C ≡ C groups while its colorometric and fluorescent response is significantly stronger than transitions detected for 10, 12-pentacosadiynoic acid (*92*), which may be of use for further development of functional adhesive interlayers. Different polymer-peptide amphiphiles with polymerizable tail and adhesive recognition groups were synthesized (Figure 1):

- **PA1:** *10,12-tricosadiynoic-glycidyl-methyl-ester (TDA-Gly-OMe)* amphiphiles were synthesized as so-called "matrix amphiphiles" by conventional solution phase organic chemistry, modifying the carboxylic head-group of the fatty acid with a glycine methyl ester (*93*).
- **PA2:** *10,12-tricosadiyonic-amido-bis-methoxy-ethoxy-Glycidyl-Arginidyl-Glycidyl-Aspartidyl-Serinyl-Proline (TDA-(EO)$_2$-GRGDSP)* amphiphiles were synthesized by implementing a short spacer molecule of polyethylene glycol (PEG) between the peptide sequence and the fatty acid, as so-called "cell-adhesive" amphiphiles. The amphiphiles were obtained by standard solid phase peptide synthesis using Fmoc chemistry protocols as previously published for non-polymerizable peptide amphiphiles (*84*, *85*), or polymerizable peptide amphiphiles (*94*).
- **PA3:** *10,12-tricosadiynoic-amido-bis(ethylene oxide)-dihydroxphenylalanine-2-hydroxyl-ethyl amide (TDA-(EO)$_2$-DOPA-C$_2$-OH)* amphiphiles were synthesized as so-called "substrate-adhesive amphiphiles" according to conventional solid-state Fmoc-peptide synthesis protocols for polymerizable peptide amphiphiles (*95*). The fatty acid is coupled to a spacer group (PEG) and a functional adhesive head-group (DOPA = 3,4-dihydroxy-L-phenylalanine) mimicking the functional groups governing adhesion of blue mussels (*M. Edulis*).

Self-Assembly of Adhesive Amphiphiles and Polymerization of Vesicles

A known drawback with the amphiphilic self-assembled architectures is their poor stability in air and solvents: e.g., bilayer films may "roll-up" into non-defined aggregates (*89*) or vesicles consisting of peptide amphiphiles may undergo thermodynamically driven aggregation, which ultimately leads to phase separation. These problems can be overcome by formation of a covalent bond between the amphiphilic molecules, i.e. chemically cross-linking the amphiphilic molecules in the assembled state. To this respect, adhesive vesicles were designed by the self-organization of polymerizable peptide-amphiphilic types and stabilized by UV-induced polymerization.

As outlined in detail in recent communications (*94*, *95*), the polymerizable peptide amphiphiles were mixed in different concentrations of the matrix amphiphiles (0 to 100 mol-%) and either one type cell-adhesive amphiphiles (0 to 100 mol-%) or another type DOPA-adhesive amphiphiles (0 to 35 mol-%), respectively. Briefly, the amphiphilic mixtures were hydrated above their melting point (T_m = 67°C) for 3 hours, and allowed for self-organization into vesicular structures by orientation of the hydrophobic tails under subsequent ultrasonification treatment. After filtration and incubation (12 hours at 4°C), the vesicle solutions were exposed to ultraviolet irradiation (4W TLC-lamp, 250 nm) in order to stabilize their structure by polymerization of the diacetylenic tails. The resulting vesicular structures are schematically shown in Figure 2, with exposure of the functional adhesive groups at their outer (and inner) surface. The vesicle solutions were stored at 4°C and vesicles remained stable for several weeks (no aggregation within the vesicular solution).

The polymerization enhances the morphological stability of the amphiphiles, as well as it produces an electron-rich polymer backbone, which consists of conjugated double and triple bonds. Therefore, interesting chromatic properties are induced by the absorption of visible light. The polymerized vesicle-solutions show blue to red colors upon UV-light irradiation, depending on the molar composition of the respective amphiphiles. Examples for UV absorption spectra of vesicles (here: mixtures of the matrix amphiphiles PA1, and PA2, respectively) after polymerization are shown in Figure 3. The change in absorption maxima from 650 nm (blue) to 550 nm (red) at higher concentrations of cell-adhesive amphiphile PA2 may be attributed to a higher number of bulky head-groups into the vesicle bilayer which can lead to a distortion in the polymer backbone. The latter can directly affect the effective conjugation length, thus changing the characteristic colorimetric properties. Also upon polymerizing different mixtures of adhesive and matrix amphiphiles, a blue color was observed at low concentrations of adhesive amphiphiles and red color was seen at the maximum of polymerizable content of 35 mol-% adhesive amphiphiles. However, amphiphilic mixtures that are non-polymerizable due to incompatibilities by steric hinderance, did not result in a color change of the clear aqueous solution. The colorimetric response serves as a direct assessment tool to verify the copolymerization ability of the respective amphiphilic mixtures.

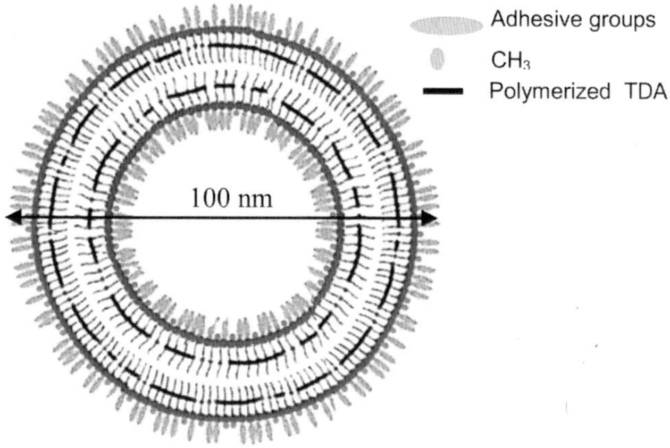

Figure 2. Schematic representation of vesicles with polymerized amphiphiles, presenting bio-inspired adhesive functionalities at the outer surface.

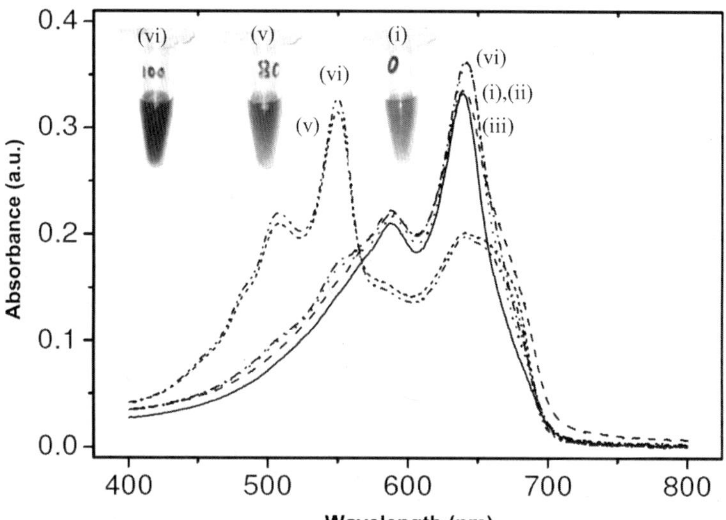

Figure 3. Colorimetric response of solutions with polymerized matrix and cell-adhesive amphiphiles in different molar ratios of PA1 and PA2, respectively: (i) 100:0, (ii) 80:20, (iii) 60:40, (iv) 40:60, (v) 20: 80, and (vi) 0:100 mol%.

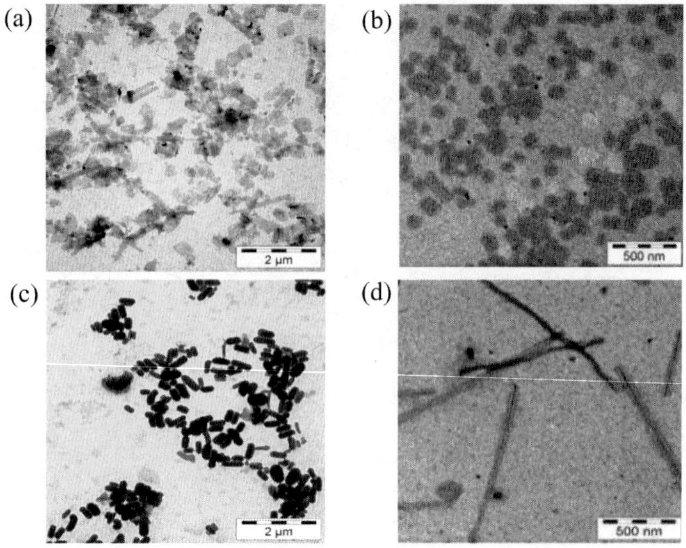

Figure 4. Cryogenic TEM observation of (a) non-polymerized PA1 amphiphile, and polymerized vesicles, containing (b) 10 mol-% PA2, (c) 35 mol-% PA2, and (d) 100 mol-% PA2, respectively.

The morphology of the polymerized vesicles was characterized by cryogenic TEM measurements, as illustrated in Figure 4. The non-polymerized fractions aggregate into instable films with parallel alignments of amphiphiles (Figure 4a). The polymerized vesicles have well-defined structures, depending on the amount of functional adhesive amphiphiles relatively to the amount of matrix amphiphiles: e.g., spherical structures are observed with diameters of 100 to 150 nm in presence of up to 20 mol-% adhesive amphiphile PA2 (Figure 4b), or bilayer structures in presence of 30 mol-% PA2 (Figure 4c). A theory describing the packing-directed assembly of amphiphiles with hydrophobic chains in the fluid state (*96*), assumes an optimal surface area per polar headgroup can be predicted from the "packing parameter" *P*, depending on the volume of the hydrophobic chain, the polar hear surface area at the critical micellar concentration and the chain length. However, the self-assembly of polymerizable amphiphiles with different head-groups may become more difficult, as it strongly depends on the concentration and mutual interactions of both constituents. E.g., the co-assembly of the matrix and DOPA-adhesive amphiphile PA3 was limited to a concentration of 35 mol-%, while the cell-adhesive amphiphile PA2 could be assembled even in 100 mol-% concentrations (*94*). Mixtures of PA1, and PA2 self-assemble into spherical vesicles for up to 80 mol-% cell-adhesive amphiphiles, while they form fibrous structures for 100 mol-% PA2 (Figure 4d).

Attachment of Adhesive Vesicles to a Substrate

Polymerized vesicles can be adsorbed to a substrate after deposition, using a surface-attached polymer gel containing photo-reactive benzophenone

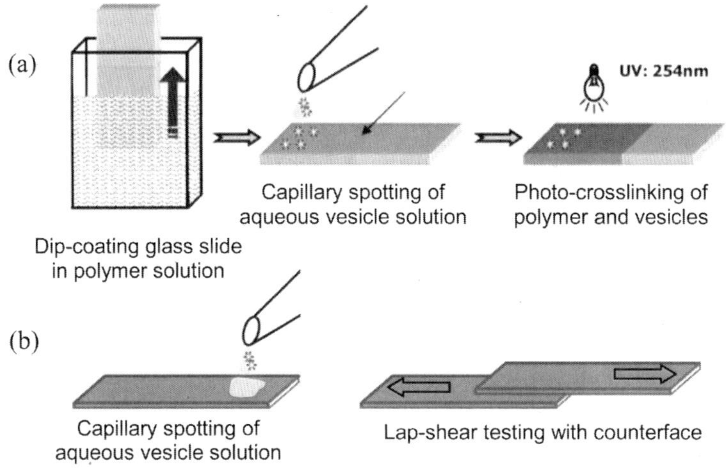

Figure 5. Immobilization of vesicles onto substrates, (a) attachment of vesicles (PA1/PA2) with cell-adhesive amphiphiles onto hydrogel support, and (b) application of vesicles (PA1/PA3) with substrate-adhesive amphiphiles into adhesive interface.

moieties. The latter hydrogel films have been extensively studied by the group of Rühe, and for details of the preparation, characterization of such surface-bound reactive polymer layers, as well as benzophenone photo-chemistry the reader is referred to the literature (*77–80*). In brief, a statistical copolymer of poly (dimethylacrylamide-benzophenone-methacrylate) or PDMAA-co-MABP (2 mol% MAPB, Mw = 10^6 g/mol) was dip-coated onto polymeric (PMMA) substrates to yield a polymer film of approximately 65 nm dry thickness. After deposition of the polymer film, a two step protocol was followed for in situ formation of a polymer network as well as to immobilize deposited PA1/PA2 vesicles within/on the polymer gel (Figure 5a): (i) the polymer film was first partially cross-linked with 25 mJ/cm^2 of UV light (254 nm), and (ii) subsequently the vesicle solutions were spotted onto the polymer layer and exposed to about 375 mJ/cm^2 of UV light to cross-link most of the remaining benzophenone within the polymer layer, as well as to chemically link the vesicles to the benzophenone moieties present in the polymer. The spot size was about 600 µm in diameter, as given by the applied volume and the spreading of the drop on the hydrophilic support. Similarly, polymerized vesicles of mixtures PA1/PA3 were spotted onto different susbtrates (glass-substrates, Si-wafer, and PMMA). As we aimed for a direct contact of these vesicles with the respective surface-chemistry of rather hard materials, we did not deposit any soft polymer gel prior to the deposition of the vesicles (*95*). A fixed volume of the aqueous dispersion of polymerized PA1/PA3 vesicles with a known concentration of vesicles was spotted onto a given surface area of 25 x 50 mm^2 (Figure 5b).

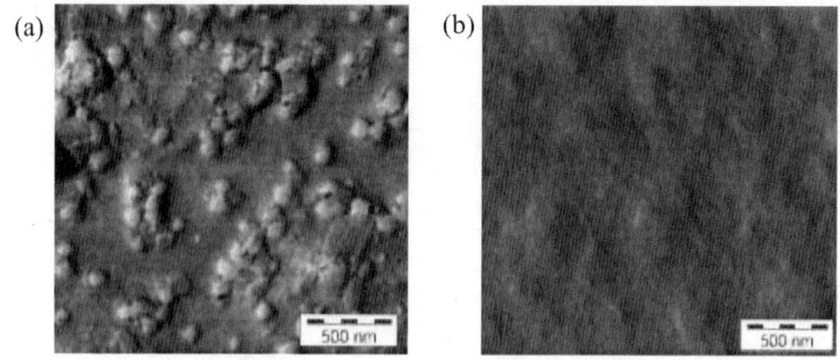

Figure 6. Detailed topographical characterization of micro-array surfaces by AFM (here: phase images), (a) surface-immobilized, polymerized PA1/PA2 vesicles on a PDMAA hydrogel support (AFM taken inside a vesicle spot), (b) surface-attached polymer hydrogel support (AFM taken outside a vesicle spot on the same substrate).

Results and Discussion: Adhesive Performance

Adhesive Performance of Surface-Attached Peptide-Polymer Vesicles in Interface between Functionalized Substrates and Living Cells

The topography of prepared micro-arrays with spotted vesicles adsorbed to the polymer gel support was characterized by AFM. Figure 6 shows the AFM phase micrographs taken within a vesicle spot, as well as outside of a vesicle spot (i.e. of the surface-attached background polymer gel). In the phase image showing the vesicle spot (Figure 6a), spherical objects are observed with a diameter of about 100-200 nm in size. The latter can be well differentiated from the background (Figure 6b). The size of these objects correlates well with the size of polymerized vesicles, as characterized by dynamic light scattering and TEM (*94*). In order to remove any non-bound vesicles, the substrates were thoroughly rinsed with water, and subsequently dried under ambient conditions.

Because the peptide amphiphile PA2 was designed to specifically interact with living cells, we were next interested in investigating the interaction of living cells with the prepared vesicle-based micro-arrays. To this respect, the substrates with micro-arrays of immobilized peptide-exposing polymerized vesicles were first sterilized by dipping in non-denatured ethanol and dried in a bio-safety cabinet. The substrates were then rinsed with PBS and placed in well-plates with pre-heated (37°C) basal medium (without serum). Human skin fibroblast cells were released from culture plates and washed with PBS. The cells were suspended in known amount of basal medium, counted and diluted to a final concentration of 10^6 cells/ml. 1 ml of this cell suspension was added to each well that contained a test substrate. The cells were mixed by gently swirling the plates on the work-bench and then cells were allowed to adhere to the substrates for one hour at 37°C (5% CO_2) under serum-free conditions. After this time, the substrates were removed from the well plates, washed 2 times with PBS, and placed in serum-containing media for longer time.

(a)

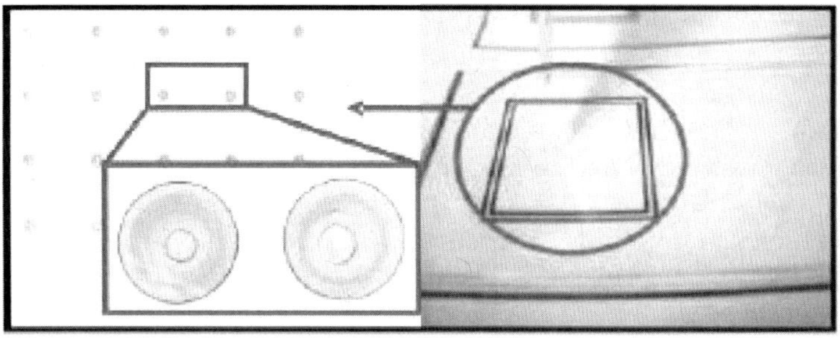

(b)

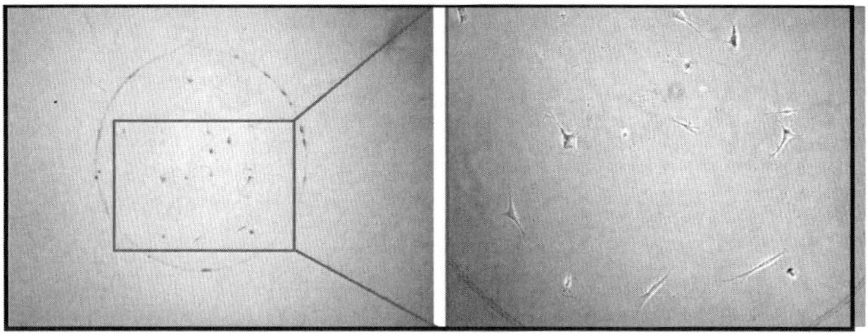

Figure 7. Optical micrographs of vesicle-based micro-arrays and adhesion of living cells, (a) micro-printing of polymerized vesicles and surface-immobilization, (b) human skin fibroplasts adhesion on vesicle spots (diameter 600 µm).

The design and effectiveness of vesicle-based micro-arrays for living cell interactions is illustrated in Figure 7, preparing micro-arrays with vesicles containing 20 mol-% of cell-adhesive PA2 amphiphiles and 80 mol-% of PA1 matrix amphiphiles. This composition has been shown as an optimum for cell-adhesion on planar surfaces that consist of the same molecular constituents (*93*). The substrates were coated with a photo-reactive polymer and polymerized vesicles were immobilized in a subsequent step, as described above. Next, living cells are being cultured on the so-prepared vesicle micro-arrays from buffered solution, as described in previous paragraph. The cells can be grown on the printed arrays via normal cell culture techniques.

Figure 7a illustrates a micro-array of 4 x 5 spots with surface-attached polymerized vesicles and Figure 7b shows the result of a subsequent cell culture experiment, using Human skin fibroblast cells to investigate the bioactivity of the prepared vesicle-based micro-arrays. However, many different cell types are known to bind to the presented GRGDSP ligand, and may therefore be used for this purpose. It is found that cells exclusively attach to the vesicle-spots, whereas outside of the spots, i.e. on the background polymer film, no cells adhere. To this respect, the chosen PDMAA hydrogels (background polymer support) are excellent to circumvent non-specific cell adhesion because it is well-known that such surface-attached (water swollen) polymer films are suited to prevent non-specific protein adsorption, that may otherwise facilitate cell-attachment outside of the vesicle-spots. As the anchoring of the transmembrane integrin receptors to the appropriate peptide-ligands is crucial for cell adhesion, proliferation, and metabolism, the cells can only adhere and spread on the vesicle spots. Cells located on the background polymer do not adhere, and thus can be washed-off from the surface. The advantage of this approach is that it allows screening cells based on specific interactions with the modified substrates as opposed to substrates immobilized with whole proteins that can support growth of numerous cell types. The cells do spread in serum-free medium, and they followed the imprinted surface-chemistry (i.e. laterally structured peptide-ligand exposure), without leaving the vesicle-spots.

Adhesive Performance of Surface-Adsorbed Polymerized Vesicles Exposing Mussel-Adhesive DOPA-Groups

The adhesive properties of vesicles with DOPA-functional amphiphiles towards oxidic silica interfaces were evaluated by drop casting a given volume of the aqueous vesicle solution (50 μL) with given vesicle concentration (0.01 to 0.125 mg/ml) onto a given surface area. As such, the number of vesicles in the interface directly correlates to the concentration of vesicles in aqueous solution (*95*). An AFM height image of the adhesive interface with low concentration (0.01 to 0.03 mg/ml) and high concentration (0.06 to 0.12 mg/ml) of vesicles absorbed onto the silica surfaces is shown in Figure 8. The vesicles form a monolayer conformation with a thickness of about 30 nm at low concentrations, while they aggregate into a multilayer with a thickness of up to300 nm at high concentrations. The local arrangement of the vesicles is very important, as it seems to determine the accessibility of the DOPA-functitonal sites in respect to the adhesive interface (*95*). In case of monolayer conformations, each single vesicle is able to establish contact between the substrate and a counter-face. The adhesive strength of each individual vesicle contributes to the overall interfacial adhesion and the adhesive strength increases with an increasing number of contact points. In case of multilayer conformation, individual vesicles in the bulk of the adhesive film may not make direct contact with the substrate, but rather make contact between each other. This results in the formation of a plane with lowest mechanical resistance into the internal adhesive film, where the interfacial shear is concentrated.

The macro-scale shear strength of interfaces with DOPA-functional vesicles was evaluated by placing a similar oxidic silica counterface in contact with the substrate on which the DOPA vesicles were drop cast. The samples were vertically clamped into a Zwick tensile tester and sheared at 5 mm/min. The adhesive bonding strength was measured as the shear force at the moment of bond rupture, using a load cell with maximum capacity of 1 kN and accuracy of 0.1 N. All test data apply for vesicles in an aqueous film between the silica substrates, and the adhesive bonding was set after about 5 minutes contact time. All data apply to polymerized vesicles that are contained in an adhesive aqueous film between the substrates.

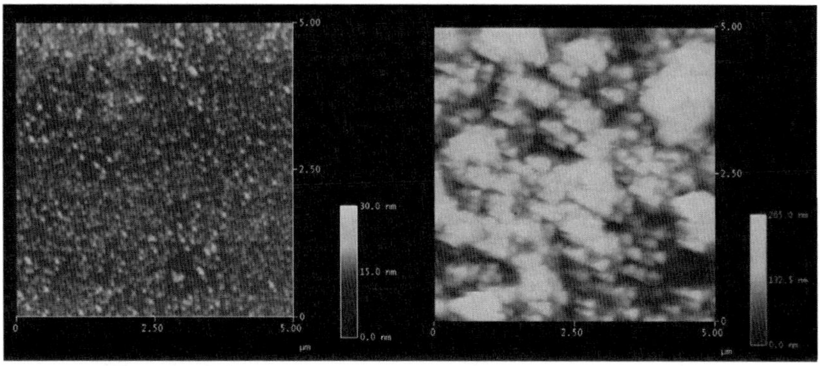

(a) (b)

Figure 8. Topography of the adhesive interlayer by AFM height scanning images, depending on the concentration of suface-adsorbed vesicles, (a) monolayer at low concentrations, (b) multilayer at high concentrations. (adapted from (95))

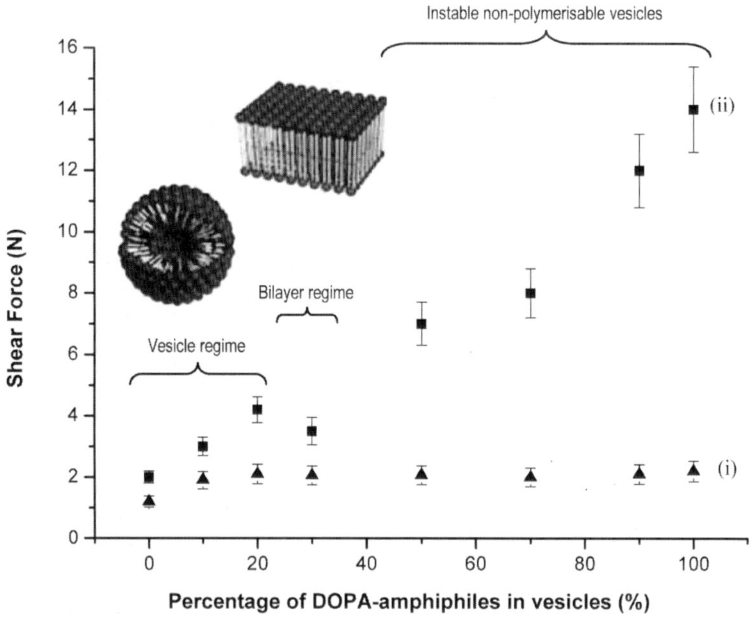

Figure 9. Influence of percentage adhesive amphiphiles relatively to matrix amphiphiles on adhesive strength (maximum shear force in lap-shear test), (i) non-assembled and non-polymerized amphiphilic mixtures, (ii) self-assembled and polymerized (< 40 mol-%) amphiphilic mixtures.

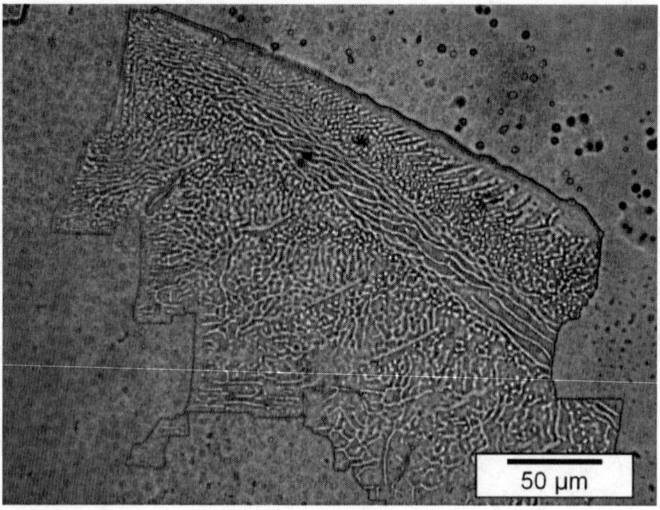

Figure 10. Adhesive interlayer of non-polymerized polymer-peptide amphiphiles with orientation of fibrous amphiphiles into instable and cracked film.

First, the effect on shear strength of UV-induced vesicle polymerization and their molar composition is represented in Figure 9. The adhesive strength of polymerized vesicles is higher and much more stable compared to disordered, non-polymerized polymer-peptide amphiphilic monolayers with a similar mol-% adhesive amphiphiles. The organization and shapes of non-polymerized amphiphiles is obviously not controlled and the assemblies may spontaneously reorganize under mechanical stress, consequently forming a discontinuous film with fiber-like structures. The fibers orient under 45° relatively to the shear direction and form inhomogeneous domains, while cracks in between these domains lead to premature failure of the adhesive bond (detailed in Figure 10). The adhesive performance of non-polymerised amphiphiles is dynamically instable over long sliding distances: the adhesive force increases over the sliding path in parallel with a spontaneous organisation of the amphiphiles. The adhesive strength of polymerized vesicles with different molar percentages of adhesive amphiphiles indicates that a higher amount of adhesive groups does not *a priori* imply higher shear forces (Figure 9). The adhesive strength increases non-linearly with the molar percentage of adhesive amphiphiles in the vesicles, and a detailed observations learns that the adhesive strength of vesicles with 20 mol-% adhesive amphiphiles was superior to vesicles containing 25 to 35 mol-% adhesive amphiphiles (*95*). This can be understood by considering the materials shapes, which are spherical when containing up to 25 mol-% adhesive amphiphiles and rather have a bilayer structure when containing higher amounts of adhesive amphiphiles. The situation for amphiphiles with consisting of more than 40 mol-% adhesive amphiphile (PA3) is different, as they could not be successfully polymerized and do not form a nanostructured adhesive interlayer. The polymerized vesicles remain stable under mechanical stress and arrange into discontinuous assemblies (detailed in Figure 11).

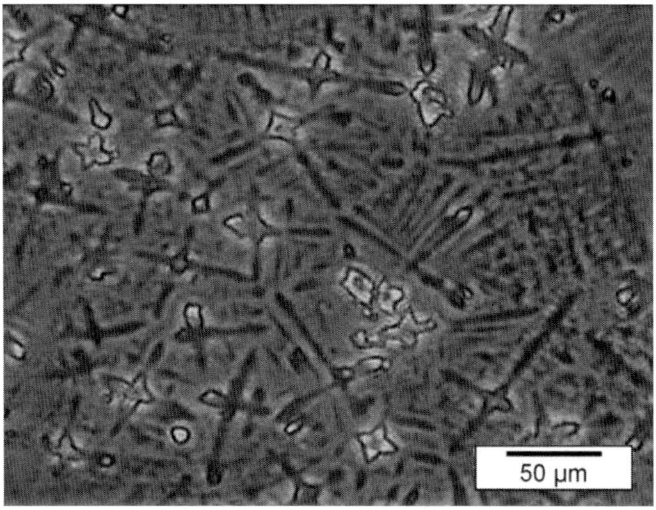

Figure 11. Adhesive interlayer of polymerized polymer-peptide amphiphiles that organize into specific dentritic structures under mechanical shear stress.

The above observations indicate important interactions between the number of adhesive points, vesicle morphology and substrate characteristics (i.e. morphology and topography), which determine the accessability and effectiveness of the adhesive groups in the interface presented to the substrates.

Second, the influence of polymerized vesicle concentrations in the interlayer on the adhesive strength is presented in Figure 12 (data in part replotted from Reference (*95*)). The concentration of polymerized vesicles was varied between 0.01 and 0.12 mg/ml for compositions with 10 mol-%, 20 mol-% and 100 mol-% adhesive amphiphiles. A higher percentage of adhesive amphiphiles does not necessarily imply higher adhesive strength for every concentration. The different trends observed at various contents of adhesive amphiphiles directly relate to the vesicle morphology (10 and 20 mol-%) or continuous interlayer film (100 mol-%). The adhesive strength of vesicles with 10 or 20 mol-% adhesive amphiphiles rises towards a maximum of about 32 to 47 N at vesicle concentrations of 0.03 mg/ml, and decreases at higher concentrations while becoming almost similar to that of a pure water film at concentrations of 0.125 mg/ml. Thus, a maximum adhesive strength occurs at intermediate vesicle concentrations: this behaviour relates to the formation of a nanoparticle monolayer at low concentrations and a nanoparticle multilayer at high concentrations (Figure 8). The adhesive strength of amphiphiles organized into a continuous interlayer is inferior and linearly increases with concentration. This behaviour was expected for a continuous film as a lower number of adhesive points decreases the overal interfacial strength.

Conclusions and Outlook

In this chapter, we demonstrated a method for the nano-scale controlled presentation of adhesive functional groups on a substrate by incorporating them

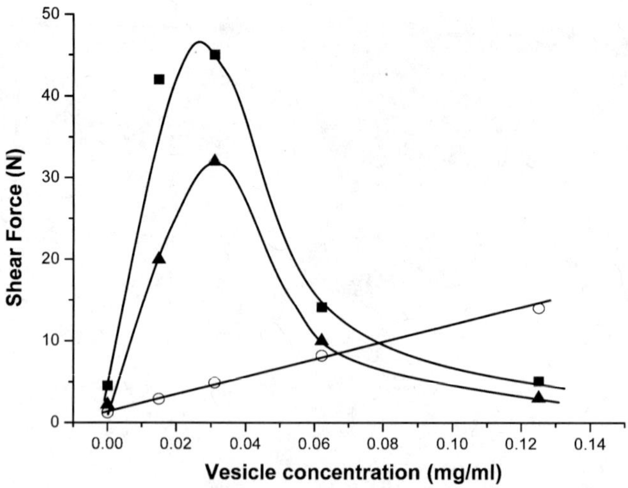

Figure 12. Influence of vesicle concentration on adhesive strength (maximum shear force in lap-shear test) for polymerized amphiphilic mixtures with 10 mol-% (▴), 20 mol-% (▪) and 100 mol-% (○) adhesive PA3 amphiphiles.

into well-defined, self-assembled vesicle architectures that can be stabilized by polymerization (i.e. polymerized vesicles exposing adhesive functional groups at the surface). The immobilization of polymerized peptide-amphiphile vesicles onto soft structured substrates allows them to be incorporated in microsystem applications. Depending on the chemical composition of the adhesive groups, adhesion towards living cells or artificial surface interfaces was established.

In order to present either cell-adhesion promoting or mussel-adhesive peptide functions in a controlled fashion at an interface, the adhesive groups were coupled to a photo-polymerizable fatty acid (tricosadiyonic acid) through standard peptide synthesis protocols. So-prepared polymerizable peptide amphiphiles were co-assembled with matrix amphiphiles into vesicular structures. The concentration of adhesive groups is determined by the molar composition of the co-assemblies, and their presentation at the vesicle surface is determined by the morphology of the polymerized vesicles. The UV-light induced polymerization of the self-assembled architectures allows for the formation of dimensionally, chemically and mechanically stable vesicles if the ratio of polymerizable groups to the other groups was chosen properly. The shape of the latter vesicles is inherent to the co-assembly process.

In a first study, micro-arrays of surface-immobilized adhesive vesicles that expose a bio-active RGD-peptide ligand at the surface were used to attract and confine living cells in a directed and spatially resolved manner onto distinct surface areas, which are defined by the volume of the applied solution during contactless deposition. In order to avoid non-specific interactions of proteins with the respective background surface, we used a surface-attached hydrogel as an underlying matrix for the attachment of the bioactive vesicles. The approach introduced here may allow for a simple, yet very versatile preparation of so-called cell-chips, i.e. cell micro-arrays, which are of interest for high throughput analysis in biomedical technology, drug development, and as biophysical model-systems.

In a second study, polymerized vesicles were prepared that expose bio-mimetic mussel-adhesive DOPA-ligands at their surface. When deposited between two chemically identical surfaces (glass-substrates), the polymerized vesicles offer much more stable and representative adhesive properties compared to the non-polymerized peptide amphiphiles. The adhesive performance strongly depends on the shape of the polymerized materials (vesicles or bilayer structures) as well as their concentration at the interface (monolayer, multilayer conformation). With respect to the latter, the accessibility of the DOPA-functional sites impacts the overall interfacial adhesion strongly, and this parameter may need to be carefully adjusted for the design of bio-inspired adhesives (i.e. bio-inspired glue). Because it is known that DOPA-adhesion functions even in water and considering high salt as well as varying pH, these materials may serve as an interesting alternative for conventional glues (typically synthetic polymers applied either from organic or aqueous solution), where the viscosity is usually rather high and increases strongly with increasing polymer content. The present aqueous-based system has a low viscosity that is almost independent from the concentration of adhesive vesicles supplied in the solvent, hence one may think of very small glueing pads to be established in the future using the polymerized DOPA-functional vesicles.

Acknowledgments

The authors thank Profs. Rühe and Wilde, Dr. Prucker and Dipl.-Ing. J. Dalin at IMTEK for fruitful and stimulating interactions within the joint "Nanopad" project. We greatly acknowledge the German Federal Ministry for Eduction and Research (BMBF) for funding within the program "Micro- Nano Integration for Microsystems Engineering (MNI-mst): Nanopad", as well as the DFG, Emmy Noether Programme (grant BI 738/1-4). Dr. Ralph Thomann at the FMF of Freiburg University assisted us in taking TEM measurements. The technical support by Mr. Matthias Preisendanz and Mr. Stefan Schwarz is most appreciated.

References

1. Drury, J. L.; Mooney, D. J. *Biomaterials* **2003**, *24* (24), 4337.
2. Langer, R. *AIChE J.* **2000**, *46*, 1286.
3. Mammen, M.; Choi, S. K.; Whitesides, G. M. *Angew. Chem., Int. Ed.* **1998**, *37*, 2754.
4. Zieziulewicz, T. J.; Unfricht, D. W.; Hadjout, N.; Lynes, M. A.; Lawrence, D. A. *Toxicol. Sci.* **2003**, *74* (2), 235.
5. Hartwell, L. H.; Hopfield, J. J.; Leibler, S.; Murray, A. W. *Nature* **1999**, *402*, C47.
6. Hayes, D.; Chen, T. *Proceedings of OFC 2005*, Anaheim CA, March 6-11, 2005; p 1.
7. Evans, E.; Ritchie, K. *Biophys. J.* **1997**, *72*, 1541.
8. Geleshuber, I. C.; Thompson, J. B.; Del Arno, Y.; Stachelberger, H.; Kindt, J. H. *Mater. Sci. Technol.* **2002**, *18*, 763.

9. Jelinek, R.; Kolusheva, S. *Biotechnol. Adv.* **2001**, *19*, 109.
10. Jenkins, A. T.; Boden, N.; Bushby, R. J.; Evans, S. D.; Knowles, P. F.; Miles, R. E.; Ogier, S. D.; Schönherr, H.; Vancso, G. J. *J. Am. Chem. Soc.* **1999**, *121*, 5274.
11. Bloor, D.; Chance, R. R. *Polydiacetylenes*; Series E: Applied sciences No. 102; Martinus Nijhoff Publishers: 1985.
12. Sasaki, D. Y.; Carpick, R. W.; Burns, A. R. J. *J. Colloid. Interface Sci.* **2000**, *229*, 490.
13. Mowery, M. D.; Menzel, H.; Cai, M.; Evans, C. E. *Langmuir* **1998**, *14*, 5594.
14. Spevak, W.; Nagy, J. O.; Charych, D. H. *Adv. Mater.* **1995**, *7*, 85.
15. Ozaki, H.; Funaki, T.; Mazaki, Y.; Masuda, S.; Harada, Y. *J. Am. Chem. Soc.* **1995**, *117*, 5596.
16. Okawa, Y.; Aono, M. *Nature* **2001**, *409*, 683.
17. Lu, Y. F.; Yang, Y.; Sellingern, A.; Lu, M. C.; Huang, J. M.; Fan, H. Y.; Haddad, R.; Lopez, G.; Burns, A. R.; Sasaki, D. Y.; Shelnutt, J.; Brinker, C. J. *Nature* **2001**, *410*, 913.
18. Wegner, G. *Z. Naturforsch., B* **1969**, *24*, 824.
19. Wilson, T. E.; Bednarski, M. D. *Langmuir* **1992**, *8*, 2361.
20. Wegner, G. *Pure Appl. Chem.* **1977**, *49*, 443.
21. Okada, S.; Peng, S.; Spevak, W.; Charych, D. *Acc. Chem. Res.* **1998**, *31*, 229.
22. Huo, Q.; Russell, K. C.; Leblanc, R. M. *Langmuir* **1999**, *15*, 3972.
23. Cheng, Q.; Stevens, R. C. *Langmuir* **1998**, *14*, 1974.
24. Lio, A.; Reichert, A.; Ahn, D. J.; Nagy, J. O.; Salmeron, M.; Charych, D. H. *Langmuir* **1997**, *13*, 6524.
25. Carpick, R. W.; Burns, A. R.; Sasaki, D. Y.; Eriksson, M. A.; Marcus, M. S. *ACS Symp. Ser.* **2005**, *888*, 82.
26. Folkman, J.; Mascona, A. *Nature* **1978**, *273*, 345.
27. Lodish, H.; Berk, A.; Zipursky, S. L.; Matsudaira, P.; Baltimore, D.; Darnell, J. *Molecular Cell Biology*, 4th ed.; W.H. Freeman and Company: New York, 2000.
28. Baszkin, A.; Norder, W. *Physical chemistry of biological interfaces*; Marcel Dekker Inc.: New York, 2000.
29. Jan, K. M.; Chien, S. *J. Gen. Physiol.* **1973**, *61*, 638.
30. Capo, C.; Bongrad, P.; Benoliel, A. M.; Ryter, A.; Depieds, R. *Ann. Immunol. (Inst. Pasteur)* **1981**, *132D*, 165.
31. Alberts, B.; Bray, D.; Lewis, J.; Roberts, K.; Watson, J. D. *Molecular Biology of the Cell*; Garland Publishers: 2002.
32. Humphries, M. J. *Biochem. Soc. Trans.* **2000**, *28*, 311.
33. Hubbell, J. A. *Curr. Opin. Biotechnol.* **1999**, *10*, 123.
34. Lin, H. B.; Lewis, K. B.; Scampavia, D.; Rater, B.; Cooper, S. L. *J. Biomater. Sci., Polym. Ed.* **1993**, *4*, 183.
35. Massia, S. P.; Hubbell, J. A. *J. Biol. Chem.* **1992**, *267*, 14019.
36. Massia, S. P.; Hubbell, J. A. *J. Cell Biol.* **1991**, *114*, 1089.
37. Brandley, B.; Schnaar, R. *Anal. Biochem.* **1988**, *172*, 270.
38. Matsuda, T.; Condo, A. *Trans. Am. Soc. Artif. Intern. Organs* **1989**, *35*, 677.

39. Massia, S. P.; Hubbell, J. A. *J. Biomed. Mater. Res.* **1991**, *25*, 223.
40. Lin, H. B.; Zhou, Z. C. *J. Biomater. Sci. Polym. Ed.* **1992**, *3*, 217.
41. Hersel, U.; Dahmen, C.; Kessler, H. *Biomaterials* **2003**, *24*, 4385.
42. Pierschbacher, M. D.; Ruoslahti, E. *Nature* **1984**, *309*, 30.
43. Pierschbacher, M. D.; Ruoslahti, E. *J. Biol. Chem.* **1987**, *262*, 17294.
44. Hautanen, A.; Gailit, J.; Mann, D. M.; Rouslahti, E. *J. Biol. Chem.* **1989**, *264*, 1437.
45. Zhao, H.; Waite, J. H. *J. Biol. Chem.* **2006**, *281*, 26150.
46. Silverman, H. G.; Roberto, F. F. *Biotechnology* **2007**, *9*, 661.
47. Wiegemann, M. *Aquatic Sci.* **2005**, *67*, 166.
48. Waite, J. H. *Int. J. Biol. Macromol.* **1990**, *12*, 139.
49. Papov, V. V.; Diamond, T. V.; Biemann, K.; Waite, J. H. *J. Biol. Chem.* **1995**, *270*, 20183.
50. Ninan, L.; Monahan, J.; Stroshine, R. L.; Wilker, J. J.; Shi, R. *Biomaterials* **2003**, *24*, 4091.
51. Kanyalkar, M.; Srivastava, S.; Coutinho, E. *Biomaterials* **2002**, *23*, 389.
52. Waite, J. H.; Holten-Andersen, N.; Jewhurst, S.; Sun, C. J. *J. Adhesion* **2005**, *81*, 297.
53. Kitamura, M.; Kawakama, K.; Nakamura, N.; Tsumoto, K.; Uchiyama, H.; Ueda, Y.; Kumagai, I.; Nakaya, T. *J. Polym. Sci.* **1999**, *A37*, 729.
54. Waite, J. H. *J. Biol. Chem.* **1983**, *258*, 2911.
55. Hwang, D. S.; Yoo, H. J.; Jun, J. H.; Moon, W. K.; Cha, H. J. *Appl. Environ. Microbiol.* **2004**, *70*, 3352.
56. Taylor, S. W. *Anal. Biochem.* **2002**, *302*, 70.
57. Loizou, E.; Weisser, J. T.; Dudigalla, A.; Procar, L.; Schmidt, G.; Wilker, J. J. *Macromol. Biosci.* **2006**, *6*, 711.
58. Haemers, S.; van der Leeden, M. C.; Koper, G. J. M.; Frens, G. *Langmuir* **2002**, *18*, 4903.
59. Burzio, L. A.; Waite, J. H. *Protein Sci.* **2001**, *10*, 735.
60. Yu, M. E.; Hwang, J. Y.; Deming, T. J. *J. Am. Chem. Sci.* **1999**, *121*, 5825.
61. Guvendiren, M.; Messersmith, P. B.; Shull, K. R. *Biomacromolecules* **2008**, *9*, 122.
62. Lee, B. P.; Chao, C. Y.; Nunalee, F. N.; Motan, E.; Shull, K. R.; Messersmith, P. B. *Macromolecules* **2006**, *39*, 1740.
63. Lee, H.; Lee, B. P.; Messersmith, P. B. *Nat. Mater.* **2008**, *7*, 8.
64. Dalsin, J. L.; Lin, L.; Tosatti, S.; Vörös, J.; Textor, M.; Messersmith, P. B. *Langmuir* **2005**, *21*, 640.
65. Catron, N. D.; Lee, H.; Messersmith, P. B. *Biointerphases* **2006**, *1*, 134.
66. Doraiswamy, A.; Dinu, C.; Cristescu, R.; Messersmith, P. B.; Chisholm, B. J.; Stafslien, S. J.; Chrisey, D. B.; Narayan, R. J. *J. Adhes. Sci. Technol.* **2007**, *13*, 1.
67. Burke, S. A.; Jones, M. R.; Lee, B. P.; Messersmith, P. B. *Biomed. Mater.* **2007**, *2*, 203.
68. Statz, A. R.; Meagher, R. J.; Barron, A. E.; Messersmith, P. B. *J. Am. Chem. Soc.* **2005**, *127*, 7972.
69. Jennissen, H. P.; Laub, M. *Materialwiss. Werkstofftech.* **2007**, *38*, 1035.
70. Lee, H.; Lee, B. P.; Messersmith, P. B. *Nature* **2007**, *448*, 338.

71. *Comprehesive Microsystems*; Zappe, H., Ed.; Elsevier: Amsterdam, 2008.
72. Stanish, I.; Santos, J. P.; Singh, A. *J. Am. Chem. Soc.* **2001**, *123*, 1008.
73. Kim, J. M.; Ji, E. K.; Woo, S.; Lee, H.; Ahn, D. J. *Adv. Mater.* **2003**, *15*, 1118.
74. Shim, H. Y.; Lee, S. H.; Ahn, D. J.; Ahn, K. D.; Kim, J. M. *Mater. Sci. Eng.* **2004**, *24*, 157.
75. Kim, J. M.; Lee, Y. B.; Yang, D. H.; Lee, J. S.; Lee, G. S.; Ahn, D. J. *J. Am. Chem. Soc.* **2005**, *127*, 17580.
76. Jelinek, R.; Kolusheva, S. *Bull. Israel Chem. Soc.* **2008**, *23*, 6.
77. Prucker, O.; Naumann, C. A.; Rühe, J.; Knoll, W.; Frank, C. W. *J. Am. Chem. Soc.* **1999**, *121*, 8766.
78. Pahnke, J.; Rühe, J. *Macromol. Rapid Commun.* **2004**, *25*, 1396.
79. Dorrer, C.; Rühe, J. *Langmuir* **2007**, *23*, 3820.
80. Toomey, R.; Freidank, D.; Rühe, J. *Macromolecules* **2004**, *37*, 882.
81. Loschonsky, S.; Shroff, K.; Wörz, A.; Prucker, O.; Rühe, J.; Biesalski, M. *Biomacromolecules* **2008**, *9*, 543.
82. Petersen, S.; Loschonsky, S.; Prucker, O.; Rühe, J.; Biesalski, M. *Phys. Status Solidi* **2009**, *A3*, 468.
83. Tu, R. S.; Tirrell, M. *Adv. Drug Delivery Rev.* **2004**, *56*, 1537.
84. Berndt, P.; Fields, G. B.; Tirrell, M. *J. Am. Chem. Soc.* **1995**, *117*, 9515.
85. Yu, Y. C.; Berndt, P.; Tirrell, M.; Fields, G. B. *J. Am. Chem. Soc.* **1996**, *118*, 12515.
86. Löwick, D. W.; van Hest, J. C. M. *Chem. Soc. Rev.* **2004**, *33*, 234.
87. Yu, Y. C.; Pakalns, T.; Dori, Y.; McCarthy, J. B.; Tirrell, M.; Fields, G. B. *Methods Enzymol.* **1997**, *289*, 571.
88. Fields, G. B.; Lauer, J. L.; Dori, Y.; Forns, P.; Yu, Y. C.; Tirrell, M. *Biopolymers* **1998**, *47*, 143.
89. Dori, Y.; Bianco-Peled, H.; Satija, S. K.; Fields, G. B.; McCarthy, J. B.; Tirrell, M. *J. Biomed. Mater. Res.* **2000**, *50*, 75.
90. Tu, R.; Biesalski, M.; Tirrell, M. *Biophys. J.* **2002**, *82*, 791.
91. Arnold, M.; Cavalcanti-Adam, E. A.; Glass, R.; Blümmel, J.; Eck, W.; Kantlehner, M.; Kessler, H.; Spatz, J. P. *Chem. Phys. Chem.* **2004**, *5*, 383.
92. Wang, S.; Lunn, R.; Krafft, M. P.; Leblanc, R. M. *Langmuir* **2000**, *16*, 2882.
93. Biesalski, M.; Knaebel, A.; Tu, R; Tirrell, M. *Biomaterials* **2006**, *27*, 1259.
94. Biesalski, M.; Tu, R.; Tirrell, M. *Langmuir* **2005**, *21*, 5663.
95. Samyn, P.; Rühe, J.; Biesalski, M. *Langmuir* **2010**, *26*, 8573.
96. Israelachvili, J. N. *Intermolecular and surface forces*; Academic Press: 1998; p 10.

Chapter 15

Synthesis of Large-Pore Periodic Mesoporous Organosilicas Using Hexane as Swelling Agent

Manik Mandal and Michal Kruk*

Center for Engineered Polymeric Materials, Department of Chemistry, College of Staten Island, City University of New York, 2800 Victory Boulevard, Staten Island, New York 10314, and Graduate Center, City University of New York, 365 Fifth Avenue, New York, New York 10016
***E-mail: michal.kruk@csi.cuny.edu**

Large-pore organosilicas with methylene and ethylene bridges in the framework were synthesized at low initial temperature (15 °C) using hexane as a swelling agent and Pluronic P123 ($EO_{20}PO_{70}EO_{20}$) block copolymers as template. Organosilica materials with the nominal (KJS) pore sizes of 18 and 10 nm were obtained in case of methylene and ethylene bridges, respectively. The variation of amount of precursor was found to have almost negligible effect on the interplanar spacing and pore size. The use of swelling agents with different chain length allowed us to tailor the pore size from 9 to 11 nm in case of ethylene-bridged organosilicas, when the first step of the synthesis was carried out at room temperature.

Introduction

Periodic mesoporous organosilicas (PMOs) were first reported in 1999 by three independent research groups (*1–4*). Initially reported syntheses involved the use of alkylammonium surfactants as micellar templates, and bis(trialkoxysilyl)organic organosilica precursors or their mixtures with a silica precursor (tetraalkoxysilane). After discovery of PMOs, a lot of research (*5–10*) has been carried out to further explore these materials. PMOs are useful from the point of view that they have highly tunable organic groups and mesopores of adjustable size, which could be used as space for loading materials (for heterogeneous catalysis, drug delivery, biomolecules encapsulation, adsorption).

© 2011 American Chemical Society

PMOs can also be used in electronics (as low-dielectric-constant materials) (*11*), and as precursors for new materials (such as ceramics and aminosilicas) (*12*, *13*).

The pore size of PMOs prepared using alkylammonium surfactants as templates was limited to ~7 nm (*14–17*). Subsequently, oligomers (*11*) and block copolymers (*18–20*) were used in order to enlarge the pore size. Also, swelling agents were used in combination with the surfactant templates (*21*). In initial studies, the use of block copolymers increased the pore size to ~10 nm with retention of an ordered structure (*19*, *20*). These syntheses was carried out mostly at or above room temperature. The combination of block copolymers with swelling agents in some cases led to pore sizes up to 20 nm with disordered structure (*21*), which could be due to uncontrolled swelling of the block copolymers by the swelling agents. So it remained a great challenge for quite some time to enlarge pore size of PMOs beyond 10 nm limit with retention of an ordered structure.

Recently, the pore size of PMOs with ethylene bridging groups was further increased (beyond 10 nm) (*22*) at sub-ambient initial synthesis temperature in the presence of a common swelling agent, 1,3,5-trimethylbenzene (TMB). Further extension of this strategy included the synthesis of methylene-bridged PMO with spherical mesopores forming weakly ordered structure (presumably Fm3m) with pore size ~13 nm (*23*). Recently, we developed routes to synthesize large-pore PMO materials with both 2-D hexagonal structures of cylindrical mesopores and face-centered cubic structures of spherical mesopores and with pore diameters up to at least ~20 nm for the 2-D hexagonal structure (*24*) and 16-17 nm for the cubic structure (*25*). The basis of our strategy (*24–27*) was the use of the understanding of the solubilization (swelling) behavior of different hydrophobic substances (swelling agent candidates) in solutions of block copolymers (*28*, *29*) to select appropriate swelling agents. The development of our strategy was stimulated by the finding of Sun et al. (*30*) that the pore diameter of SBA-15 prepared in the presence of aliphatic hydrocarbons (nonane, heptanes, hexane) as swelling agents follows the extent of solubilization of alkanes in Pluronic copolymer micelles (*29*). We envisioned that the selection of appropriate swelling agents on the basis of the experimental data and predictions for solubilization of organic compounds in Pluronics will lead to advances in the surfactant-templated synthesis of ultra-large-pore mesoporous materials. Later, we have shown that aliphatic hydrocarbons (cyclohexane) and aromatic hydrocarbons (benzene derivatives, such as toluene, xylene, 1,3,5-triisopropylbenzene) can be used as superior swelling agents in combination with Pluronic block copolymers to produce large-pore ordered mesoporous silicas (*26*, *27*) and PMOs (*24*, *25*) with 2-D hexagonal structure of cylindrical mesopores (*24*, *26*) and face-centered cubic structure of spherical mesopores (*25*, *27*). Herein, we show that hexane, an aliphatic hydrocarbon that has been successfully used to synthesize large-pore SBA-15 silicas with 2-D hexagonal structure (*30–32*), can be used under similar synthesis conditions (15 °C) to induce formation of large-pore PMO materials with pore size of about 10 nm or larger. PMOs with quite large pores can even form at ambient initial temperature and their pore diameter followed a sequence of the extent of solubilization of the swelling agents in solutions of Pluronics.

Materials and Methods

Synthesis of Methylene-Bridged PMO

PMOs with methylene bridges were synthesized using bis-(triethoxysilyl)methane (BTEM) as a precursor and hexane as a swelling agent. The molar composition of the synthesis mixture was BTEM/P123/hexane/HCl/H_2O = 0.50+x/0.0168/4.88/4.42/186, with x = 0.2, 0.3 and 0.4. It should be noted that x = 0 corresponds to Si/P123 molar ratio equal to that commonly used in SBA-15 synthesis (*26*, *30*). This composition was similar to that used in our recent study (*24*) involving cyclohexane as a swelling agent. In a typical experiment, 2.40 g of Pluronic P123 was dissolved in 84 ml 1.30 M HCl solution followed by mechanical stirring until the whole polymer dissolved completely at 15 °C. Then, premixed BTEM (different amounts) and 13 ml hexane were added at once. The whole mixture was stirred for 1 day. Then the mixture was treated hydrothermally at 100 °C for 2 days. The resulting as-synthesized materials were then filtered, washed with deionized water and dried at ~ 60 °C under vacuum. Finally, the materials were calcined under air at 300 °C for 5 hr (heating ramp 2 °C min^{-1}).

Synthesis of Ethylene-Bridged PMO

The synthesis of ethylene-bridged PMOs was performed similarly to the synthesis of methylene-bridged PMOs, except for the fact that the precursor used was bis(triethoxysilyl)ethane, BTEE. This composition was also similar to that in our recent study (*24*) involving cyclohexane as a swelling agent. The molar composition of the synthesis mixture was BTEE/P123/hexane/HCl/H_2O = 0.50+x/ 0.0168/4.88/4.42/186, with x=0.2, 0.3 and 0.4. The surfactant was removed by ethanol extraction in Soxhlet apparatus.

Measurements

The small-angle X-ray scattering patterns were recorded on Bruker Nanostar U equipped with Cu Kα radiation source (rotating anode operated at 50 kV, 24 mA) and Vantec 2000 area detector. Samples were placed in the hole of an aluminum sample holder and secured both sides using a tape. The nitrogen adsorption measurements were performed at -196 °C on a Micromeritics ASAP 2020 volumetric adsorption analyzer. The samples were outgassed at 140 °C in the port of the adsorption analyzer before analysis. Transmission electron microscopy (TEM) was carried out using FEI Tecnai G2 Twin microscope at an accelerating voltage of 120 kV. The samples were first sonicated in ethanol, then drop casted on carbon-coated copper grid and the solvent was dried in air before analysis. The specific surface area was calculated using the BET method (*33*). The total pore volume was calculated from the amount adsorbed at a relative pressure of 0.99 (*33*). The pore size distributions (PSDs) were calculated using the BJH method (*34*) with the KJS correction for the pore diameter (*35*). It is expected that the method will overestimate the pore diameters of cylindrical pores by 1-2 nm in the considered pore size range (*26*).

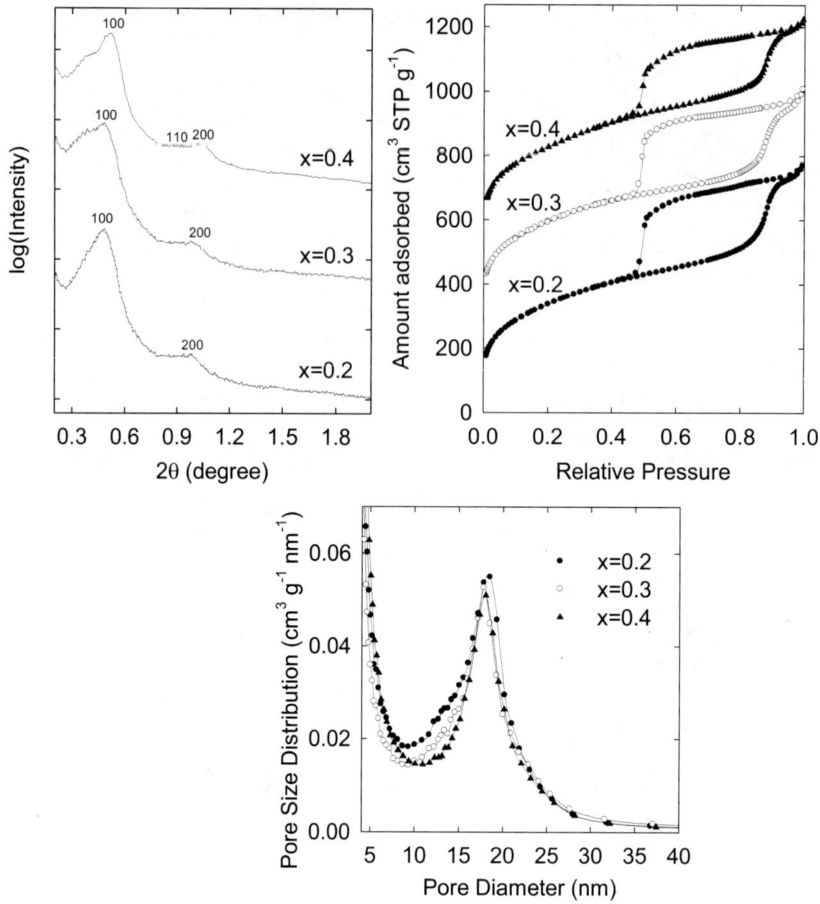

Figure 1. SAXS patterns (top left), nitrogen adsorption isotherms (top right) and pore size distributions (calculated from adsorption branches of the isotherms) (bottom) of calcined methylene-bridged PMOs synthesized with different amounts of organosilica precursor using hexane as a micelle expander at initial temperature of 15 °C followed by the hydrothermal treatment at 100 °C for 2 days. The isotherms were offset vertically by 250 and 500 cm³ STP g⁻¹ for x = 0.3 and 0.4, respectively.

Results and Discussion

Methylene-Bridged PMO

The SAXS patterns for calcined methylene-bridged PMOs synthesized at initial temperature of 15 °C using hexane as a swelling agent with different amounts of bis(triethoxysilyl)methane (BTEM) precursor are shown in Figure 1. All the materials showed peaks which can be assigned as (100) and (200) peaks. With increasing the amount of the precursor, there appeared a shoulder on left side of the (100) peak. The origin of the shoulder is not clear, but it could be related to an unidentified impurity present in the materials. The first peak position remained almost constant indicating the independence of (100) interplanar spacing on the amount of the precursor. In the case of large-pore 2-D hexagonal methylene-bridged PMOs synthesized using cyclohexane as a swelling agent, the (100) interplanar spacing was found to be strongly dependent on the relative amount of the precursor, and the interplanar spacing reached a maximum of ~22 nm at an intermediate amount of the precursor (x = 0.3) (*24*). On the other hand, when x was equal to 0.2 and 0.4, the interplanar spacing was much smaller, unlike in the case of hexane used as a swelling agent. It should be noted that the amount of the precursor used in all of these studies (x from 0.2 to 0.4) corresponded to higher Si/P123 ratio than that used for synthesizing ordered pure-silica materials under similar conditions (*26*, *31*, *36*). During calcination, the PMOs essentially did not shrink. The (100) interplanar spacing of the calcined methylene-bridged PMOs was very large, ranging from 17.2 to 18.2 nm (Table 1), which is smaller only from the (100) interplanar spacing of 2-D hexagonal methylene-bridged PMOs prepared in the presence of cyclohexane. Although the SAXS patterns featured quite broad peaks, making the structural assignment less convincing, TEM images of the considered PMOs showed stripes attributable to a 2-dimensional hexagonal structure (as a side projection of a honeycomb), indicating that the materials were fairly well ordered (Figure 2). However, TEM also revealed some regions, which may have been less well organized. The weakness of (110) peak, which was also observed for cyclohexane-swollen PMOs (*24*), could be related to a large pore wall thickness relative to the unit-cell parameter, because the ratio of the pore wall thickness to the unit-cell parameter is known to influence the relative peak intensity on X-ray diffraction patterns for ordered mesoporous materials (*37*).

The materials were further analyzed by the nitrogen physisorption technique. All the materials showed isotherms with broad hysteresis loops having steep capillary condensation and evaporation steps. The capillary condensation steps appeared at relatively high relative pressure of ~ 0.88, indicating the presence of large mesopores. The capillary evaporation step appeared at a lower limit of adsorption-desorption hysteresis, indicating the presence of narrow entrances (of size below 5 nm) (*38*) through which the uniform mesopore channels were accessible from the surrounding. The specific surface area of 1200-1250 m^2 g^{-1}, and pore volume of 1.11-1.17 cm^3 g^{-1} were obtained for the considered materials, being very similar in all cases. The pore size distributions (PSDs) showed maxima at the pore diameter of ~ 18 nm, which is very large compared to PMOs synthesized so far using different synthetic conditions, except for our recent report of the synthesis of 2-D hexagonal large-pore PMOs using cyclohexane as a swelling agent (*24*). Another interesting feature to notice is that the capillary condensation pressure remained almost constant over a wide range of relative amounts of bis(triethoxysilyl)methane precursor. This indicates that the pore size remained constant upon variation in the amount of precursor from x=0.2 to x=0.4, which corresponds to the Si : P123 ratio 40-80% higher than that in the standard SBA-15 silica synthesis. This invariance is different from the significant sensitivity to the synthesis conditions observed for methylene-bridged PMOs prepared in the presence of cyclohexane (*24*), but resembled the behavior

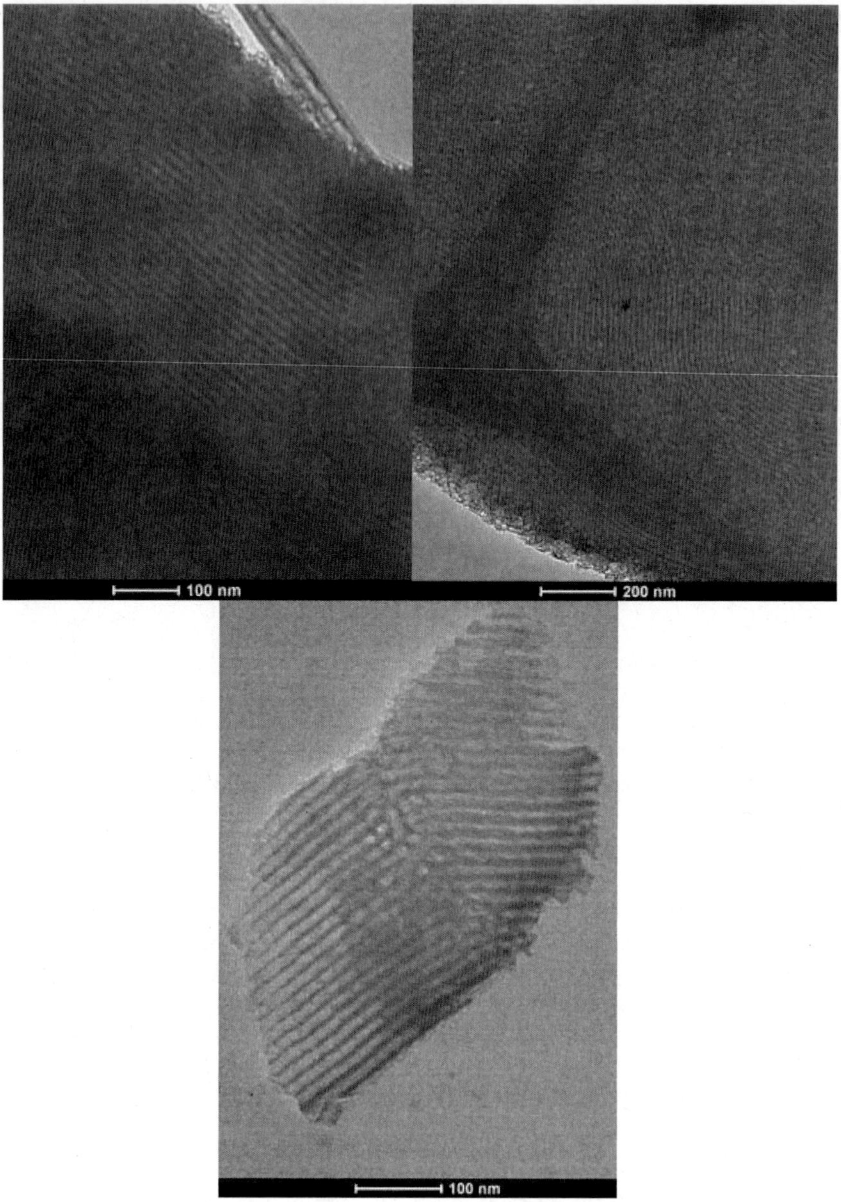

Figure 2. TEM images of methylene-bridged PMOs synthesized using hexane as a swelling agent at x = 0.2 (top left), 0.3 (top right) and 0.4 (bottom).

of ethylene-bridged PMOs synthesized in the presence of cyclohexane. These findings suggest that the response to the variations in composition of the synthesis mixture varies from one combination of the framework precursor and swelling agent to another. As we discussed elsewhere (*24*), both the swelling agent and the framework precursor (or its hydrolysis products) are expected to be present in the cores of the Pluronic P123 micelles and thus the uptake of the swelling agent may

Table 1. Structural properties of organosilicas[a]

Conditions: Bridging group, initial synthesis temperature, precursor amount, swelling agent	d_{100} *(nm)*	$d_{100,AS}$ *(nm)*	S_{BET} *(m²/g)*	w_{KJS} *(nm)*	V_t *(cm³/g)*
Methylene, 15°C, x=0.2, Hexane	18.2	18.2	1223	18.2	1.17
Methylene, 15°C, x=0.3, Hexane	18.3	18.2	1252	18.0	1.14
Methylene, 15°C, x=0.4, Hexane	17.2	17.2	1198	18.1	1.11
Ethylene, 15 °C, x=0.2, Hexane	15.0	14.4	497	10.9	0.62
Ethylene, 15 °C, x=0.3, Hexane	14.1	14.2	185	10.4	0.27
Ethylene, 15 °C, x=0.4, Hexane	14.1	14.2	77	10.0	0.12
Ethylene, 24 °C, x=0.4, Hexane	13.3[b]	13.3[b]	520	11.0	0.92
Ethylene, 24 °C, x=0.4, Octane	12.8[b]	12.8[b]	416	9.6	0.62
Ethylene, 24 °C, x=0.4, Decane	11.9[b]	12.3[b]	465	8.9	0.57

[a] Notation: d_{100}, interplanar spacing for calcined or extracted sample; $d_{100,AS}$, interplanar spacing for as-synthesized sample; S_{BET}, BET specific surface area; w_{KJS}, KJS pore diameter; V_t, total pore volume. [b] Tentative assignment.

be influenced by the identity and the relative amount the framework precursor and the swelling agent. It is also noteworthy that the swelling agent in the present case is used in large excess (*24*, *31*), and apparently its uptake by the micelles is limited. The amount of precursor when x = 0.4, corresponds to 1.8 times the amount of precursor (on Si : P123 molar ratio basis) used in the synthesis of ordered pure-silica materials with 2-dimensional hexagonal structure (commonly referred as SBA-15) under similar conditions (*26*, *31*). Upon further increasing the amount of the precursor (which corresponded to its double amount), the height of PSD decreased significantly (data not shown).

Ethylene-Bridged PMO

A series of large-pore ethylene-bridged PMOs was synthesized using hexane as a swelling agent and different amounts of bis(triethoxysilyl)ethane (BTEE) precursor at initial synthesis temperature of 15 °C followed by the hydrothermal treatment at 100 °C for 2 days. The SAXS patterns of the extracted materials (Figure 3) showed a strong, but somewhat broad main peak and several weak peaks, which can be tentatively assigned as (100), (200), and (210) reflections, indicating the formation of 2-dimensional hexagonal structure (p6mm symmetry). The (100) interplanar spacings of 14.1-15.0 nm were obtained for the considered materials (see Table 1). It should be noted that the methylene- and ethylene-bridged PMOs were synthesized under similar conditions, and yet the (100) interplanar spacing was appreciably larger for methylene-bridged PMOs.

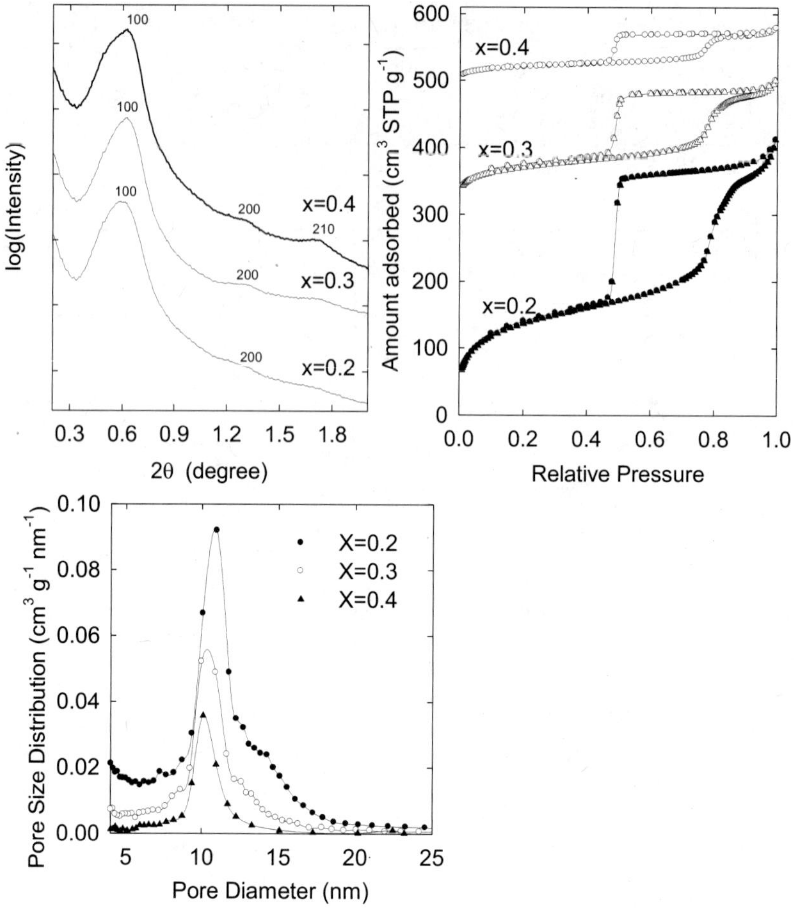

Figure 3. SAXS patterns (top left), nitrogen adsorption isotherms (top right) and pore size distributions (calculated from adsorption branches of the isotherms) (bottom) of extracted ethylene-bridged PMO synthesized with different amounts of organosilica precursor using hexane as a micelle expander at an initial synthesis temperature of 15 °C, followed by hydrothermal treatment at 100 °C for 2 days. The isotherms are offset vertically by 320, and 500 cm³ STP g⁻¹ for x = 0.3, and 0.4, respectively.

Nitrogen adsorption isotherms (Figure 3) showed broad hysteresis loops similar to those observed for the methylene-bridged PMOs. Both capillary condensation and evaporation steps were steep. The capillary condensation steps slightly shifted to a lower relative pressure (p/p_0) with increasing the amount of the precursor. When the amount of the precursor was low (i.e., x = 0.2), the capillary condensation step did not level off completely. After increasing the amount of the precursor, the capillary condensation steps leveled off to a much more significant extent, suggesting a lower interparticular porosity. The samples showed a delayed capillary evaporation step at the lower limit of adsorption-desorption hysteresis, which is due to the presence of "plugs" or constrictions in the main mesopore

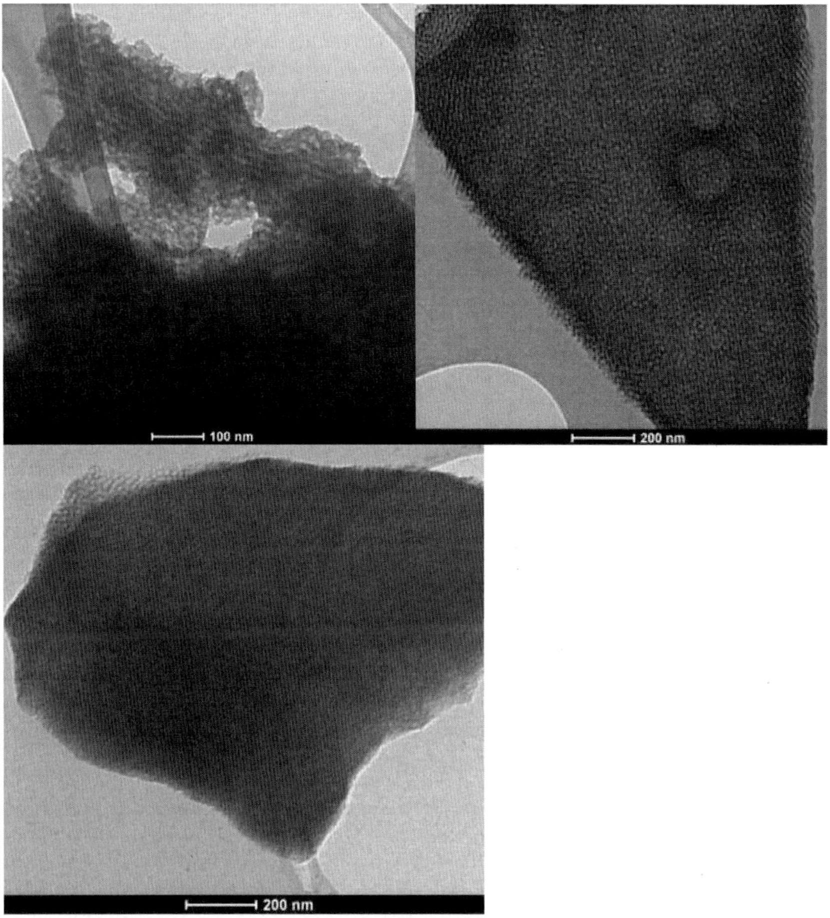

Figure 4. TEM images of ethylene-bridged PMO synthesized using hexane as a swelling agent with the BTEE/P123 relative amounts corresponding to x = 0.2 (top left), 0.3(top right) and 0.4 (bottom).

channels (*39*, *40*). The "plugs" were observed first in the case of templated pure-silica materials (*39*), which was explained to be due to the solubilization of excess silica precursor in the hydrophobic cores of the block copolymer micelles (*40*). In the case of organosilica materials, the delayed capillary evaporation step could be related to a similar behavior (*24*, *41*). The "plugs" reside in the main mesopores or at the entrances to the mesopores (*24*). The PSDs indicated that with increasing the amount of the precursor, the pore size decreased to some extent, because pore sizes of 10.9, 10.4, and 10.0 nm were obtained when x = 0.2, 0.3, and 0.4 respectively. With the increase in the amount of the precursor, the mesopore volume and the height of the capillary condensation step decreased, which may be in part due to inefficient surfactant extraction. A lower particle size was observed by TEM at lower relative amount of the precursor (Figure 4). The TEM images also showed the presence of stripes, which is typical for hexagonal structures viewed from the side, although honeycomb projections were

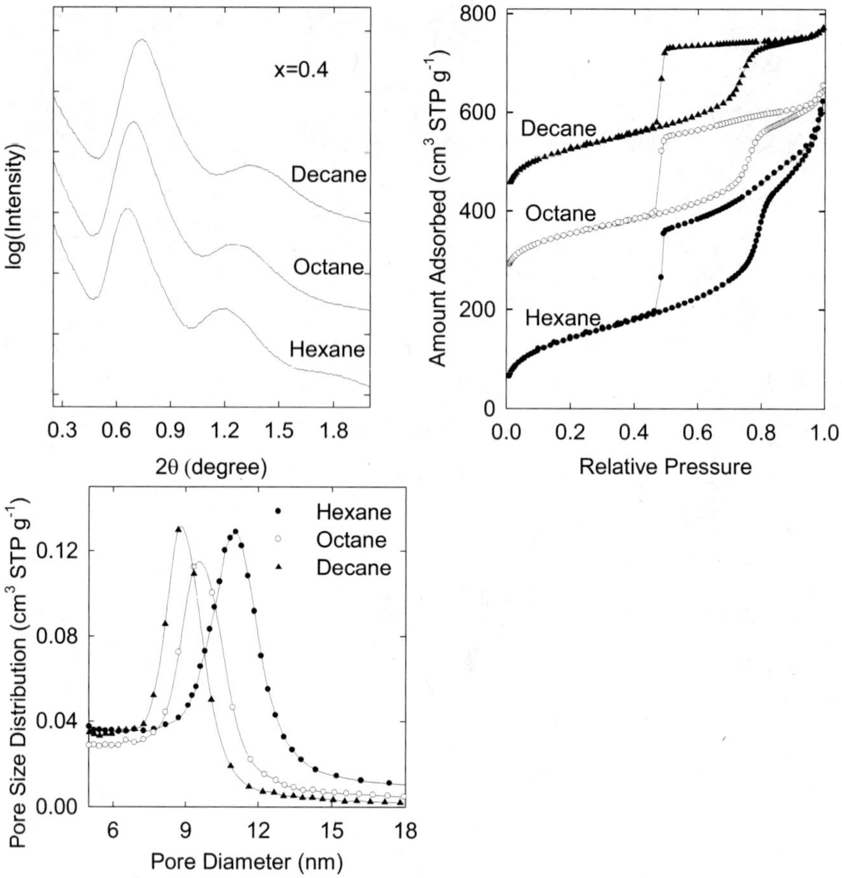

Figure 5. SAXS patterns (top left), nitrogen adsorption isotherms (top right) and pore size distributions (calculated from adsorption branches of the isotherms) (bottom) of extracted ethylene-bridged PMOs synthesized with different micelle expanders at an initial synthesis temperature of 24 °C followed by hydrothermal treatment at 100 °C for 2 days. The isotherms are offset vertically by 240, and 400 cm^3 STP g^{-1} for materials synthesized using octane and decane as swelling agents, respectively.

not observed (Figure 4). On the other hand, some areas on TEM images appeared to be quite weakly ordered, perhaps worm-like.

In the case of ethylene-bridged PMO, aliphatic swelling agents of different chain length (hexane, octane, and decane) were used to tailor the pore size. In this case, the materials were synthesized at an initial synthesis temperature of 24 °C followed by the hydrothermal treatment at 100 °C for 2 days. While the SAXS patterns could not be unequivocally assigned to 2-d hexagonal structure, it was clear that with the increase in the chain length of the hydrocarbon swelling agent, the interplanar spacing appeared to decrease, as inferred from the position of the main peak at SAXS patterns (Figure 5). This can be attributed to the fact that solubilization of aliphatic hydrocarbons in micelles of Pluronics decreases with increase in the chain length (*28*, *29*). Bao et al. proposed that this solubilization behavior is a reason why the unit-cell size and pore diameter of SBA-15 silica can be increased by decreasing the chain length of the linear hydrocarbon swelling agent (*30*, *32*). Our observation for PMOs follows their finding for pure-silica materials, but is even a more direct evidence of the considered behavior, because all our samples were prepared at the same initial synthesis temperature, whereas Bao et al. varied the initial temperature. The nitrogen adsorption isotherms and pore size distributions for the organosilicas synthesized using different linear hydrocarbon swelling agents (Figure 5) show that with the increase in the chain length from hexane to octane and decane, the pore diameter decreased from 11.0, to 9.6, and 8.9 nm. The pore volumes showed similar increasing trend (Table 1), but this may be coincidental, as the development of secondary mesoporosity could be inferred from nitrogen adsorption isotherms as the hydrocarbons with shorter chain length were employed in the synthesis.

Conclusions

Large-pore organosilicas with methylene and ethylene bridges in the framework were synthesized at low initial temperature (15 °C) using hexane as a swelling agent. KJS (nominal) pore diameters of 18 nm was obtained in the case of methylene-bridged PMO, while for ethylene-bridged PMO, the pore size of 10-11 nm was obtained. The obtained materials were moderately ordered, as seen from SAXS patterns with rather broad peaks possibly featuring shoulders. The pore diameters achieved for methylene-bridged PMOs were unusually large and they were only matched by those reported in our study involving cyclohexane. For ethylene-bridged PMOs, the synthesis was also successful when the initial part of it was performed at room temperature, but the structure of the resulting materials was not evident from SAXS patterns. In this case, octane and decane swelling agents were found to afford lower pore sizes than of the pore size induced by hexane. All the materials showed considerable constrictions either inside the main mesopores or at the entrances to them.

Acknowledgments

Imaging facility at CSI is acknowledged for providing access to TEM. The NSF is acknowledged for supporting the acquisition of SAXS/WAXS instrument through grant CHE-0723028 and for partial support of this research (award DMR-0907487). BASF is acknowledged for providing Pluronic P123 block copolymers.

References

1. Inagaki, S.; Guan, S.; Fukushima, Y.; Ohsuna, T.; Terasaki, O. *J. Am. Chem. Soc.* **1999**, *121*, 9611–9614.
2. Melde, B. J.; Holland, B. T.; Blanford, C. F.; Stein, A. *Chem. Mater.* **1999**, *11*, 3302–3308.

3. Asefa, T.; MacLachlan, M. J.; Coombs, N.; Ozin, G. A. *Nature* **1999**, *402*, 867–871.
4. Yoshina-Ishii, C.; Asefa, T.; Coombs, N.; MacLachlan, M. J.; Ozin, G. A. *Chem. Commun.* **1999**, 2539–2540.
5. Fujita, S.; Inagaki, S. *Chem. Mater.* **2008**, *20*, 891–908.
6. Asefa, T.; Yoshina-Ishii, C.; MacLachlan, M. J.; Ozin, G. A. *J. Mater. Chem.* **2000**, *10*, 1751–1755.
7. MacLachlan, M. J.; Asefa, T.; Ozin, G. A. *Chem. Eur. J.* **2000**, *6*, 2507–2511.
8. Hatton, B.; Landskron, K.; Whitnall, W.; Perovic, D.; Ozin, G. A. *Acc. Chem. Res.* **2005**, *38*, 305–312.
9. Hoffmann, F.; Cornelius, M.; Morell, J.; Froba, M. *Angew. Chem., Int. Ed.* **2006**, *45*, 3216–3251.
10. Stein, A.; Melde, B. J.; Schroden, R. C. *Adv. Mater.* **2000**, *12*, 1403–1419.
11. Lu, Y. F.; Fan, H. Y.; Doke, N.; Loy, D. A.; Assink, R. A.; LaVan, D. A.; Brinker, C. J. *J. Am. Chem. Soc.* **2000**, *122*, 5258–5261.
12. Asefa, T.; Kruk, M.; Coombs, N.; Grondey, H.; MacLachlan, M. J.; Jaroniec, M.; Ozin, G. A. *J. Am. Chem. Soc.* **2003**, *125*, 11662–11673.
13. Toury, B.; Babonneau, F. *J. Eur. Ceram. Soc.* **2005**, *25*, 265–270.
14. Sayari, A.; Hamoudi, S.; Yang, Y.; Moudrakovski, I. L.; Ripmeester, J. R. *Chem. Mater.* **2000**, *12*, 3857–3863.
15. Liang, Y. C.; Hanzlik, M.; Anwander, R. *Chem. Commun.* **2005**, 525–527.
16. Liang, Y.; Hanzlik, M.; Anwander, R. *J. Mater. Chem.* **2006**, *16*, 1238–1253.
17. Liang, Y.; Erichsen, E. S.; Hanzlik, M.; Anwander, R. *Chem. Mater.* **2008**, *20*, 1451–1458.
18. Muth, O.; Schellbach, C.; Froba, M. *Chem. Commun.* **2001**, 2032–2033.
19. Zhu, H. G.; Jones, D. J.; Zajac, J.; Roziere, J.; Dutartre, R. *Chem. Commun.* **2001**, 2568–2569.
20. Matos, J. R.; Kruk, M.; Mercuri, L. P.; Jaroniec, M.; Asefa, T.; Coombs, N.; Ozin, G. A.; Kamiyama, T.; Terasaki, O. *Chem. Mater.* **2002**, *14*, 1903–1905.
21. Burleigh, M. C.; Markowitz, M. A.; Wong, E. M.; Lin, J. S.; Gaber, B. P. *Chem. Mater.* **2001**, *13*, 4411–4412.
22. Zhou, X.; Qiao, S.; Hao, N.; Wang, X.; Yu, C.; Wang, L.; Zhao, D.; Lu, G. Q. *Chem. Mater.* **2007**, *19*, 1870–1876.
23. Kruk, M.; Hui, C. M. *J. Am. Chem. Soc.* **2008**, *130*, 1528–1529.
24. Mandal, M.; Kruk, M. *J. Mater. Chem.* **2010**, *20*, 7506–7516.
25. Mandal, M.; Kruk, M. *J. Phys. Chem. C* **2010**, *114*, 20091–20099.
26. Cao, L.; Man, T.; Kruk, M. *Chem. Mater.* **2009**, *21*, 1144–1153.
27. Huang, L.; Yan, X.; Kruk, M. *Langmuir* **2010**, *26*, 14871–14878.
28. Nagarajan, R.; Barry, M.; Ruckenstein, E. *Langmuir* **1986**, *2*, 210–215.
29. Nagarajan, R. *Colloids Surf., B* **1999**, *16*, 55–72.
30. Sun, J.; Zhang, H.; Ma, D.; Chen, Y.; Bao, X.; Klein-Hoffmann, A.; Pfaender, N.; Su, D. S. *Chem. Commun.* **2005**, 5343–5345.
31. Kruk, M.; Cao, L. *Langmuir* **2007**, *23*, 7247–7254.
32. Zhang, H.; Sun, J.; Ma, D.; Weinberg, G.; Su, D. S.; Bao, X. *J. Phys. Chem. B* **2006**, *110*, 25908–25915.
33. Sing, K. S. W.; Everett, D. H.; Haul, R. A. W.; Moscou, L.; Pierotti, R. A.; Rouquerol, J.; Siemieniewska, T. *Pure Appl. Chem.* **1985**, *57*, 603–619.

34. Barrett, E. P.; Joyner, L. G.; Halenda, P. P. *J. Am. Chem. Soc.* **1951**, *73*, 373–380.
35. Kruk, M.; Jaroniec, M.; Sayari, A. *Langmuir* **1997**, *13*, 6267–6273.
36. Cao, L.; Kruk, M. *Colloids Surf., A* **2010**, *357*, 91–96.
37. Schmidt, W. *Microporous Mesoporous Mater.* **2009**, *117*, 372–379.
38. Kruk, M.; Jaroniec, M. *Chem. Mater.* **2003**, *15*, 2942–2949.
39. Van Der Voort, P.; Ravikovitch, P. I.; De Jong, K. P.; Benjelloun, M.; Van Bavel, E.; Janseen, A. H.; Neimark, A. V.; Weckhuysen, B. M.; Vansant, E. F. *J. Phys. Chem. B* **2002**, *106*, 5873–5877.
40. Kruk, M.; Jaroniec, M.; Joo, S. H.; Ryoo, R. *J. Phys. Chem. B* **2003**, *107*, 2205–2213.
41. Bao, X.; Zhao, X. S.; Li, X.; Li, J. *Appl. Surf. Sci.* **2004**, *237*, 380–386.

Chapter 16

Template-Free Synthesis and *in Situ* Functionalization of Nanocapsules

Ramjee Balasubramanian* and Zaharoula M. Kalaitzis

Department of Chemistry and Biochemistry, 4541 Hampton Blvd, Old Dominion University, Norfolk, VA 23529
***bramjee@odu.edu**

A single component, multifunctional resorcinarene thiol-ene monomer, resorcinarene tetra alkene tetra thiol (RTATT), has been used as a photopolymerizable building block, for the synthesis of hollow nanocapsules, and other morphologically distinct polymeric architectures. Photopolymerization of RTATT in chloroform resulted in the direct formation of nanocapsules, in the absence of any template or preorganization. We show that the polymerization media plays a crucial role in determining the polymer morphology, as a variety of architectures, like lattices, fibrous networks, and nanoparticles were produced by varying the solvent. Further, we have developed a one-pot, two-stage method for the synthesis and *in situ* functionalization of resorcinarene nanocapsules.

Introduction

Polymeric nanocapsules, i.e., spherical polymeric architectures of nanoscopic dimensions with a hollow interior, have attracted enormous attention in recent years (*1*). Nanocapsules are capable of encapsulating a wide variety of hydrophilic and lipophilic guest molecules and offer potential applications as drug delivery vehicles, imaging agents, nanoreactors, and in catalysis (*1*, *2*). The challenging task of fabricating hollow nanocapsules is typically accomplished by sacrificial template based approaches or emulsion polymerization (*1*, *2*). Typically, in a template based approach an organic/polymeric shell is either grown or self-assembled around the template by a variety of means, following which the sacrificial template is selectively removed, resulting in the formation

© 2011 American Chemical Society

of nanocapsule. A number of templates, including gold (*3*) and silica (*2c*) nanoparticles, dendrimers (*4*), self-assembled amphiphilic block copolymers (*2b–2d*), liposomes (*5*), and vesicles (*6*), have been employed for the synthesis of nanocapsules. A hybrid approach involving the layer-by-layer self-assembly of oppositely charged polymers on sacrificial templates such as polystyrene, melamine-formaldehyde, silica, carbonate etc. leading to the formation of polymeric multilayer capsules has been developed and investigated over the years (*7*). Conventional emulsion, miniemulsion and interfacial polymerizations have also been employed for the fabrication of nanocapsules (*1*, *8*). Overall, the current strategies for nanocapsule fabrication have certain intrinsic limitations, such as tedious procedures for the removal of sacrificial template or surfactants, lack of nanocapsule robustness, low efficiency, etc. (*2c*, *8b*). Hence, there exists a strong need to develop new, simpler and direct routes to fabricate polymeric nanocapsules with well-defined morphology and surface composition.

In recent years, thiol-ene photoreaction has emerged as a powerful polymerization tool (*9*). Thiol-ene "click" chemistry – the radical addition of thiol to a non-activated double bond – is a rapid, simple, inexpensive and orthogonal reaction, offering other advantages like little or no oxygen inhibition, self-initiation, and tolerance to a variety of functional groups and solvents (*9*). Recently, Kim and coworkers showed that the solution thiol-ene photopolymerization of a cucurbituril-ene with various alkyl dithiols directly resulted in the formation of nanocapsules (*10*).

Resorcinarenes are a class of macrocyclic compounds, typically obtained by the acid catalyzed condensation between resorcinol and an aliphatic or aromatic aldehyde (*11*). These bowl-shaped molecules, along with their conformationally rigid derivatives with enforced cavities known as "cavitands", are well-established building blocks in supramolecular chemistry (*11*, *12*). Pioneering work by Cram and coworkers generated carcerand, the first closed molecular container compound formed by the covalent linking of two cavitands in a head-to-head fashion (*13*). Further, the groups of Sherman (*14*), Sherburn (*15*) and Warmuth (*16*) have reported the formation of larger container molecules formed by the covalent linking of 5 or 6 resorcinarene molecules. Simultaneously, supramolecular assembly of resorcinarenes mediated by non-covalent interactions has also been explored over the years. For example, MacGillivray and Atwood demonstrated the spectacular self-assembly of resorcinarenes held together by 60 hydrogen bonds, which resembled an Archimedian snub cube (*17*). From resorcinarenes and related calixarene derivatives, a wide-variety of self-assembled architectures like vesicles (*18*), nanocapsules (*19*), solid-lipid nanoparticles (*20*), and nanotubes (*21*) have also been reported.

Here we describe the photopolymerization of a single component, multifunctional resorcinarene thiol-ene monomer, resorcinarene tetra alkene tetra thiol (RTATT, Figure 1), into nanocapsules, in the absence of self-assembly, or sacrificial template (*22*). Further morphologically distinct architectures like nanocapsules, lattices, fibrous networks, and nanoparticles were prepared by varying the polymerization solvent (*22*). Also, we have developed a direct, one-pot method for the synthesis and surface functionalization of resorcinarene nanocapsules. The synthesis of RTATT was recently reported by Wei and

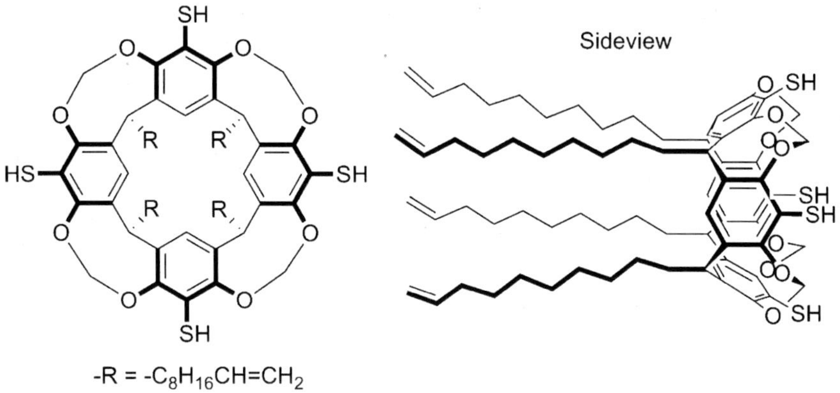

Figure 1. Resorcinarene tetra alkene tetra thiol (RTATT).

coworkers, who employed RTATT as crosslinkable surfactants around gold nanoparticles using olefin-metathesis (*23*).

Experimental

General Remarks

All photopolymerization reactions were carried out in quartz glassware. Synthesis of RTATT has been described in detail elsewhere (*23*). Synthesis of the resorcinarene-ene cavitand (Figure 6) was carried out from the corresponding resorcinarene, by following Kaifer's sealed tube method (*24*), and its spectral data was consistent with those reported in the literature (*25*).

Photopolymerization

In a typical photopolymerization experiment, a degassed RTATT solution (1.5 mM in dried and freshly distilled solvent), was irradiated for 3 h in a Rayonet reactor equipped with four 254 nm and 300 nm lamps. In synthesis and surface functionalization experiments, the polymerization was briefly interrupted after 3 h of UV irradiation, and the functionalizing agent (MPA or PEGMA; 1 equiv.) was introduced into the quartz vessel under argon atmosphere. UV irradiation was then continued for an additional 10 minutes. The reaction mixture was either subjected to an aqueous workup or dialysis. Dialysis of surface functionalized nanoparticles was performed using a Spectra/Por RC membrane (MWCO: 25,000) with a 3:1 chloroform/methanol mixed solvent system. The solvent was changed once after 16 h and the dialysis continued for another day. Contents were used as such for further characterization.

Characterization Methods

All data reported in this study are of as prepared samples, unless otherwise mentioned. IR analysis of solution drop cast polymer samples was carried out on a

Thermo Electron Nicolet 370 DTGS spectrophotometer operating in transmission mode. TEM and EDS analysis were carried out in a JEOL JEM-2100F field emission microscope operating at 200 kV equipped with an Oxford INCAx-sight EDS detector and a Gatan SC1000 ORIUS CCD camera (11 megapixel). Samples for TEM analysis were stained by mixing photopolymers prepared in organic solvent with an equal volume of 0.1% aqueous OsO_4 solution for at least 1 h with occasional shaking and deposited on a carbon coated copper grid. SEM images were obtained using a JEOL JSM-6060LV scanning electron microscope operating at 5-30 kV. SEM samples were prepared by depositing dilute solutions of photopolymers on a freshly cleaned glass slide followed by deposition of a thin layer of gold to prevent charging. Dynamic light scattering experiments were performed on a Zetasizer nano series model: ZEN 3200 supplied by Malvern Instruments.

Results and Discussion

Solvent Dependent Morphologies in Thiol-ene Photopolymerization

Thiol-ene photopolymerizations of RTATT solutions (1.5 mM) were carried out by irradiating them with 254 nm and 300 nm lamps for 3 h. Although thiol-ene polymerization can proceed with irradiation at longer wavelengths, the polymerization rate is known to be significantly faster at 254 nm (*26*). A variety of solvents, such as chloroform, dichloromethane, ethyl acetate and tetrahydrofuran, with a wide range of UV radiation transmission efficiency at 254 nm, were employed in this study (*27*). Even after 3 h of photopolymerization, we did not notice any visible phase separation in these polymerization reactions. The proof that thiol-ene polymerization indeed took place under these reaction conditions was provided by IR spectroscopy (Figure 2). IR spectra (Figure 2a) of RTATT photoproducts obtained in various solvents, though similar to RTATT monomer, showed reduced peak intensities at 2583 cm^{-1} and 1640 cm^{-1}, corresponding to S-H and C=C stretching (*28*). To quantify the extent of reduction, the areas of the thiol (~ 2620 – 2520 cm^{-1}) and alkene (~ 1660 – 1610 cm^{-1}) peaks were normalized by the area of the acetal peak (~ 1010 – 920 cm^{-1}) and plotted in Figure 2b. The extent of reduction of the thiol and ene peaks was significantly different for polymers prepared in various solvents, suggesting differential polymerization progress, perhaps reflecting differential polymerization rates of RTATT in various solvents.

Interestingly, polymerization of RTATT in chloroform led to the formation of nanocapsules (Figure 3), in the absence of any template, pre-organization or emulsifiers. The formation of a darker rim with a lighter core in TEM (Figure 3a and 3b) suggested that they were indeed hollow nanocapsules. The hollow nature of the nanocapsule could be further confirmed by energy dispersive spectroscopic analysis of the center (Figure 3c) and the rim (Figure 3d) of a single nanocapsule, which showed enhanced S content (as measured by S+Os/Cu+Os ratio) in the latter when compared to the former. SEM analysis (Figure 3e) confirmed the formation of an almost spherical particle. Size analysis from SEM images provided an average nanocapsule dimension of 106 ± 18 nm (Figure 3f),

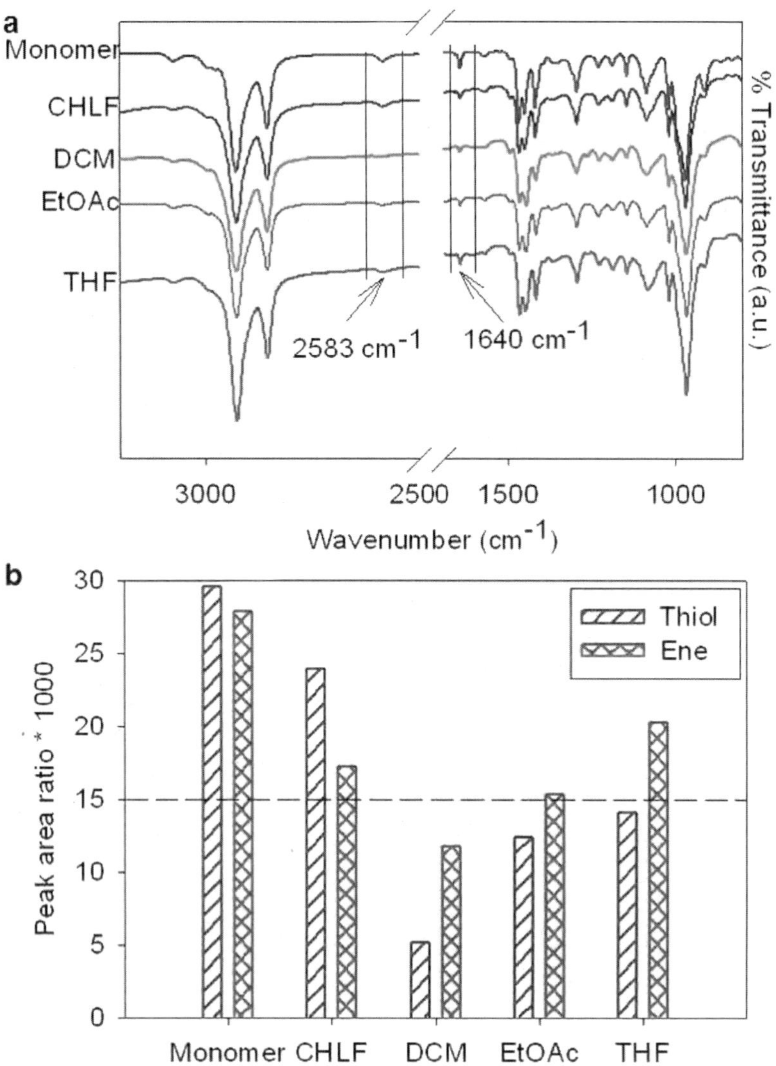

Figure 2. IR spectra (a) and peak area analysis (b) of RTATT monomer and RTATT polymers prepared in chloroform (CHLF), dichloromethane (DCM), ethyl acetate (EtOAc), and tetrahydrofuran (THF).

which was further confirmed by light scattering based particle size measurements (Figure 3g).

Notably, RTATT polymers synthesized in different solvents exhibited substantial differences in morphology. The TEM (Figure 4a) and SEM (Figure 4b) analysis of the polymer prepared in dichloromethane showed the formation of network lattice structures that extended several microns. Ethyl acetate as polymerization media led to fibrous network structures (Figure 4c), and polymerization in tetrahydrofuran yielded nanoparticles (Figure 4d). Such dramatic differences in morphological features are impressive in the light of a recent report of solvent variation in thiol-ene photopolymerization, where nanocapsules of different sizes were observed (*10a*).

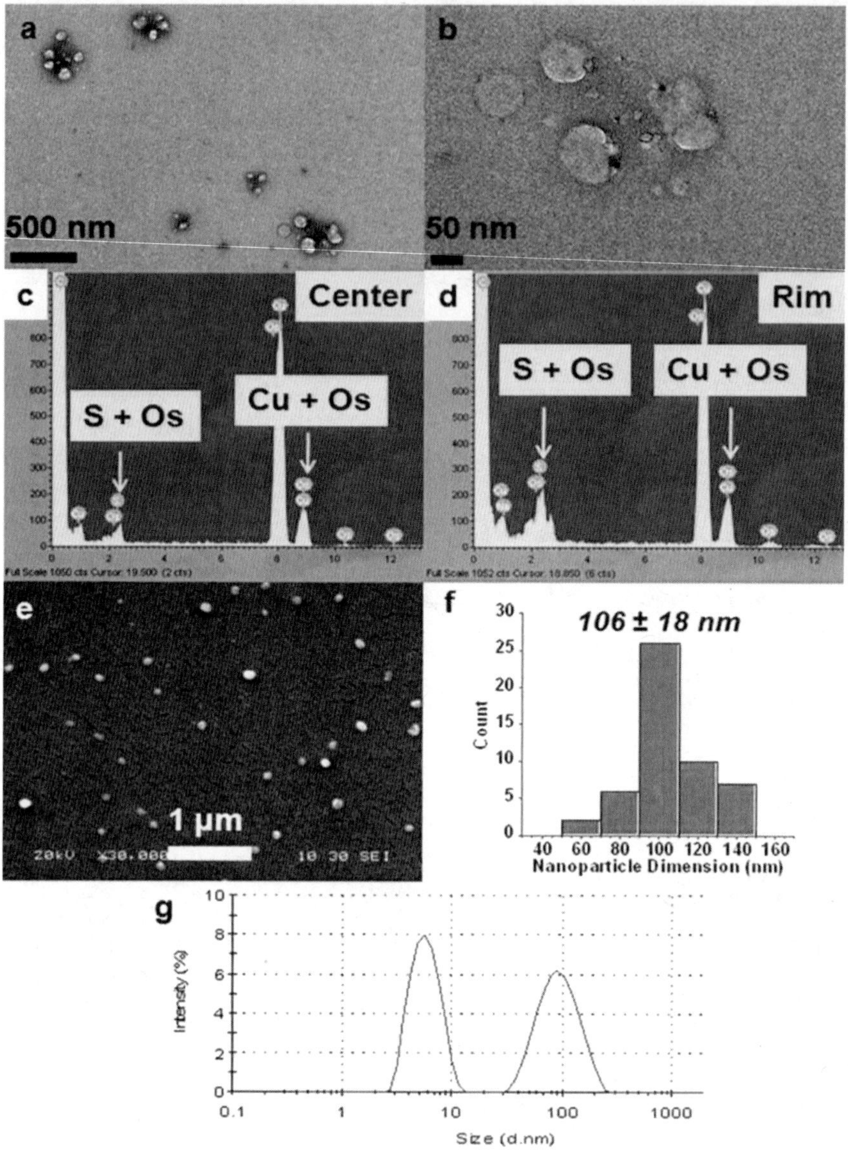

Figure 3. Analysis of the RTATT polymer synthesized in chloroform: TEM (a-b), EDS of the nanocapsule center (c) and nanocapsule rim (d), SEM (e), and particle size analysis from SEM (f) and dynamic light scattering (g) data.

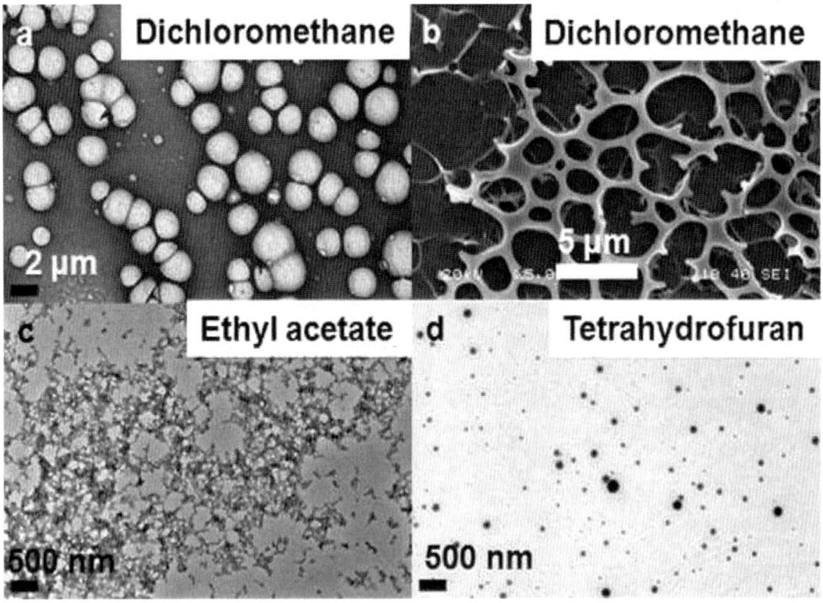

Figure 4. TEM (a) and SEM (b) images of RTATT lattices formed in dichloromethane. TEM images of RTATT polymers prepared in ethyl acetate (c) and tetrahydrofuran (d).

We believe that the formation of morphologically distinct polymeric structures in various solvents could be a function of two antagonistic effects, thiol-ene polymerization rate or reaction extent and phase separation effects, as the rate of the crosslinking reaction is known to suppress the phase separation by slowing down the polymer diffusion (*29*). As shown in Scheme 1, initially a dimer is expected to be formed by the addition of a thiol of RTATT, to a C=C double bond of another RTATT. The dimer is expected to further react with RTATT monomers to form oligomers and polymers, although the rate of oligomerization/polymerization may be solvent dependent. We have already confirmed the differential polymerization extent of RTATT in various solvents from the IR data. Note that the IR spectra of RTATT polymers synthesized in dichloromethane and ethyl acetate contain much reduced intensities of thiol (2583 cm^{-1}) and ene (1640 cm^{-1}) stretching when compared to those obtained in tetrahydrofuran and chloroform (Figure 2). It is documented in literature that the early onset of phase separation at lower conversion is known to result in formation of spherical particles, whereas the late onset of phase separation with high conversion produces web-like, sparse networks (*30*). The RTATT polymerization reactions with higher reaction extent (with dichloromethane and ethyl acetate as reaction media) led to lattice or fibrous networks, in contrast to those with lower conversion (from tetrahydrofuran and chloroform), which led to the formation of spherical nanoparticles and nanocapsules. We believe that the formation of hollow covalent polymeric nanocapsules could be related to the non-covalent self-assembly of lipids (*22*). It is worth noting that Kim and coworkers have invoked the initial formation of a two dimensional disk like oligomeric intermediate, and its bending to reduce its total energy to explain the covalent synthesis of hollow cucurbituril nanocapsules in a somewhat similar thiol-ene photopolymerization reaction (*10a*).

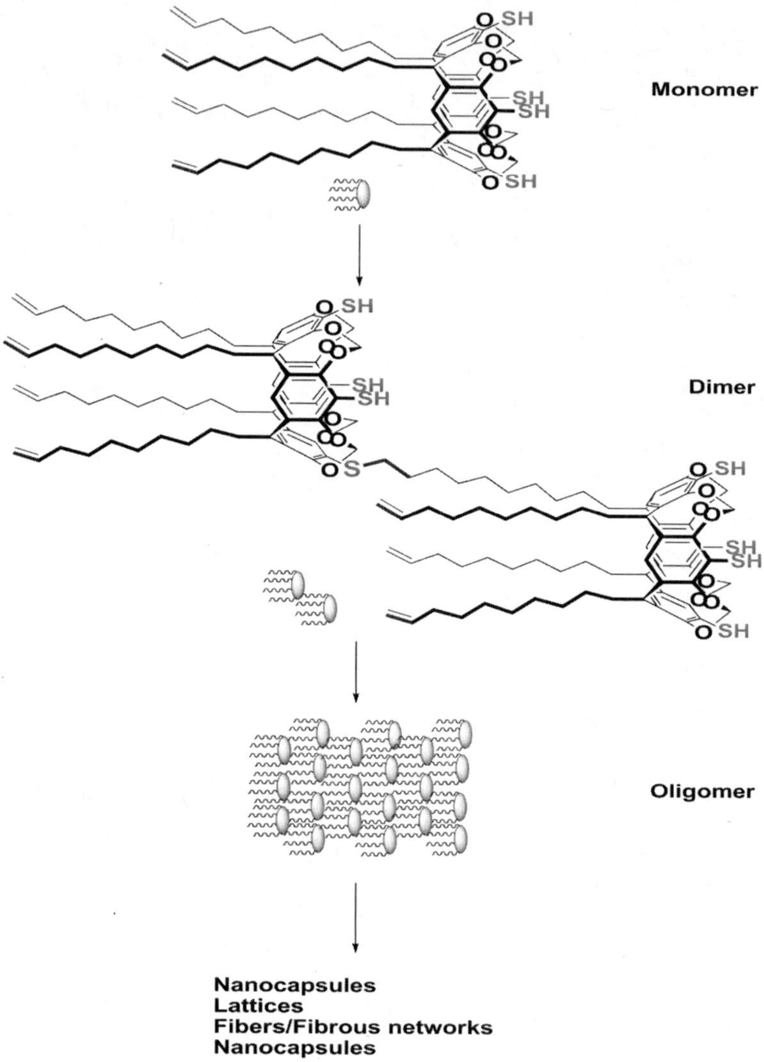

Scheme 1. Formation of nanocapsule and other morphologically distinct polymers.

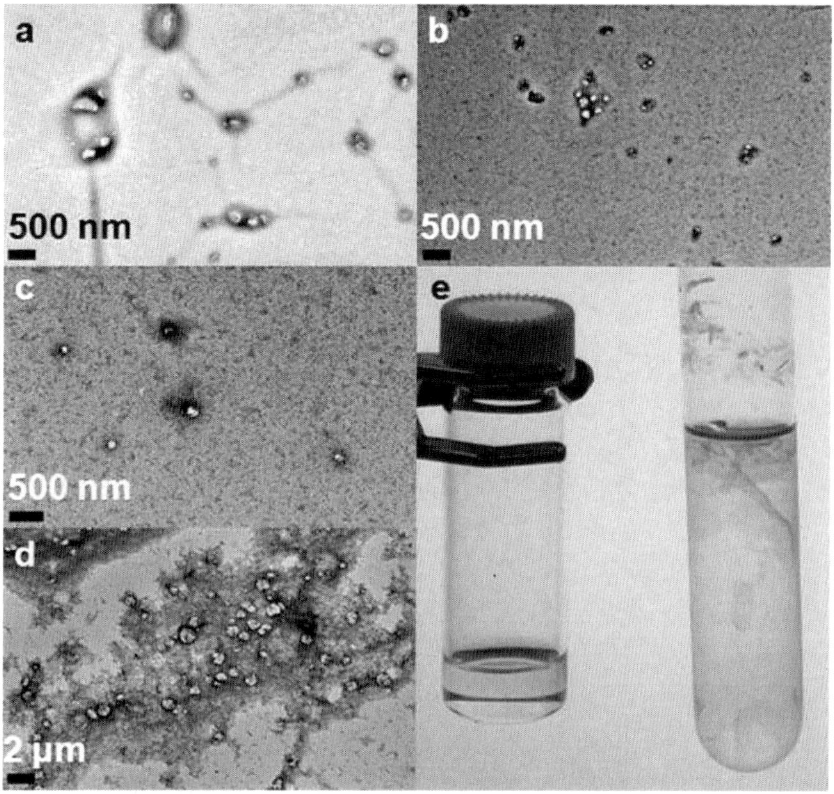

Figure 5. TEM images of various RTATT polymerizations. RTATT concentration and polymerization duration were a) 0.75 mM, 3 h b) 1.5 mM, 15 min c)1.5 mM, 20 h d) 1.5 mM, 20 h (e) Photograph of 5c (left) and 5d (right).

Insights into the Formation of Nanocapsules

Having established the importance of solvent in nanocapsule formation we probed the role of monomer concentration and polymerization time in determining the formation of nanocapsules. RTATT concentration did play a crucial role (*31*), as lowering it from 1.5 mM to 0.75 mM resulted in the formation of elongated linear structures along with polydisperse nanocapsules (Figure 5a). Remarkably, nanocapsules were formed after just 15 minutes of UV irradiation (Figure 5b). Increasing the polymerization duration to 20 h did not alter the shape of the nanocapsules formed (Figure 5c). However, photopolymerization reactions carried out for longer durations occasionally led to the formation of macroscopic sheets (Figure 5e), accompanied by solvent evaporation. Their microscopic analysis revealed the presence of much larger capsule-like structures embedded in a featureless polymer (Figure 5d).

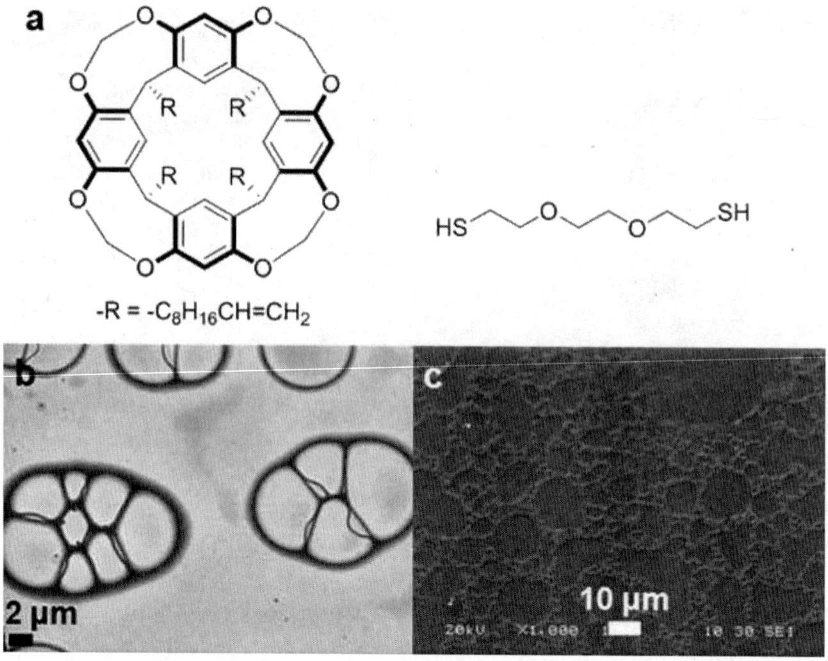

Figure 6. Photopolymerization of resorcinarene-ene cavitand and triethyleneglycol-dithiol (a) and its TEM (b) and SEM (c) images.

The photopolymerization of a structurally related resorcinarene-ene cavitand (Figure 6a) in the presence of diethyleneglycol-dithiol did not result in the formation of nanocapsules (Figure 6). This result suggests that resorcinarene-ene molecules do not behave like Kim's "flat core" monomers (*10a*). This also underlines the importance of structural features present in RTATT, leading to the formation of nanocapsules.

One-Pot Synthesis and Functionalization of Nanocapsules

Beyond morphological control of size and shape of nanomaterials, appropriate engineering of their surface composition is vital for various applications. While numerous approaches are available for the fabrication of nanocapsules and related nanocontainers (*1–8*), functionalization strategies are still in their infancy (*32*). In particular, fabrication of multifunctional nanocapsules poses a formidable challenge (*32*, *33*). Recently, Matyjaszewski and coworkers have introduced the concept of dual reactive surfactant, where one of the functional groups is involved in the miniemulsion polymerization reaction leading to the formation of nanocapsules and the other functional group is later utilized for functionalization (*33*).

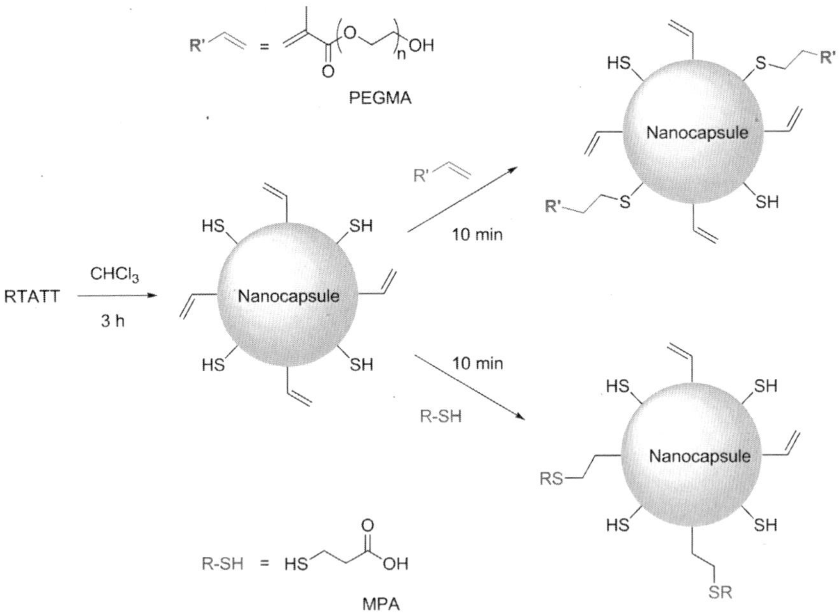

Scheme 2. One-pot synthesis and functionalization of nanocapsules.

Nanocapsules with residual thiol and ene groups offered exciting possibilities for orthogonal functionalization under photopolymerization reaction conditions by further reaction with other thiol or ene groups. We envisaged a simple one-pot, two-stage photopolymerization reaction for the synthesis and functionalization of nanocapsules as illustrated in Scheme 2. Note that the location of the residual thiol and ene functional groups in the schematic cartoon of nanocapsules (Scheme 2) is for illustrative purposes only and could as well be located on the inner periphery of the nanocapsules. Proof-of-the-concept experiments were performed by functionalizing with representative thiol and ene compounds like 3-mercaptopropionic acid (MPA) and poly (ethylene glycol) methacrylate (PEGMA, MW ~ 360).

In a typical surface functionalization experiment, RTATT solution in chloroform (1.5 mM) was irradiated for 3 h to generate nanocapsules, to which MPA or PEGMA (1 equiv) was added and UV irradiation continued for an additional 10 min. Unreacted MPA or PEGMA was removed by either dialysis or an aqueous workup. IR spectra (Figure 7) of functionalized nanocapsules clearly showed the presence of carboxyl (1700 cm^{-1}) and hydroxyl groups (3400 cm^{-1}), confirming the incorporation of MPA or PEGMA on the nanocapsules. It must be noted that under these reaction conditions, the formation of side products like disulfide linkages cannot be completely ruled out, but nevertheless they do lead to covalently functionalized nanocapsules. Detailed structural characterization of such functionalized nanocapsules is currently underway in our laboratories.

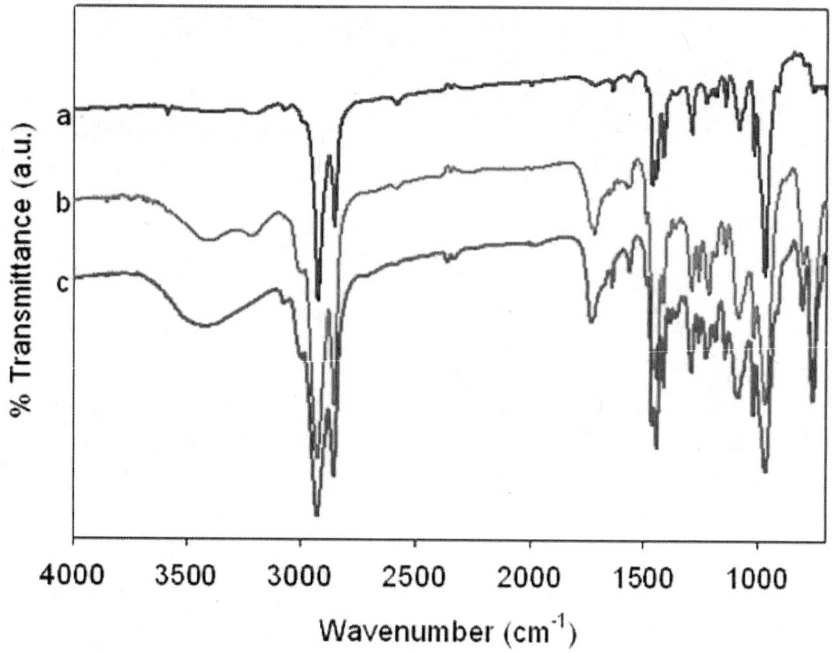

Figure 7. IR spectra of nanocapsules (a) before functionalization and dialyzed samples of (b) MPA and (c) PEGMA functionalized nanocapsules.

Conclusions

We have shown that the photopolymerization of RTATT in different solvents resulted in the formation of morphologically distinct architectures. Notably, photopolymerization of RTATT in chloroform resulted in the template-free fabrication of nanocapsules. In addition to nanocapsules, a variety of nanostructures, including lattices, fibrous structures, and nanoparticles, were obtained by varying the reaction media. A simple one-pot, two-stage photopolymerization based approach has been developed for the synthesis and surface functionalization of nanocapsules, which can be extended to other polymeric nanostructures and for the introduction of a variety of functional groups. These morphologically distinct sulfur containing polymers can act as natural templates for the self-assembly of metal nanostructures. Also, a variety of guest molecules can be encapsulated by introducing them in the polymerization medium during the formation of these nanocapsules (*10a*). Such encapsulated guest molecules can be potentially released by exploiting the solvent responsive behavior of nanocapsules (*10b*) or other external stimuli (*7*). Given their inherent guest encapsulation capabilities, further development of this one-pot fabrication and functionalization approach for nanocapsules will enable their applications in theranostic nanomedicine (*34*).

Acknowledgments

The authors gratefully acknowledge Old Dominion University (ODU) for financial support. TEM and SEM analysis were carried out at the Applied Research Center, ODU. The authors thank Dr. Wei Cao, Applied Research Center, ODU for help with TEM and SEM measurements. The authors also thank Prof. Anna H. Jeng, ODU, for access to Malvern Zetasizer and Professors Alexander Wei (Purdue University) and Richard V. Gregory (ODU) for insightful discussions.

References

1. Meier, W. *Chem. Soc. Rev.* **2000**, *29*, 295–303.
2. (a) Shchukin, D. G.; Sukhorukov, G. B. *Adv. Mater.* **2004**, *16*, 671−682. (b) O'Reilly, R. K.; Hawker, C. J.; Wooley, K. L. *Chem. Soc. Rev.* **2006**, *35*, 1068−1083. (c) Lou, X. W.; Archer, L. A.; Yang, Z. C. *Adv. Mater.* **2008**, *20*, 3987−4019. (d) Lensen, D.; Vriezema, D. M.; van Hest, J. C. M. *Macromol. Biosci.* **2008**, *8*, 991−1005.
3. Boyer, C.; Whittaker, M. R.; Nouvel, C.; Davis, T. P. *Macromolecules* **2010**, *43*, 1792–1799.
4. Wendland, M. S.; Zimmerman, S. C. *J. Am. Chem. Soc.* **1999**, *121*, 1389–1390.
5. Fukui, Y.; Fujimoto, K. *Langmuir* **2009**, *25*, 10020–10025.
6. Ali, S. I.; Heuts, J. P. A.; van Herk, A. M. *Langmuir* **2010**, *26*, 7848–7858.
7. (a) del Mercato, L. L.; Rivera-Gil, P.; Abbasi, A. Z.; Ochs, M.; Ganas, C.; Zins, I.; Sonnichsen, C.; Parak, W. J. *Nanoscale* **2010**, *2*, 458−467. (b) De Cock, L. J.; De Koker, S.; De Geest, B. G.; Grooten, J.; Vervaet, C.; Remon, J. P.; Sukhorukov, G. B.; Antipina, M. N. *Angew. Chem., Int. Ed.* **2010**, *49*, 6954−6973. (c) Angus, P. R. J.; Cortez, C.; Angelatos, A. S.; Caruso, F. *Curr. Opin. Colloid Interface Sci.* **2006**, *11*, 203−209. (d) Antipov, A. A.; Shchukin, D.; Fedutik, Y.; Petrov, A. I.; Sukhorukov, G. B.; Mohwald, H. *Colloids Surf., A.* **2003**, *224*, 175−183.
8. (a) Luo, Y. W.; Gu, H. Y. *Polymer* **2007**, *48*, 3262−3272. (b) Li, W. W.; Matyjaszewski, K.; Albrecht, K.; Moller, M. *Macromolecules* **2009**, *42*, 8228−8233.
9. (a) Kade, M. J.; Burke, D. J.; Hawker, C. J. *J. Polym. Sci., Polym. Chem.* **2010**, *48*, 743−750. (b) Dondoni, A. *Angew. Chem., Int. Ed.* **2008**, *47*, 8995−8997. (c) Hoyle, C. E.; Lee, T. Y.; Roper, T. *J. Polym. Sci., Polym. Chem.* **2004**, *42*, 5301−5338.
10. (a) Kim, D.; Kim, E.; Kim, J.; Park, K. M.; Baek, K.; Jung, M.; Ko, Y. H.; Sung, W.; Kim, H. S.; Suh, J. H.; Park, C. G.; Na, O. S.; Lee, D. K.; Lee, K. E.; Han, S. S.; Kim, K. *Angew. Chem., Int. Ed.* **2007**, *46*, 3471−3474. (b) Park, K. M.; Suh, K.; Jung, H.; Lee, D. W.; Ahn, Y.; Kim, J.; Baek, K.; Kim, K. *Chem. Commun.* **2009**, 71−73.
11. Timmerman, P.; Verboom, W.; Reinhoudt, D. N. *Tetrahedron* **1996**, *52*, 2663–2704.
12. Hooley, R. J.; Rebek, J. *Chem. Biol.* **2009**, *16*, 255–264.

13. Cram, D. J.; Karbach, S.; Kim, Y. H.; Baczynskyj, L.; Kalleymeyn, G. W. *J. Am. Chem. Soc.* **1985**, *107*, 2575–2576.
14. Makeiff, D. A.; Sherman, J. C. *J. Am. Chem. Soc.* **2005**, *127*, 12363–12367.
15. Barrett, E. S.; Irwin, J. L.; Edwards, A. J.; Sherburn, M. S. *J. Am. Chem. Soc.* **2004**, *126*, 16747–16749.
16. Liu, Y.; Liu, X.; Warmuth, R. *Chem.-Eur. J.* **2007**, *13*, 8953–8959.
17. MacGillivray, L. R.; Atwood, J. L. *Nature* **1997**, *389*, 469–472.
18. Houmadi, S.; Coquiere, D.; Legrand, L.; Faure, M. C.; Goldmann, M.; Reinaud, O.; Remita, S. *Langmuir* **2007**, *23*, 4849–4855.
19. Lee, M.; Lee, S. J.; Jiang, L. H. *J. Am. Chem. Soc.* **2004**, *126*, 12724–12725.
20. Shahgaldian, P.; Sclotti, M. A.; Pieles, U. *Langmuir* **2008**, *24*, 8522–8526.
21. Mansikkamaki, H.; Busi, S.; Nissinen, M.; Ahman, A.; Rissanen, K. *Chem.-Eur. J.* **2006**, *12*, 4289–4296.
22. Balasubramanian, R.; Kalaitzis, Z. M.; Wei, C. *J. Mater. Chem.* **2010**, *20*, 6539−6543.
23. Balasubramanian, R.; Kwon, Y. G.; Wei, A. *J. Mater. Chem.* **2007**, *17*, 105–112.
24. Roman, E.; Peinador, C.; Mendoza, S.; Kaifer, A. E. *J. Org. Chem.* **1999**, *64*, 2577–2578.
25. Pirondini, L.; Bonifazi, D.; Menozzi, E.; Wegelius, E.; Rissanen, K.; Massera, C.; Dalcanale, E. *Eur. J. Org. Chem.* **2001**, 2311–2320.
26. Cramer, N. B.; Reddy, S. K.; Cole, M.; Hoyle, C.; Bowman, C. N. *J. Polym. Sci., Polym. Chem.* **2004**, *42*, 5817–5826.
27. Montalti, M.; Credi, A.; Prodi, L.; Gandolfi, M. T. *Handbook of Photochemistry*, 3rd edition; Taylor and Francis: Boca Raton, 2006; p 541.
28. Crews, P.; Rodriguez, J.; Jaspars, M. *Organic Structure Analysis*; Oxford University Press, New York, **1998**; p 332.
29. (a) Inoue, T. *Prog. Polym. Sci.* **1995**, *20*, 119−153. (b) Li, G.; Huang, Z. B.; Xin, C. L.; Li, P.; Jia, X. L.; Wang, B. H.; He, Y. D.; Ryu, S.; Yang, X. P. *Mater. Chem. Phys.* **2009**, *118*, 398−404.
30. (a) Tanaka, H. *Adv. Mater.* **2009**, *21*, 1872−1880. (b) Serbutoviez, C.; Kloosterboer, J. G.; Boots, H. M. J.; Touwslager, F. J. *Macromolecules* **1996**, *29*, 7690−7698. (c) Guenthner, A. J.; Hess, D. M.; Cash, J. J. *Polymer* **2008**, *49*, 5533−5540.
31. Kimura, K.; Zhuang, J. H.; Wakabayashi, K.; Yamashita, Y. *Macromolecules* **2003**, *36*, 6292–6294.
32. (a) Liu, X.; Basu, A. *J. Am. Chem. Soc.* **2009**, *131*, 5718−5719. (b) Ievins, A. D.; Moughton, A. O.; O'Reilly, R. K. *Macromolecules* **2008**, *41*, 3571−3578.
33. Li, W. W.; Yoon, J. A.; Matyjaszewski, K. *J. Am. Chem. Soc.* **2010**, *132*, 7823–7825.
34. Khemtong, C.; Kessinger, C. W.; Gao, J. *Chem. Commun.* **2009**, 3497–3510.

Chapter 17

Synthesis of Oil Core/Polymer-Shell Particles via Miniemulsion Templating

Thorsteinn Adalsteinsson,[1,2,*] E. V. Fette,[1] A. Pham,[1] J. K. Black,[1] L. E. Tracy,[1] C. P. Roche,[1] and J. B. Pesavento[3]

[1]Department of Chemistry & Biochemistry, Santa Clara University, Santa Clara, CA 95053
[2]Center for Nanostructures, Santa Clara University, Santa Clara, CA 95053
[3]Life Science Division, Lawrence Berkeley National Laboratory, Berkeley, CA 94720, USA
[*]Tadalsteinsson@scu.edu

Studies of sub-micron-sized polymer-core/shell particles, or polymer nanocapsules, where the capsule core consists of a liquid hydrocarbon are discussed. The capsules are prepared via a one-step, miniemulsion polymerization procedure. The capsule stability and the effect of droplet size and the polymer/oil interfacial tension are investigated using differential scanning calorimetry. Changes in the integrity of the polymer shell can be observed by measuring the solidus temperature of the encapsulated n-hexadecane. The type of polymer and volume of the oil droplet also affect the solidus temperature. We observe a limited agreement with the Gibbs-Thomson equation.

Introduction

Oil-filled, polymer nanocapsules have the potential for wide-ranging applications. These capsules are sub-micron-sized particles in which a polymer shell, a few nanometers thick, encapsulates an oil phase. These capsules have a high weight fraction of oil, and a low weight fraction of polymer. This maximizes the payload the capsules carry. The overall size of these particles is also smaller than common microfluidic channels, which is an attractive feature for many applications such as using capsules as carriers for inks or fragrances. For this

© 2011 American Chemical Society

application, oils are contained until mechanical stress releases them, raising the question of if oil solutions can be encapsulated before they separate.

Another proposed application is using the capsule as a semi-permeable sheath around a reaction site, thus producing a nano-reactor system. This requires a well-controlled solution environment inside the capsule, where the catalytic molecule is isolated from the continuous phase. Reactant molecules should freely enter and exit the capsule. Small solvent volumes have different solvent behavior than large volumes, since the liquid structure in the capsule may be different from the bulk. The outcome may be that the encapsulated phase does not dissolve the reactant, even when the same solvent does so in the bulk phase.

Yet another popular proposal is to use nanocapsules as drug carriers for a magic bullet, where an encapsulated drug is released when an infection is detected. This requires the capsules to be delivered in a stable form to the infection where they then become unstable and release the enclosed drug. In addition to this opening of the capsule, the encapsulated phase must be destabilized as well in order for it to leave the capsule.

Before any such applications can be reliably designed, we must thoroughly understand physical properties that differentiate encapsulated volumes from bulk phase systems. Although physical principles that hold colloidal particles such as these nanocapsules in dispersion are well understood (*1*), the physical nature of the encapsulated oil volume has been studied less. A number of studies of dispersed oil phases have been done in emulsion systems (*2–5*), in aerosols (*6*, *7*) or in micro-porous media (*8*). Methods for obtaining polymeric nanocapsules were reported by Tiarks et al. (*9*) in 1999. The resulting particles provide new systems where the physical chemistry of microscopic oil droplets can be studied (*10*, *11*).

Possible Differences between Bulk and Small Particles

It is worthwhile to consider the changes in the physical environment when a liquid system is reduced from having a characteristic diameter of several microns to a diameter of several tens of nanometers. One change that is commonly pointed out is the rapid increase in surface area per unit of volume, which can be expressed as the surface area to volume ratio (S/V). S/V varies inversely with the radius of the particles, (1/r). With increasing S/V ratio, the system can be described as becoming surface-like, in that the thermodynamics of the system are governed by the surface behavior. The S/V ratio is, however, of limited use, since the surface and bulk have different dimensionalities. It is more informative to consider how the fraction of molecules that are *close* to the interface varies with the droplet volume. Here, the term *close* is assumed to include molecules that are 2 – 4 nm away from the interface, as this is where the slipping plane (Debye length) is in the oil. Molecules inside this slipping plane are in a very different dynamic environment than molecules in the bulk. The fraction of molecules within this distance increases rapidly as the particles become smaller. Figure 1 shows calculated values for volume fraction of molecules that are within 4 nm from a surface as a function of the diameter of the particle.

The figure shows that nearly 20% of 100 nm droplets are within this 4 nm shell, for 500 nm capsules, the fraction is less than 5%. The 4 nm fraction drops

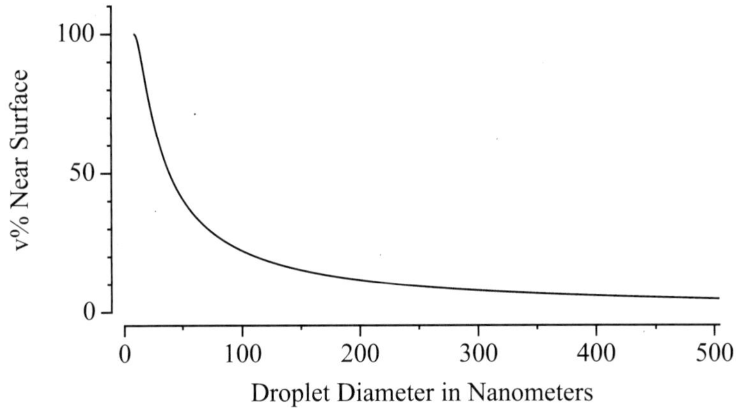

Figure 1. Calculated fraction of molecules that are "close to" or within 4-nm distances from a particle/bulk interface as a function of the diameter of the particle.

below 1% once the particle diameter exceeds 2.5 µm. This would imply that there is little, or no difference in the overall system behavior between particles that are 2.5 µm and particles that are 500 nm in size, but below this limit, the system becomes rapidly more dominated by the surface behavior. Furthermore, the effect that concave curvature has on the thickness of the electrical double layer and the Debye length is not well understood, and surface dominated properties may extend to larger capsules.

Also worthwhile is considering the behavior of the lowest nominal solute concentration as a function of the size of the encapsulated volume. If we assume that the most dilute solution has one solute molecule in the entire volume, the lowest concentration in a 150 nm droplet is in the micro-molar (µM) range. This value decreases with increasing droplet size so when the particle is 1.5 µm in diameter, the lowest possible concentration is in the nano-molar (nM) range. Figure 2 shows the variation of this one-per capsule concentration as a function of particle diameter.

The lowest possible concentration lowers the value of Gibbs free energy of the encapsulated phase, since the Gibbs free energy of a solution is closely connected with concentration of the solute in the solution. By breaking a bulk phase with a low solute concentration into sufficiently small droplets, the end-point will be that some capsules have a single solute molecule, whereas other volumes have no solute. This results in a different thermal behavior in the capsules that have a solute molecule and capsules that do not.

Observing the effect that volume confinement has on thermodynamic behavior hinges on being able to isolate the droplet and prevent droplet-to-droplet exchange, at least over moderate time intervals. Such slow dynamics are possible where a continuous polymer shell, rather than a dynamic surfactant layer, stabilizes the oil droplet. These polymer capsules should also have a well-defined size and low distribution in capsules sizes.

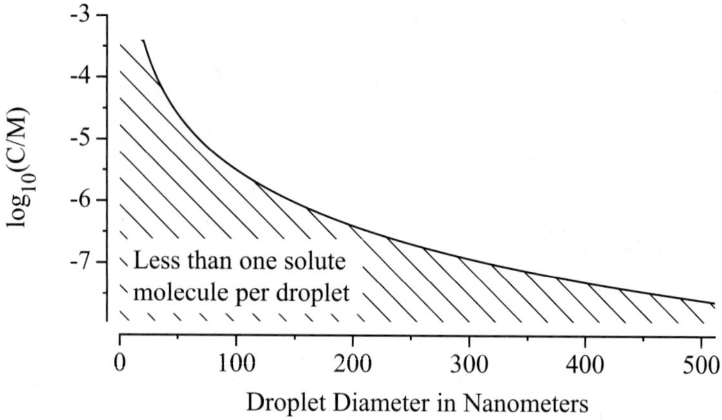

Figure 2. Calculation of the boundary between a single solute molecule per droplet and less than one solute molecule per droplet.

Synthesis of Polymer Nanocapsules with Liquid-Oil Core via Miniemulsion Polymerization

Synthetic approaches for nanocapsules are still being developed in the literature, where a variety of approaches have been suggested. The approaches range in complexity and control of the polymer shell and particle size distribution. Methods that have high control over the polymer shell composition, such as the layer-by-layer polymer assembly followed by core dissolution (*12*, *13*), tend to be ill suited to large-scale production. Scalable methods, such as polymerization nucleation from a seed particle (a capsule can be formed by core-swelling (*14*)), tend to come at a considerable cost in control of the capsule architecture. An alternative method that falls in between these two extremes of scalability is the capsule formation via miniemulsion polymerization (*9*). This approach provides a route for encapsulating a variety of oil phases and solutions while keeping some options open for the nature of the polymer. Such formulation extends over oil-in-water (*9*, *15*), water-in-oil (*16*, *17*) and oil-in-oil emulsions (*18–20*).

The synthesis of the oil filled nanocapsules is based on establishing a meta-stable precursor system via miniemulsion process (*21*). The key in establishing a miniemulsion is to use a high shearing force, commonly achieved by tip-sonication or by repeated extrusion through microporous glass membrane. The first method, which we use, is more applicable to research laboratories. The tip sonicator oscillates a metal rod (horn) at frequencies of 10 – 20 kHz (ultra sound) using pieso-electric stacks. A laboratory sized sonicator can efficiently process samples of a few-milliliters. The extrusion technique is more suitable for larger scale productions of miniemulsions as used by the cosmetic industry to produce creams, lotions and other similar emulsions.

Most miniemulsion formulations include a surfactant and a co-surfactant. The exact requirement for a co-surfactant is not clearly specified, but in general, the term refers to a molecule that is not a surfactant on itself, but partitions into a surfactant layer (*21*). An example, which we used in this work, is dodecanol in combination with sodium dodecyl sulfate (SDS). Dodecanol is not considered to be particularly efficient co-surfactant, although it is a convenient choice because it is a common impurity in SDS (*22*).

Miniemulsion stability can be further enhanced by adding an ultra-hydrophobe (for standard emulsions) (*23*) or an ultra-lyophobe, or salts, (for inversed miniemulsion) (*24*) in the formulation. Ultra hydrophobes are generally long chain, or partially fluorinated hydrocarbons.

Miniemulsions of SDS, dodecanol and vinyl monomers are extremely stable when a small amount of *n*-hexadecane is included in the formulation as an ultra-hydrophobe. After sonication the dispersion has the characteristic behavior of being non–foaming, likely due to depletion of surfactant micelles from the aqueous phase. Once depleted, the dispersion can be centrifuged to isolate the emulsion to the top of the vial, and the cream re-dispersed in fresh water. The particle size, according to dynamic light scattering (DLS), remains unchanged. This emulsion can also be heated up to 90°C, and cooled back down without a noticeable change in particle size. This stability is the same for miniemulsions where the majority of the oil phase is *n*-hexadecane. The heat stable miniemulsion, where the majority of the encapsulated phase is inert to free-radical polymerization is the precursor system for nanocapsule synthesis.

The theories describing the physical principles for this extreme stability are somewhat divergent. One explanation is based on the low size distribution (poly-dispersity) that miniemulsion droplets have after sonication. With truly mono-dispersed particles, the effect of Ostwald ripening process is small. It is still possible to make poly-dispersed miniemulsions, especially by mixing two emulsions of different sizes. The particle size remains unchanged despite this mixing, which indicates that this explanation may be too simplistic. Another explanation is based on the presence of the co-surfactant in the surfactant layer. The assumed action of the co-surfactant is to increase the density of the interfacial layer whereby a barrier for oil/water exchange is established. It was commonly supposed that miniemulsions have negligible exchange dynamics. This assumption is not strictly correct since micelle to micelle exchange has been verified in miniemulsions, although the exchange rate is longer than in standard emulsions (*25*). Katarina Landfester (*23*) proposed that the stability of miniemulsions is due to the surface of the droplet being under-covered by surfactant. In this condition, the Laplace pressure on the droplets is connected to the actual surface tension of the oil/water interface, or:

$$p_{\mathrm{Laplace}} = \frac{3\gamma_{ij}}{R}$$

The presence of the ultra-hydrophobe also creates osmotic pressure between the oil phase and the water phase, or:

$$\Pi = \frac{RT}{M}C$$

In a system at equilibrium, the Laplace pressure on the particles is equal to the osmotic pressure out of the particles. But in miniemulsions, the Laplace pressure is greater than the osmotic pressure, so the system has net excess pressure on the droplets. The Laplace pressure increases with 1/R, whereas the osmotic pressure increases with solute concentration, or 1/V. The lowest state of average pressure per droplet is reached when all the droplets have the same size and changing the droplet sizes is uphill in Gibbs free energy.

The most stable miniemulsions are indeed formed when oil droplets have less than full coverage in the final dispersed stage and the emulsion does not foam. These emulsions are stable for several days or longer. Even better stability is obtained if the final volume of the emulsion after sonication is then doubled by adding 1 CMC concentrated surfactant solution. The dispersion foams when it is shaken, indicating that there is excess surfactant in the continuous phase. The miniemulsion micelles are most likely fully covered, but no free surfactant micelles should be in the continuous phase since the concentration is below 1 CMC.

Methods for Varying Particle Size

The particle size is first and foremost determined by the mechanical method of making the emulsion. The output power and the duration of the sonication step are the primary variables. The temperature during the sonication, as well as the viscosity of the sample has an effect to a lesser extent. Small variations can also be seen depending on the placement of the sonication tip in the sample and the total sample volume. We have also observed dependence of particle size by varying the surfactant to oil ratio, although this effect is far less pronounced than one would expect. The particle size depends on the type of surfactant used in the emulsion, where we see larger micelles formed when using polymeric surfactants such as TWEEN-80 and BRIJ types. A typical summary of DLS size after sonication are shown in Figure 3 below. Here, the sample was sonicated for a total of two minutes.

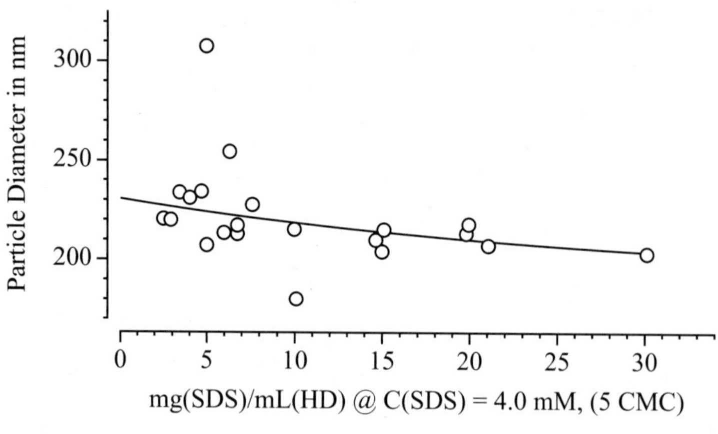

Figure 3. The final miniemulsion particle diameter as a function of the surfactant to n-hexadecane ratio.

To the best of our knowledge, the connection between particle size and the sonication process is not understood. One possible theory is that a standing wave is generated between the tip and the solution during the sonication step. The standing wave drives the system into a steady state between splitting and recombination of oil particles as the solution responds to the force delivered by the tip. This response depends on the viscosity and elasticity of the solution at the given tip frequency. The final steady state condition is therefore dependent on temperature, surface tension between the oil and water (hence, surfactant) and viscosity of the oil phase. Figure 4 shows the size of the miniemulsion droplets as a function of volume ratio of n-hexadecane and methyl methacrylate. By changing the volume ratio, the viscosity of the oil phase changes significantly. The air/water surface tension of the oil solution is roughly that of hexadecane, and remains largely unchanged.

The inability to control the frequency of the sonicator tip is a challenge for further investigation. The frequency is set by the oscillation of the pieso electric stacks in the horn, which are tuned to the eigenfrequency of the horn, or vice verse.

Method for Depositing the Polymer and Forming the Nanocapsules

Various reactive vinyl monomers, such as alkyl methacrylates and styrene are soluble in long chain hydrocarbons, such as n-hexadecane, 1-octanol, 1-dodecane and Isopar-M®. Polymers derived from these monomers are, however, insoluble in the oil phase and precipitate once a sufficient molecular weight is reached. The use of n-alkanes and vinyl-type monomers is somewhat of a limiting in terms of the types of polymers that can be deposited, since free-radical polymerization has a high likelihood to interfere with molecules solvated in the oil phase. Poly-condensation and reactions are impossible in the presence of water, but this limitation can be overcome by using oil-in-oil emulsions (*18*, *19*). If the polymers have a spreading coefficient that is lower than the spreading coefficient of the oil on the water, a polymer film will form at the oil/water interface (*23*). This has been our primary method for producing oil-core/polymer-shell capsules for thermodynamic studies. The polymer is deposited between the oil phase and the surfactant. This allows us to remove the surfactant from the capsule dispersion by dialysis, or repeated rinsing. Once the surfactant is removed, the capsules flocculate. By varying the type of polymer used, we obtain capsules that have varying oil/solid interfacial tensions. Furthermore, we can vary the size of the final oil droplet by increasing, or decreasing the volume fraction of monomer in the initial formulation of the miniemulsion. This method gives us capsules with slightly varying shell thickness. Figure 5 shows our results for poly(styrene)/n-hexadecane capsules. For this experiment, the size of both the emulsion prior to synthesis and the particle size post synthesis was measured. In general, the particle size prior to synthesis was slightly larger than the particle size post synthesis. It should also be noted that the diameter of the droplet was not measured directly, but calculated from the particle size measured using DLS.

We have also stabilized hexadecane miniemulsions and subsequently deposited poly(*N*-isopropyl acrylamide), pNIPA, onto the emulsion droplets by initiating the polymerization from the surfactant layer. This method generates particles where we can vary the polymer thickness shell while keeping the

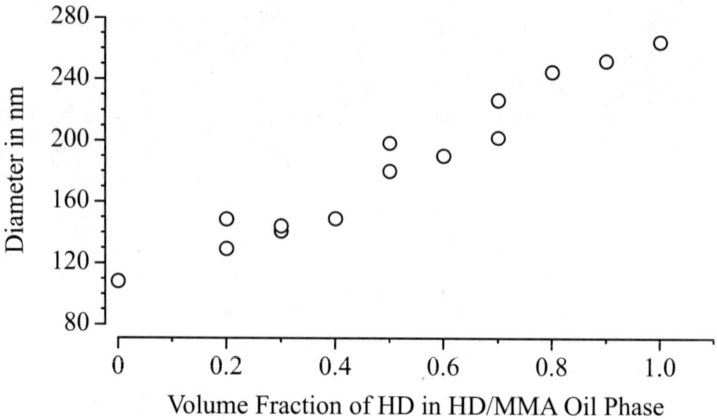

Figure 4. Variation of particle size with varying the composition of the oil phase. The samples were sonicated once for 1 minute at 25 – 28°C under argon atmosphere. Reproduced with permission from reference (11)

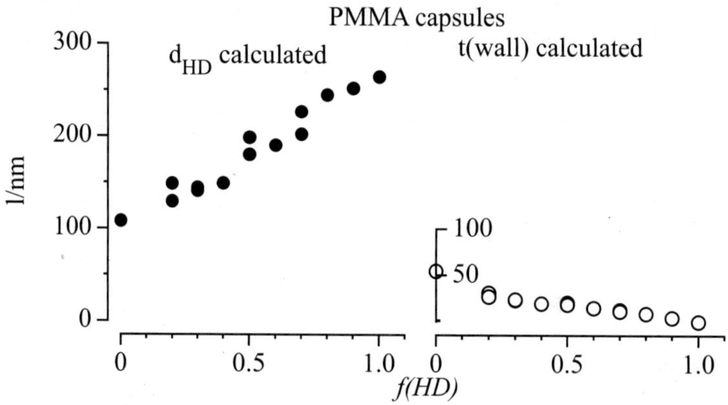

Figure 5. Oil droplet size and poly(styrene) wall thickness in capsules synthesized by polymerizing styrene within an HD/Styrene miniemulsion. The volume fraction on the horizontal axis is the volume fraction prior to synthesis. Reproduced with permission from reference (10)

volume of the encapsulated oil constant. Here the surfactant and not the polymer dominates the oil/solid interfacial tension. The pNIPA capsules are stable against washing and dialysis. Depositing the polymer from the continuous phase allows us to include solute molecules in the dispersed phase, since the synthesis avoids generating a free radical species in the oil phase and does not require a fine balance of spreading coefficients between the oil/polymer and water.

Transmission electron microscopy images of samples of poly(styrene) and poly(methyl methacrylate) capsules showed that the particles appeared to be hollow, and have similar size as indicated by the DLS measurements. Figure 6 shows TEM images (uranyl acetate stained capsules) of poly(methyl

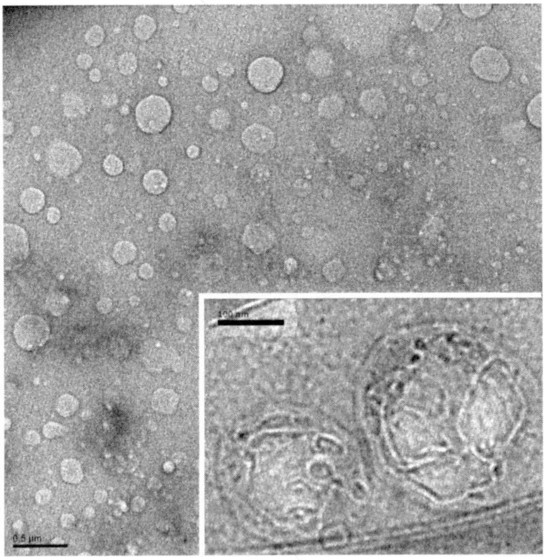

Figure 6. Main figure: TEM image of uranyl acetate stained poly(methyl methacrylate) / HD capsules. The scale bar is 0.5 µm. Inset figure: Cryo-TEM image (no stain) of the same capsule sample. The scale bar is 0.1 µm. For this sample 80% of the oil volume in the synthetic setup was HD. Reproduced with permission from reference (11)

methacrylate) capsules. The inset figure shows a Cryo-TEM image of the same capsule sample (f(HD)=0.8).

It is interesting to note that the polymer capsule appears to be uneven, or made from domains, in the CRYO-TEM image in Figure 6. This behavior is not unexpected since we do not include any specific a cross-linking monomer in the polymer reaction. The interface between the oil and the PMMA is not smooth, and there may be some contact between the surfactant and the oil, even after polymerization and deposition of the polymer from within the capsule.

Differential Scanning Calorimetry of the Encapsulated Oil Phase

Small oil droplets have phase transitions that differ significantly from the bulk phase. This observation has been explained by two theoretical approaches. The older thermodynamic approach originates from Gibbs and Thomson. In that theory, the free energy cost that is required to form an interface between two phases (γ_s) must be overcome by the phase that is to undergo the phase transition. This increase in free energy results in lowering of the transition temperature in comparison to the bulk transition. This difference $\Delta T=T_b - T_s$ is referred to as supercooling. The Gibbs-Thomson theory correlates this supercooling inversely with the radius of curvature of the surface and linearly with the surface tension. The theory is simply stated as:

$$\Delta T = T_{\text{bulk}} - T_{\text{drop}} = \frac{K}{r}$$

where K is

$$K = \frac{\gamma_{ij} v_i T_{\text{bulk}}}{\Delta H_{\text{tr}}}$$

γ_{ij} is the surface tension between the liquid and the adjacent solid, v_i is the molar volume of the liquid and ΔH is the enthalpy for the transition.

The 1/r dependence is found in systems where the transitioning phase is clearly insoluble in the surrounding phase, e.g. in glass pores (*26*). This behavior has not been verified for oil droplets in emulsions, but this may be due to some ambiguity regarding the exact radius of curvature. Montenegro et al. (*5*) did not observe curvature dependence for miniemulsions, which are the closest relatives of the particles we study.

An alternative explanation for the observation of supercooling comes from nucleation theory (*7, 27, 28*). Here, the argument is that in order for a dispersed phase to undergo a phase transition, every droplet must have a nucleation event. This event occurs in the pure phase, and is coined homogeneous nucleation. Heterogeneous nucleation is unlikely in these systems, since every particle must have an impurity where nucleation can originate. A bulk phase needs only one successful heterogeneous nucleation event and heterogeneous nucleation is therefore likely in bulk phases. Observation of supercooling in the excess of 15 degrees is considered a signature of homogeneous nucleation.

It is more challenging to study the effect of interfacial tension on supercooling than it is to study the effect of curvature. Montenegro et al. (*5*) studied phase transition as a function of surface tension in hexadecane miniemulsions by first preparing miniemulsions that depleted the dispersion of excess surfactant, and then introducing different amounts of surfactants after sonication. This method certainly alters the final surface tension of the air/water interface, but it is not clear if the surfactant gets introduced into the miniemulsion droplets. The excess surfactant may therefore have little effect on the oil/water interface, which led to the observation that the phase transition of the hexadecane was independent of the final air/water interfacial tension.

The phase transition of hexadecane in miniemulsions depends on the type of surfactant used, but the emulsion droplet size also varies with the surfactant, so it is hard to separate the effect, or even know the exact surface tension between the oil and water in miniemulsions. In our method we deposit a polymer at the hexadecane/water interface from inside the miniemulsion droplets. The final interface in question should consequently be between the hexadecane and the polymer.

Mechanical Stability of the Capsules in DSC

The liquid to solid (solidus) phase transition of hexadecane occurs in a single step, without the formation of a rotator phase, or other pre-transitional phases (*28*). The normal freezing point of hexadecane is 18.1°C. When hexadecane freezes, it undergoes a slight contraction in volume, from a density of 0.773 g/mL to 0.81 g/mL. This volume change puts stress on the capsule to contract. Figure 7 shows a typical multiple DSC scans for the cooling and heating of poly(styrene) encapsulated hexadecane droplets. The capsule diameter in the sample shown in the figure 155 nm, with an estimated 25 nm thick poly(styrene) shell around 105 nm droplet of oil. The scan rate in the experiment was 1.0 K/min.

The response of poly(styrene) shown in Figure 7 is typical for what we observed for other capsule samples. The first solidus transition occurs at ΔT values close to 16 K, when the onset temperature for the transition is used as the phase transition temperature. This transition appears as a single, broad peak in the first cooling cycle. The extent of supercooling changes significantly with repeated thermal cycling, until the transition settles to a sharp transition, at $\Delta T = 2$ K. The onset temperature and peak width vary with the construction of the sample; thicker-walled samples need more cooling cycles before the final transition temperature is reached. The cooling and heating rate also affects how many cycles are required before the final state is reached; slower cooling rates need fewer cool/heat cycles.

The first solid to liquid (liquidus) transition registers as a broadened peak with its onset slightly lower than the bulk melting point. There is a slight change in this peak as well with repeated thermal cycling. This peak is less sensitive to the construct of the capsules and the heating rate.

Our current model for this behavior is that the shift in the solidus transition is a result of the poly(styrene) not having the structural rigidity necessary to withstand multiple volume changes in the encapsulated hexadecane. This leads to point-failure in the capsules from where subsequent freezing transitions can occur via heterogeneous nucleation. We observed that thinner capsules fail more readily than thicker capsules and that slow cooling rates result in quicker capsule failure.

The extent of supercooling is in line with what is expected for homogeneous nucleation, $\Delta T = 15$ K, whereas the later transitions are more indicative of heterogeneous nucleation. We do not observe the hexadecane leaking out of the capsule, or any increase in mobility of the hydrocarbon when this change occurs.

Figure 8 shows similar cool-heat experiment done on hexadecane encapsulated within a poly(methyl methacrylate) (PMMA) shell. The PMMA polymer is more mechanically stable than poly(styrene), thus the freezing point of the encapsulated oil remains constant for multiple cool-heat cycles.

Effect of the Oil Droplet Size on the Solidus Phase Transition Temperature

The volume fraction of n-hexadecane was varied in the formulation of the synthesis mixture. This gave us a set of samples, where the diameter of the encapsulated hexadecane varied, while the capsule shell thickness remained mostly constant. Throughout, the supercooling (ΔT) was measured as a function of the inverse size of the oil droplet and compared to the Gibbs-Thomson equation. Figure 9 shows this comparison for PMMA encapsulated hexadecane. Although there may appear to be a slight variation in the onset temperature, the overall variance of the ΔT is close to the reproducibility of the experiment (see the three points in the figure at 1000 nm/R = 13.5). Furthermore, the total change in ΔT

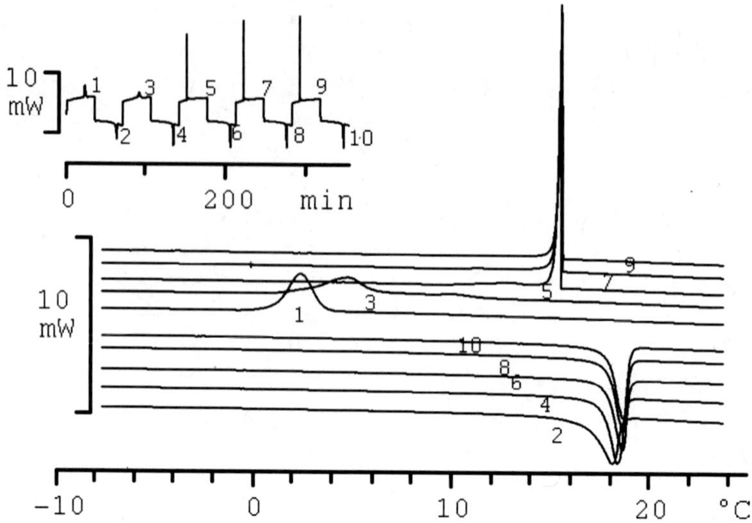

Figure 7. Stacked DSC thermographs of poly(styrene) encapsulated n-Hexadecane. Odd numbers indicate subsequent cooling runs. Reproduced with permission from reference (10)

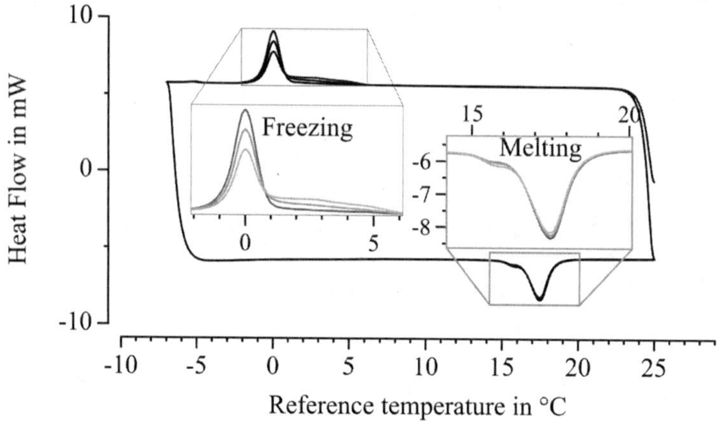

Figure 8. Cool-heat DSC cycles for poly(methyl methacrylate) encapsulate n-hexadecane. The volume fraction of PMMA in the capsule volume was 20%. Reproduced with permission from reference (11)

that we see over the entire range of 1/R is 0.2 K, which is close to the sensitivity of the DSC. Therefore, although it is tempting to state that we see agreement with the Gibbs-Thomson equation, such statement is not proper considering our experimental sensitivity.

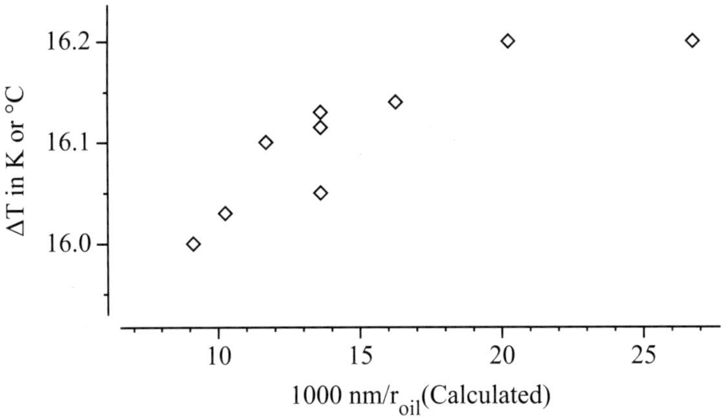

Figure 9. Supercooling as a function of inverse radius of the hexadecane encapsulated within a PMMA shell. The three points at 1000 nm/R = 13.5 demonstrates the reproducibility of the experiment. Reproduced with permission from reference (11)

The vertical intercept of a line through the data (omitting the last two points) is at 15.8 K, which typical for homogenous nucleation for the phase transition. This value should be zero, but a non-zero intercept is a confirmation of homogeneous nucleation.

Connection to Surface Tension and Droplet Size with Phase Transition Temperature

By changing the alkyl side-chain on the methyl methacrylate backbone, we obtain polymer capsules, where the interfacial tension between the hexadecane and polymer varies. This variation is coupled with a change in the glass transition temperature of the polymer and the mechanical stability of the capsule. Table 1 shows the different monomers we have worked with, along with the glass transition temperature of the polymer in melt and the air/polymer surface tension (sources indicated).

Poly(n-hexyl methacrylate)/hexadecane capsules were too unstable to be useful for long-time heat/cool thermal cycling; data for the onset temperatures varied significantly with preparation and age of the sample. Poly(ethyl methacrylate), PEMA, poly(n-butyl methacrylate), PnBMA, and poly(t-butyl methacrylate), PtBMA, capsules were stable.

We chose to study further capsules where the total volume of the precursor miniemulsion droplet contained 50% vol/vol of the monomer. Sonication and polymerization of these miniemulsion droplets gave particles that varied slightly in size (by ± 15 nm), but Figure 9 establishes that small variation in size has little, or no effect on ΔT as long as the oil is fully encapsulated by the polymer. In case of partial encapsulation, we expect to see both particle concentration dependence of ΔT, and change in the value with multiple freeze-thaw cycles. Figure 10 shows the onset temperatures for the solidus transition for n-hexadecane in the nanocapsules.

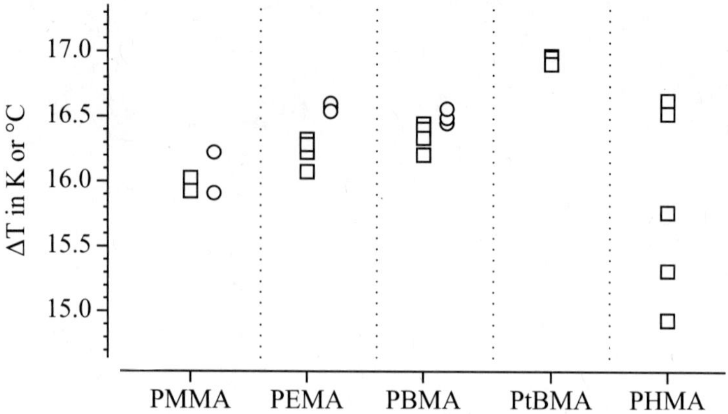

Figure 10. Supercooling for the solidus temperature of n-hexadecane encapsulated within various polymer capsules. Reproduced with permission from reference (11)

The apparent scattering of the onset values for poly(n-hexyl methacrylate) encapsulated hexadecane is due to the instability of those capsules. Each point represents the first cooling cycle for the samples and the difference in symbols represents two unrelated persons in the experiment. The ΔT values in Figure 10 are sorted according to the perceived complexity of the alkyl side chain, from methyl to n-hexyl methacrylates.

According to the Gibbs-Thomson equation, ΔT varies linearly with the interfacial tension ($\gamma_{o/p}$) at the polymer/oil interface. We do not, however, have direct access to this value with much accuracy. Considerable research work has been done on the poly(alkyl-methacrylate)/water interface. From this work it is clear that the interfacial tension depends on the method used to prepare the interface (*29*), and the surface structure of the polymer tends to change over time.

Here we can only give reasonable estimates of the polymer/oil surface tension using contact angle measurements that we did on polymers isolated from our synthesis and spin-coated onto silicon wafers. Our measurements of the air/hexadecane/polymer contact angle gave nearly identical angle for all the polymers (between, 10° and 12°). We did, however, get air/water/polymer contact angles that were in a reasonably good agreement with the literature.

We can then estimate the surface tension of the polymer/oil interface using the Young's equation:

$$\gamma_{o/p} = \gamma_{a/p} - \gamma_{a/o} \times \cos(\theta_{a/o/p})$$

the surface tension of hexadecane (γ_{ao}) is well known to be 28.12 mN/m. Since values for $\gamma_{a/p}$ vary somewhat in the literature between sources, we end up with a range for the calculated value for $\gamma_{o/p}$. These values are collected in Table 1.

Table 1. Interfacial tension of the polymer/air interface from various literature sources. Reproduced with permission from reference (*11*)

Polymer	T_g (°C)	$\gamma_{a/p}$ mN/m	*Source*	$\gamma_{o/p}$ mN/m
PMMA	108	38.3, 43.2	Kwok 2000 (*30*), Lee 1991 (*31*)	11 – 16
PEMA	65	33.6, 35.9	Kwok 2000 (*30*), Wu 1971 (*29*)	6 – 8
PnBMA	20	28.8, 34.6	Kwok 2000 (*30*), Wu 1971 (*29*)	1 – 7
PtBMA	110	18.1, 30.5	Grundke 2005, Wu 1971 (*29*)	-9.5 – 2.9
PnHMA	-5	31.1	Clarke 2006 (*29*)	3.5

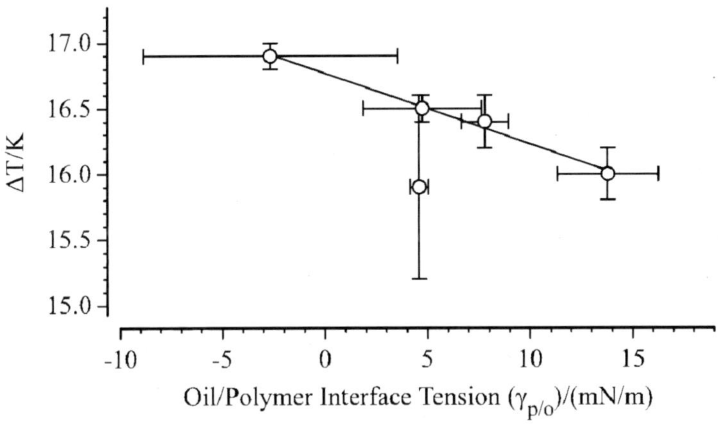

Figure 11. Supercooling as a function of polymer/n-hexadecane interfacial tension. The horizontal error bars represent variation in experimental values found in the literature, vertical error bars represent experimental scattering in this work. Reproduced with permission from reference (11)

Figure 11 shows the measured ΔT values from Figure 10 drawn against the $\gamma_{o/p}$ values in Table 1. We represent the mid-point of the calculated range of surface tension by a point in the figure, but the range is shown as error bars. The data falls somewhat fortuitously on a line, although lines drawn through the lowest $\gamma_{o/p}$ values in the range, as well as lines through the highest values also gives fairly good linear correlation. What is, however, important here is that the slope of this line is negative, rather than positive as is predicted by the Gibbs-Thomson equation.

The poor agreement we find in our work with the classical Gibbs-Thomson equation may not come as a big surprise, although the negative slope in Figure 11 was rather unexpected. Researchers studying phase transitions in emulsions have found little, or no dependence of the supercooling with particle size. This has been explained by the dominating effect of the surfactant in the system, or by the insensitivity that small changes in particle radius has on the 1/r term in the equation, once r approaches several hundred nanometers.

Our system is also not a fully equilibrated system. The miniemulsion which forms the template is a meta-stable state, and we do observe relaxation, or failure of the capsules studied here after long times (months). Finally, the Gibbs-Thomson equation is derived for a system where phase undergoing the transition has zero partition into the solid phase and the phase directly adjacent to the solid surface is bulk-like. This may be a boundary condition in the theory that does not apply to our system, the oil may orient itself next to the polymer film, forming a surface layer of either highly dense oil, or a layer of less packed oil molecules. This type of packing has been observed for water molecules in porous glasses using NMR spectroscopy. Similar work has not been done for our types of systems, but this measurement is one of our current pursuits.

References

1. Hiemenz, P. C.; Rajagopalan, R. *Principles of Colloid and Surface Chemistry*, 3rd ed.; Marcel Dekker: New York, 1997.
2. Turnbull, D.; Cormia, R. L. Kinetics of Crystal Nucleation in Some Normal Alkane Liquids. *J. Chem. Phys.* **1961**, *34* (3), 820–831.
3. Kashchiev, D.; Kaneko, N.; Sato, K. Kinetics of crystallization in polydisperse emulsions. *J. Colloid Interface Sci.* **1998**, *208* (1), 167–177.
4. Jiang, K.; Su, Y. L.; Xie, B. Q.; Jiang, S. C.; Zhao, Y.; Wang, D. J. Effect of Geometrical Confinement on the Nucleation and Crystallization Behavior of n-Alkane Mixtures. *J. Phys. Chem. B* **2008**, *112* (51), 16485–16489.
5. Montenegro, R.; Antonietti, M.; Mastai, Y.; Landfester, K. Crystallization in miniemulsion droplets. *J. Phys. Chem. B* **2003**, *107* (21), 5088–5094.
6. Kozisek, Z.; Demo, P. Homogeneous nucleation rate at various initial supersaturations in a closed system. *J. Aerosol Sci.* **2009**, *40* (9), 802–806.
7. Kozisek, Z.; Demo, P. Size distribution of nuclei formed by homogeneous nucleation in closed systems. *J. Aerosol Sci.* **2009**, *40* (1), 44–54.
8. Gelb, L. D.; Gubbins, K. E.; Radhakrishnan, R.; Sliwinska-Bartkowiak, M. Phase separation in confined systems. *Rep. Prog. Phys.* **1999**, *62* (12), 1573–1659.
9. Tiarks, F.; Landfester, K.; Antonietti, M. Preparation of polymeric nanocapsules by miniemulsion polymerization. *Langmuir* **2001**, *17* (3), 908–918.
10. Fette, E. V.; Pham, A.; Adalsteinsson, T. Crystallization and Melting Transitions of Hexadecane Droplets in Poly(styrene) Nanocapsules. *J. Phys. Chem. B* **2008**, *112* (17), 5403–5411.
11. Black, J. K.; Tracy, L. E.; Roche, C. P.; Henry, P. J.; Pesavento, J. B.; Adalsteinsson, T. Phase Transitions of Hexadecane in Poly(alkyl methacrylate) Core-Shell Microcapsules. *J. Phys. Chem. B* **2010**, *114* (12), 4130–4137.
12. Donath, E.; Sukhorukov, G. B.; Caruso, F.; Davis, S. A.; Möhwald, H. Novel hollow polymer shells by colloid-templated assembly of polyelectrolytes. *Angew. Chem., Int. Ed.* **1998**, *37* (16), 2202–2205.

13. Sukhorukov, G. B.; Donath, E.; Davis, S.; Lichtenfeld, H.; Caruso, F.; Popov, V. I.; Möhwald, H. Stepwise polyelectrolyte assembly on particle surfaces: a novel approach to colloid design. *Polym. Adv. Technol.* **1998**, *9* (10–11), 759–767.
14. Kowalski, A.; Vogel, M. Multi-stage opacifying polymer particles containing non-polymeric acid absorbed therein. U.S. Patent 4,880,842, 1989.
15. Steiert, N.; Landfester, K. Encapsulation of organic pigment particles via miniemulsion polymerization. *Macromol. Mater. Eng.* **2007**, *292*, 1111–1125.
16. Landfester, K. Synthesis of colloidal particles in miniemulsions. *Annu. Rev. Mater. Res.* **2006**, *36*, 231–279.
17. Paiphansiri, U.; Tangboriboonrat, P.; Landfester, K. Polymeric nanocapsules containing an antiseptic agent obtained by controlled nanoprecipitation onto water-in-oil miniemulsion droplets. *Macromol. Biosci.* **2006**, *6* (1), 33–40.
18. Crespy, D.; Stark, M.; Hoffmann-Richter, C.; Ziener, U.; Landfester, K. Polymeric nanoreactors for hydrophilic reagents synthesized by interfacial polycondensation on miniemulsion droplets. *Macromolecules* **2007**, *40* (9), 3122–3135.
19. Klapper, M.; Nenov, S.; Haschick, R.; Muller, K.; Mullen, K. Oil-in-oil emulsions: A unique tool for the formation of polymer nanoparticles. *Acc. Chem. Res.* **2008**, *41* (9), 1190–1201.
20. Muller, K.; Klapper, M.; Mullen, K. Preparation of high molecular weight polyurethane particles by nonaqueous emulsion polyaddition. *Colloid Polym. Sci.* **2007**, *285* (10), 1157–1161.
21. El-Aasser, M. S.; Sudol, E. D. Miniemulsions: Overview of research and applications. *JCT Res.* **2004**, *1* (1), 21–31.
22. Antonietti, M.; Landfester, K. Polyreactions in miniemulsions. *Prog. Polym. Sci.* **2002**, *27* (4), 689–757.
23. Landfester, K.; Bechthold, N.; Tiarks, F.; Antonietti, M. Formulation and stability mechanisms of polymerizable miniemulsions. *Macromolecules* **1999**, *32* (16), 5222–5228.
24. Landfester, K.; Willert, M.; Antonietti, M. Preparation of polymer particles in nonaqueous direct and inverse miniemulsions. *Macromolecules* **2000**, *33* (7), 2370–2376.
25. El-Aasser, M. S.; Sudol, E. D. In *Emulsion polymerization and emulsion polymers*; El-Aasser, M. S., Ed.; John Wiley & Sons Inc.: New York City,1997; p 826.
26. Schreiber, A.; Ketelsen, I.; Findenegg, G. H. Melting and freezing of water in ordered mesoporous silica materials. *Phys. Chem. Chem. Phys.* **2001**, *3* (7), 1185–1195.
27. Herhold, A. B.; Ertas, D.; Levine, A. J.; King, H. E. Impurity mediated nucleation in hexadecane-in-water emulsions. *Physical Review E* **1999**, *59* (6), 6946–6955.
28. Herhold, A. B.; King, H. E.; Sirota, E. B. A vanishing nucleation barrier for the n-alkane rotator-to-crystal transformation. *J. Chem. Phys.* **2002**, *116* (20), 9036–9050.

29. Wu, S. Calculations of interfacial tension in polymer systems. *J. Polym. Sci., Part C: Polym. Symp.* **1971**, *34* (1), 19–30.
30. Kwok, D. Y.; Neuman, A. W. Contact angle measrements and contact angle interpretation: Relevance to the thermodynamics of adhesion. In *Acid-Base Interactions*; Mittal, K. L., Ed.; VPS: Utrecht, The Netherlands, 2000; pp 91−166.
31. Lee, L.-H. *Fundamentals of Adhesion*; Plenum Press: New York, 1991; Vol. 1.

Chapter 18

Molecular Dynamics Simulations of Nanoparticles and Surfactants at Oil/Water Interfaces

A Qualitative Overview

R. J. K. Udayana Ranatunga,[1] Chuong T. Nguyen,[1] Chi-cheng Chiu,[1] Wataru Shinoda,[2] and Steven O. Nielsen[*,1]

[1]Department of Chemistry, University of Texas at Dallas, 800 West Campbell Road, Richardson, TX 75080
[2]Nanosystem Research Institute, National Institute of Advanced Industrial Science and Technology (AIST), 1-1-1 Umezono, Tsukuba, Ibaraki 305-8568, Japan
[*]steven.nielsen@utdallas.edu

Nanoparticles (NPs) and molecular surfactants are two classes of compounds which spontaneously localize at oil/water interfaces. Industrial and commercial applications of these systems often require precise two-dimensional organization of the localized NPs. An impediment to the realization of such systems is our under-developed understanding of the physics which governs the behavior of NPs in the presence of surfactants. Here we present a range of molecular dynamics simulation studies on non-ionic NP/surfactant systems. Analysis of the results allows us to relate the dispersive interactions of the NPs and surfactants to their physical behavior at oil/water interfaces.

Introduction

Documented study of the interfacial behavior of small colloidal particles began over a century ago, when Pickering (*1*) and Ramsden (*2*) separately investigated the spontaneous adsorption of colloidal particles to fluid interfaces (*3*). Since then the affinity of colloids to localize at fluid interfaces has been

© 2011 American Chemical Society

harnessed in a number of fields (*4*). The interfacial activity of molecular surfactants, on the other hand, has been known for far longer. For example, historical evidence suggests that surfactants have been used for over 3500 years in the production of soap (*5*).

Recently, interest has arisen in systems containing both colloids and surfactants (*6, 7*). Advances in synthetic technology have resulted in the production of particles with progressively smaller sizes, unlocking the unique properties which emerge due to quantum confinement (*8–10*). Developments in methodology have also narrowed the size distribution of synthesized particles, greatly improving the aggregate properties and the reproducibility of systems composed of these nano-scaled particles. Many current and proposed applications of nanoparticles (NPs) involve systems of oil and water which also contain surfactants as emulsifiers. Therefore, given the propensity of NPs to localize at fluid interfaces it is important to understand the behavior of NPs and surfactants in the presence of one another.

In this chapter we focus on non-ionic surfactants and uncharged NPs at oil/water interfaces, studied through molecular dynamics (MD) computer simulations. The high spatial and temporal resolution of the MD trajectories facilitates a thorough examination of the interactions between the system components, and statistical mechanics provides us with a path to evaluate thermodynamic properties. A multi-pronged approach is used whereby several simulation schemes are carried out, each corresponding to a separate avenue of investigation, the results of which are presented in three sections:

- Unfunctionalized NPs at oil/water interfaces. Simulation results of individual NPs and groups of NPs are presented.
- Surfactants and NPs at oil/water interfaces. Simulations are presented of NPs in the presence of free surfactants. The cooperative behavior of the system components in lowering the oil/water interfacial tension is explored.
- Surfactant functionalized NPs at oil/water interfaces. Simulations are presented of NPs in the presence of chemically absorbed surfactant ligands. The implication of the ‘softness’ that these coatings introduce to the composite particle is examined.

System Components and Simulation Details

Four components are essential for this study, namely nanoparticles, surfactants, water and oil. The choice of studying non-ionic species was taken because of the many advantages that these species present. The first of these advantages is that non-ionic surfactant-stabilized colloidal aggregates are known to be more robust than systems stabilized through electrostatic repulsion (*11*). Second, the dominant role that electrostatics plays in systems of large particles is mitigated at the nanoscale, and other factors such as dispersion interactions, entropy and steric interactions play equally important roles in determining the physical behavior of these systems (*12*). Apart from these reasons, technically, the calculation of electrostatics consumes a significant portion of the computational expense of a simulation, thus limiting the simulation timescale and hence the sampling available to us.

For the non-ionic surfactants, linear polyethyleneglycol (PEG)-alkane diblock copolymers were chosen, denoted by C_xE_y where 'x' is the number of carbons in the hydrophobic alkane tail and 'y' is the number of ethylene glycol units in the hydrophilic head block. It should be noted that these surfactants have an extended as opposed to a compact head group (*13*). A linear alkane (heptane) is used for the oil phase. The NP sizes in the simulations presented here are in the range of 5.6 Å – 20.0 Å in radius.

Simulations were carried out in MPDyn (*14*) under isothermal-isobaric conditions, using a Nosé-Hoover thermostat (*15*), at 300.15 K, and an Andersen barostat (*16*), at 0.1 MPa, respectively. The reversible RESPA (*17*) algorithm was applied with a timestep of 20 fs for long ranged interactions, and a 2 fs timestep for all other interactions. Simulation box sizes were varied as neccessary, with a typical unit cell side length of 100−150 Å.

All the simulated systems contained oil and water. Under constant composition, pressure, and temperature (NPT) conditions the system evolves to lower the Gibbs free energy, resulting in a phase separated box of oil and water corresponding to the minimum oil/water interfacial area, and hence the minimum free energy. These systems model oil/water interfaces of negligible curvature with respect to the dimensions of the individual NPs and surfactant molecules. Therefore, our results may be valid for macroscopically planar systems such as the water/toluene systems studied by Wang *et al.* (*18*), and also for oil/water emulsions where the droplet size is sufficiently large that the droplet boundary is essentially planar on the simulated length scale. An example of a simulation box is given in Figure 1, which shows the system components while also illustrating the periodic boundary image convention used in the simulations. The use of periodic boundaries maintains the solvent environment at the edges of the simulation box and allows for the representation of an extended oil/water interface without using an excessively large box size (*19*).

The interactions among the components must be specified to carry out simulations. This, and the representation of the species, are collectively known as the 'force field'. The representation we have used for this study is a coarse-grained (CG) force field where one CG 'bead' is representative of several atoms. For many areas of study this is an invaluable approach when the focus is on collective phenomena resulting from multitudes of molecules (*20*). By avoiding the complexity of calculating atomic movements and interactions CG simulations can attain high computationally efficiency, and hence are able to simulate larger systems for longer timescales compared to fully atomistic simulations. In principle, the effect of atomic interactions are included in coarse grained force fields, *i.e.*, the interactions between two CG beads should be the ensemble averaged interaction due to all the atoms those beads represent. For example, two CG water beads do not form hydrogen bonds (because they do not contain explicit hydrogen atoms), but should account for the additional strength of interaction between water molecules due to hydrogen bonds. However, because CG approaches essential discard fine detail it is critical to apply this technique in

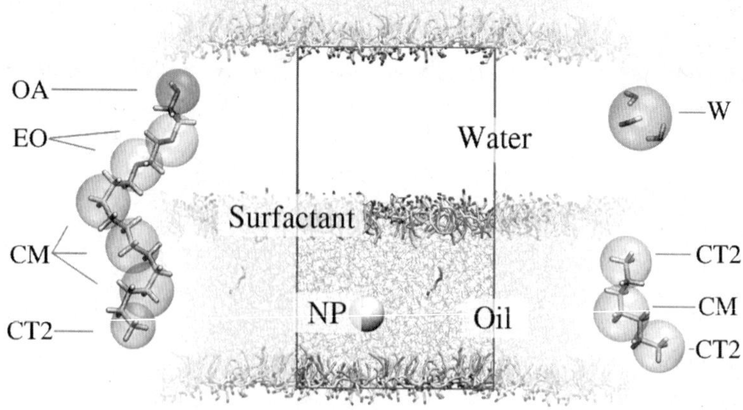

Figure 1. A side view of a representative simulation box comprising water (not shown), oil, surfactant and a single nanoparticle. The rectangular box designates the unit cell, while the periodic images are also shown. Due to the periodic image convention each simulation box contains two oil/water interfaces, which are perpendicular to the plane of the page in this example. To the left and right of the simulation box, the coarse-grained molecular representation (transparent spheres and bead labels) is superimposed on the atomistic representation (stick model).

systems where the atomistic interactions are not crucial to the phenomena under investigation (*20*). In this study we are investigating a wide range of systems in a qualitative manner, hence a strong argument can be made for the use of a CG model since it is not clear what additional knowledge could be gained through using a fully atomistic force field.

For the soft components (water, heptane and surfactant) the coarse-grained force field of Shinoda was used (*21*). This force field has been shown to accurately predict the phase behavior of PEG-alkane surfactants, and also the properties of oil/water systems including, most critically, the interfacial tension. Bond stretching is modeled through a harmonic potential, $U_{bond}(r) = k_b(r - r_e)^2$, where k_b is the bond force constant, r the separation between two covalently bound beads, and r_e their equilibrium separation. Bond bending is also modeled through a harmonic potential, $U_{bend}(\theta) = k_\theta(\theta - \theta_0)^2$, where k_θ is the force constant, θ the instantaneous angle between the three beads involved in a bond angle, and θ_0 the equilibrium angle. The relevant parameters used are given in Tables 1 and 2. Since the components we are using are uncharged, the non-bonded interactions comprise exclusively dispersive interactions which are modeled through the Lennard-Jones function,

$$U_{LJ}(r) = B \times \varepsilon \left\{ \left(\frac{\sigma}{r}\right)^m - \left(\frac{\sigma}{r}\right)^n \right\} \qquad (1)$$

Table 1. Force field parameters for bond stretching

	k_b kcal mol^{-1}Å^{-2}	r_e Å
CT2–CM	9.00	3.13
CM–CM	6.16	3.64
CM–EO	7.10	3.56
EO–EO	4.90	3.28
EO–OA	15.00	2.79

Table 2. Force field parameters for bond bending

	k_θ kcal mol^{-1}deg^{-2}	θ_0 deg
CT2–CM–CT2	1.70	173.0
CT2–CM–CM	1.60	172.0
CM–CM–CM	1.19	173.0
CM–CM–EO	1.50	172.0
CM–EO–EO	3.20	146.0
EO–EO–EO	3.40	132.0
EO–EO–OA	3.00	131.0

where σ is the exclusion volume, and m, n and B are chosen such that (1) −ε is the minimum of the potential profile, and (2) $U_{LJ}(\sigma) = 0$. Different choices for the integer powers m and n correspond to interaction potentials of different stiffness. The parameters are specific for each pair of interacting beads and are given in Table 3.

The representation of the NP and its interaction with the soft system components is based on a mean field approximation where the NP is modeled as a continuum solid sphere with a uniform interaction site density (*22*) in analogy to Hamaker's treatment of colloidal particles (*23*). The total interaction of the NP with another entity is found by integrating the site-site interaction over the volume of the NP. Hence, the interaction is governed by two parameters apart from particle geometry; (1) the density of interaction sites in the NP, ρ, and (2) the interaction strength parameter, ε, between sites on the NP and those on the interacting body. The ρ value chosen was that of graphitic carbon at 0.113 Å^{-3}. The ε values for the interaction with oil were taken directly from fully-atomistic simulation data (*22*), while the NP-water interaction strength, ε_{NP-W}, is iteratively adjusted to obtain the desired hydrophobicity, for example, to match the experimental contact angle of a water droplet on a graphene surface (*24*). For this study a base value for ε_{NP-W} was taken as 0.596 kcal mol^{-1}, which for a NP of 10.0 Å radius corresponds to a slightly hydrophobic particle. During this chapter the set of ε values for the NP interaction with the all the different beads which make up the two solvents are collectively indicated by ε_{Solv}. It should

Table 3. Force field parameters for non-bonded interactions

	CT2		CM		EO		OA		W	
	σ	ε	σ	ε	σ	ε	σ	ε	σ	ε
W	4.296	0.290	4.438	0.340	4.310	0.570	3.950	0.700	4.371	0.895
OA	3.840	0.380	4.274	0.377	3.890	0.440	3.713	0.4491		
EO	4.140	0.370	4.274	0.377	4.250	0.405				
CM	4.364	0.362	4.506	0.420						
CT2	4.221	0.312								

The σ and ε parameters for pairs of interacting types of CG beads given in units of Å and kcal mol^{-1}, respectively. For interactions involving water, m=12 and n=4 while for all other interactions m=9 and n=6.

be noted that the curvature of the NP-water interface (and hence the NP size) also strongly affects the NP hydrophobicity/hydrophilicity (*25*).

For the interaction between two NPs the same approach is used. The only difference is that the total interaction is found through integrating over the volumes of both NPs. The interaction strength between two interaction sites on different particles, ε_{NP-NP}, is 0.070 kcal mol^{-1}, which is the value for graphitic carbon.

Unfunctionalized NPs at Oil/Water Interfaces

In this section we discuss the ability of unfunctionalized NPs to lower the oil/water surface energy, beginning with the conceptual difference of NPs compared to molecular surface active agents (surfactants).

Interfacial Behavior of Surfactants versus Nanoparticles

For an isothermal-isobaric system (NPT ensemble), the interfacial localization of any discrete species is determined by the change in Gibbs free energy resulting from its adsorption to the interface (*3*). The excess free energy for an oil/water interface can be given as,

$$G = A_{ow}\gamma_{ow} \qquad (2)$$

where A_{ow} is the surface area of the oil/water interface and γ_{ow} is the oil/water surface tension. The thermodynamic definition of the surface tension follows from the above equation,

$$\gamma_{ow} = \left(\frac{\partial G}{\partial A_{ow}}\right)_{N,P,T} \qquad (3)$$

It is clear from Equation 2 that at the simplest level two factors affect the interfacial free energy, and hence that two mechanisms exist to lower this free energy. These two mechanisms are the ability of a species to reduce the oil/water contact area or to reduce the tension of the interface. The difference between molecular surfactants versus NPs at oil/water interfaces is essentially based on this observation: surfactants lower γ_{ow} while NPs lower A_{ow}.

Some surfactants can lower the surface tension of the oil/water interface to the stability limit of zero. Beyond this limit addition of surfactants to the interface causes γ_{ow} to be lowered below zero, resulting in an instability which is released by an increase in the interfacial area corresponding to buckling or folding of the interface. In general the performance of surfactants in changing an interfacial property is measured through their efficiency and effectivity. In this context the efficiency of a surfactant is the equilibrium concentration of surfactant needed to lower the interfacial tension by a specific amount, and the effectivity is the maximum change the surfactant can induce in the interfacial tension (*26*). Due to their extensive use in industry, the behavior of many surfactant species (cationic, anionic, zwiterionic and neutral) at oil/water interfaces is well known. Many of the non-ionic surfactants, including those with polyethyleneglycol (PEG) heads, are effective at lowering the oil/water surface tension to zero. We refer any reader interested in learning more about the intrinsic activity of surfactants to the wide body of literature available on this subject (*26*, *27*).

Conversely, theoretical treatments of NPs assume there is no change in the surface tension with adsorption of the NP to a fluid interface. The free energy change upon adsorption of a NP from the oil phase to the oil/water interface can be derived by considering the total surface excess free energy from all the interfaces present (*3*, *28*). For a spherical particle of radius R, this leads to an expression as a function of η, the particle position measured from the interfacial plane (i.e. in the direction normal to the interface),

$$\Delta G(\eta) = \begin{cases} 0 & \eta < -R \\ 2\pi R^2(1+\eta/R)(\gamma_{pw}-\gamma_{po}) - \gamma_{ow}\pi(R^2-\eta^2) & -R \le \eta \le R \\ 4\pi R^2(\gamma_{pw}-\gamma_{po}) & \eta > R \end{cases} \tag{4}$$

Here the reference state ($\eta = -R$) corresponds to the particle being completely submerged in the oil phase and γ_{pw} and γ_{po} are the particle/water and particle/oil surface tensions, respectively. For the $-R \le \eta \le R$ range of $\Delta G(\eta)$, the first term is due to the gain or loss of particle/solvent surface area. The second term is the free energy change due to the removal of oil/water contact arising from the excluded volume of the NP. By considering the equilibrium condition (i.e. minimizing ΔG with respect to η), Young's equation can be derived,

$$\frac{\eta}{R} = \frac{\gamma_{po} - \gamma_{pw}}{\gamma_{ow}} \tag{5}$$

Here the quantity η/R is related to the equilibrium contact angle, θ, through $\cos\theta = \eta/R$, which is an experimental observable (*29*).

Some thermodynamic approaches suggest that accumulation of any solute at a fluid interface results in a lowering of the fluid/fluid interfacial tension (*30*), but this is at odds with both experimental and computational results which often show the surface tension of fluid interfaces does not change upon particle adsorption (*6, 7*).

Transfer Free Energy Profile of Individual NPs

From molecular dynamics simulations the free energy profiles of NPs in the vicinity of an oil/water interface can be determined using the method of thermodynamic integration (*31*), in which a series of simulations are run varying the position η of the NP. In each simulation, η is kept constant through application of a harmonic biasing potential, from which the force of constraint is tracked. The mean value of the force of constraint for each simulation can be integrated with respect to η to obtain the free energy profile, $\Delta G(\eta)$ (*32*).

The transfer free energy profiles of two individual NPs at a heptane/water interface are shown in Figure 2. The general features predicted by Equation 4 are clearly visible, where $\Delta G(\eta)$ shows a parabolic minimum close to the interface ($\eta = 0$). The stability of the NP at the interface is seen to increase with particle size, in line with Equation 2 which predicts this stability increase to scale as R^2. This is seen experimentally in polydisperse NPs at fluid interfaces, where the larger NPs displace the smaller NPs over time (*33*). Polydispersity can also lead to phase separation of the NPs into small islands at the fluid interface.

Dimerization Free Energy Profile of NPs Localized at the Interface

The spatial organization of NPs at fluid interfaces is a problem given much attention due to the implications born to many industrial and commercial fields (*3, 34*). An exhaustive survey of this topic is beyond the scope of this chapter, but conventional approaches to treat aggregation behavior are based on the Derjaguin-Landau-Verwey-Overbeek (DLVO) theory (*3*). DLVO theory treats charged particles in electrolyte solution by decomposing the net interaction between particles into an electrostatic double-layer repulsion and a dispersive attraction. Adoption of this theory to interfacial systems requires modifying the contributions to account for particle immersion in media of different permittivity and the distribution of charges which accompanies this (*3*). Furthermore, treatment of particle interactions at the interface needs to address phenomena not seen in the bulk such as capillary forces.

The work we have carried out is on small uncharged NPs and hence the forces due directly or indirectly to electrostatics and gravity are avoided. In this section we focus on the dimerization free energy of NPs at an oil/water interface evaluated through MD simulation data and we attempt to give a conceptual reasoning for the observed features.

In an ideal gas reference state the NPs, interacting only through dispersion forces, show a simple form for their interaction free energy as a function of their separation. A single minimum exists corresponding to the physical contact of the particles. Separation values less than the minimum are strongly disfavored due to steric overlap.

When the NPs are localized at an oil/water interface, their mutual interaction is more complicated (*35*) as seen in the dimerization free energy profiles shown

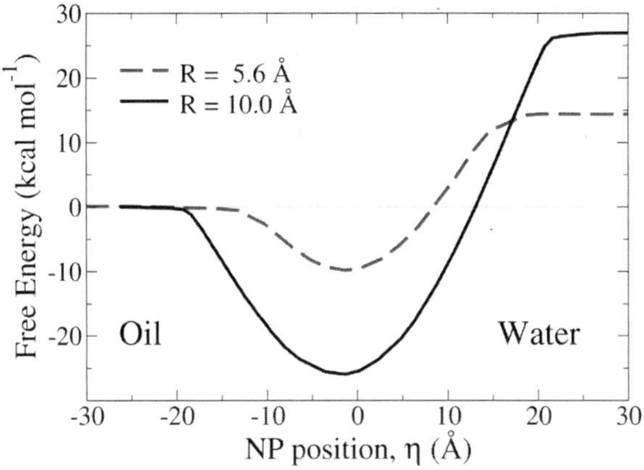

Figure 2. Transfer free energy profiles of two unfunctionalized NPs, from a reference state in bulk heptane to a bulk water environment. These NPs are considered hydrophobic due to their preference to solvate in the oil phase. Note that the active η range is broader than predicted by Equation 2 due to dispersion forces.

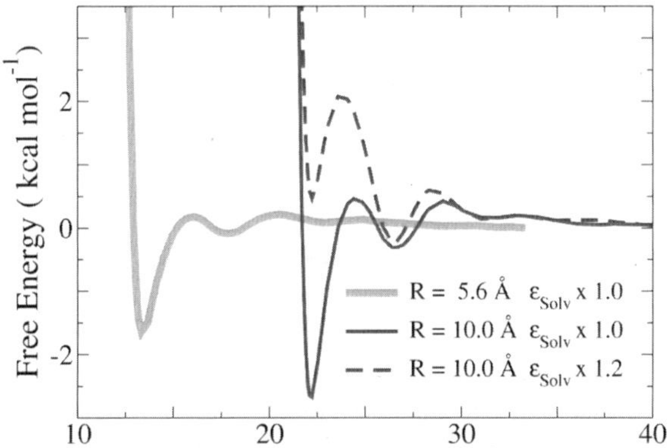

Figure 3. Dimerization free energy profiles of three pairs of unfunctionalized NPs as a function of their interparticle separation at an oil/water interface. The increase in interaction energy with particle size is seen between the two profiles shown in solid lines (radius 5.6 Å versus radius 10.0 Å; ε_{NP-W} = 0.596 kcal mol^{-1}). The effect of solvation is seen through the two radius 10.0 Å profiles.

in Figure 3. Dimerization profiles for two sizes of NPs were calculated using the previously described techniques, namely constrained molecular dynamics and thermodynamic integration of the mean force of constraint with respect to the inter-particle separation.

The profiles in Figure 3 show the undulations in free energy which arise due to the solvation of the NPs at the interface. Individual particles tend to order solvent molecules in layers around them. The dimerization of two such particles involves the restructuring and desolvation of these solvent layers giving the undulations in the dimerization free energy profile. Solvent interaction with the particle diminishes with distance and hence the undulations in the profile decrease in magnitude the further away the particles are from one another. Conceptually the disruption of a solvation shell entails an enthalpic loss of the solvent-particle interaction which is countered by the enthalpic gain of solvent-solvent interactions, particle-particle interaction and an entropic gain through disruption of the ordered solvent layer.

The scaling of interaction magnitude with particle size is also exemplified in Figure 3 (profiles in solid lines). Here the deepening of the free energy minimum at particle contact is clear, as well as the increase in the amplitude of undulations caused by solvent restructuring. Both these effects are caused by size (curvature) dependence of the NP/solvent and NP/NP interactions.

The dimerization free energy profiles of Figure 3 also show the effect of increasing the particle-solvent interaction strength. Through scaling ε_{Solv} by a factor of 1.2 the global minimum of the free energy profile can be shifted from being at close-contact (Figure 3, thin solid line) to a minimum corresponding to a solvent layer separating the two particles (Figure 3, dashed line). This is thought to be due to the particle-solvent enthalpic contribution overtaking the contributions which stabilize the system when the particles are in contact.

Organization of NPs at an Oil/Water Interface

The dimerization profiles shown Figure 3 are for two particles at an oil/water interface. However, the behabior of a NP in the presence of several others (localized at an oil/water interface) depends on the relative positions of all particles within their effective interaction range. Hence, predicting the two-dimensional spatial organization of NPs is a challenging task. Although the mutual interaction of particles is a manifold landscape it can be expected that there is a strong correlation between the features of the dimerization free energy profile and the overall interfacial structure. From simulations of different NPs at an oil/water interface we observed this relationship as shown in Figure 4. Specifically, simulations were carried out for the three types of NPs shown in the dimerization free energy profiles of Figure 3 and the resulting 2D structures are shown in Figure 4. As expected from the dimerization free energy profiles the particles with unscaled ε_{Solv} values form clusters with the particles in contact with one another. The radius 10.0 Å NPs form extended close-packed clusters while the radius 5.6 Å particles form smaller clusters. This may be due to the differences in free energy at particle contact which for the radius 5.6 Å particles is comparable to the thermal energy and hence subject to thermal fluctuations.

The effect of increasing the solvent interaction strength is also seen in Figure 4 where the radius 10.0 Å with ε_{Solv} scaled by 1.2 shows no NP clustering. This behavior is commensurate with the largely repulsive dimerization profile of Figure 3.

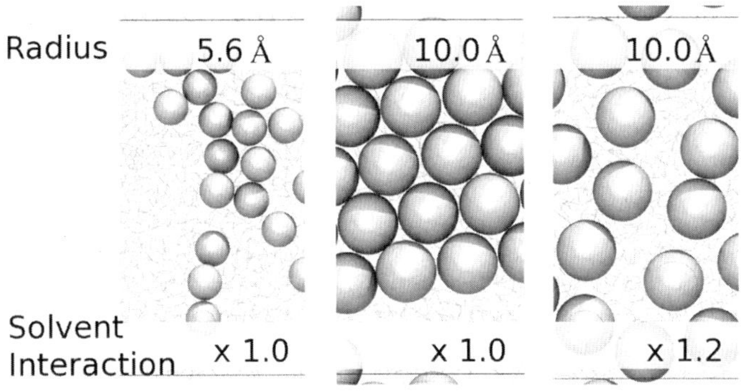

Figure 4. Equilibrium structure of NPs localized at an oil/water interface. Radius 5.6 Å particles (left) form small clusters while 10.0 Å radius NPs (center) form extended close packed structures. By increasing the particle-solvent interaction strength by a factor of 1.2 (right) the two-dimensional spatial organization of the radius 10.0 Å NPs is completely different.

Synergistic Interactions of Surfactants and NPs at Oil/Water Interfaces

The co-existence of surfactants and NPs can be broadly classified into two categories based on their interaction; (1) NPs in the presence of free surfactants; (2) chemically adsorbed surfactants.

First we consider NPs in the presence of free surfactants. At low coverage of surfactants (~65 Å^2 per surfactant) and NPs (~651 Å^2 per NP) both species are predominantly found at the interface. Representative systems are shown in Figure 5, for 10.0 Å radius particles with different ε_{Solv} values.

Previous simulation studies have shown that $C_{12}E_3$ surfactants are successful at lowering the oil/water surface tension, γ_{ow}, to zero at a surface concentration of ~32.5 Å^2 per surfactant. For a simulation cell of surface area 125 Å × 125 Å this corresponds to ~480 surfactants. To investigate how the presence of NPs affects the behavior of the surfactants a system of the above surface area was prepared with half the maximum coverage, i.e. 240 $C_{12}E_3$ molecules localized at each oil/water interface. In addition to the surfactants, varying numbers of NPs are positioned at the interface. The systems are allowed to equilibrate for 10 ns, after which data is collected from longer equilibrium simulations. Conformations of these systems after 50 ns are shown in Figure 6.

From Figure 6, it is clear that the surfactants are sparingly soluble in oil. It is also clear that at higher NP concentrations the NP interfacial distribution is shifted towards the oil phase and some NPs are dislodged from the interface towards the oil phase. To further discuss this trend let us quantify the effectivity of the surfactants through their effect on γ_{ow}, the oil/water interfacial tension.

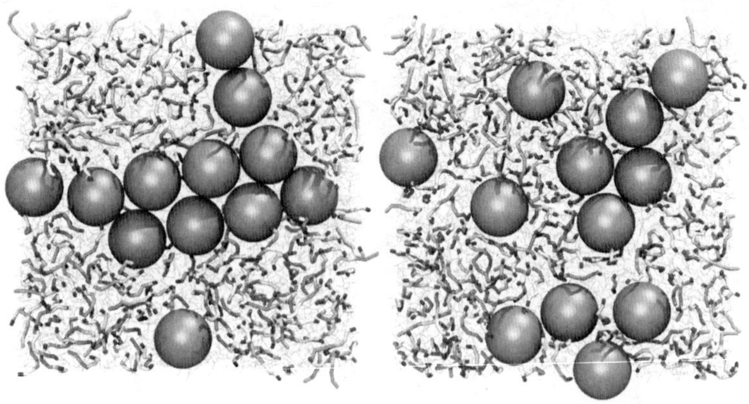

Figure 5. (left) 24 radius 10.0 Å particles with 240 $C_{12}E_3$ surfactants at an oil/water interface. (right) The same system with the particle-solvent interactions scaled by a factor of 1.2.

Surface Tension Measurements and Cooperativity Between NPs and Surfactants

The synergy of surfactants and NPs in lowering the surface tension of the oil/water interface has received interest recently (*6*, *7*, *36*, *37*) and research into this phenomena is ongoing. However, most of the observed synergy is caused by electrostatic interactions between charged particles and ionic surfactants, whereas we report synergistic effects in systems of uncharged NPs and non-ionic surfactants.

Since both surfactants and NPs are localized at the oil/water interface some cooperativity in lowering the oil/water surface tension may be expected. A simple hypothesis is that NPs, through removal of oil/water contact, decrease the effective interfacial surface area over which the surfactants act. This would effectively lower γ_{ow} compared to when the same quantity of surfactants were acting on the interface in the absence of NPs.

The series of simulations shown in Figure 6 was designed to test this hypothesis. When localized at the interface, a single NP of radius 10 Å removes over 300 Å^2 of oil/water surface area. Therefore by using 24 NPs in a 125 Å × 125 Å surface area system, more than half of A_{ow} is effectively removed. If our hypothesis holds true throughout this composition range, the surface tension would be expected to show a monotone decrease to a final value of close to zero for the four systems of Figure 6. However, the actual change of γ_{ow} is given in Figure 7 and invalidates our hypothesis.

Initially, when no NPs are present, the system has a surface tension of 29.3 mN m^{-1} which is roughly half of the pure oil/water surface tension of 50.1 mN m^{-1}. The surface tension drops upon addition of NPs, indicating that there is synergy between the two species at low NP concentrations. However, at higher concentrations the expected lowering of surface tension is not seen, and instead the surface tension plateaus at ~12 mN m^{-1}.

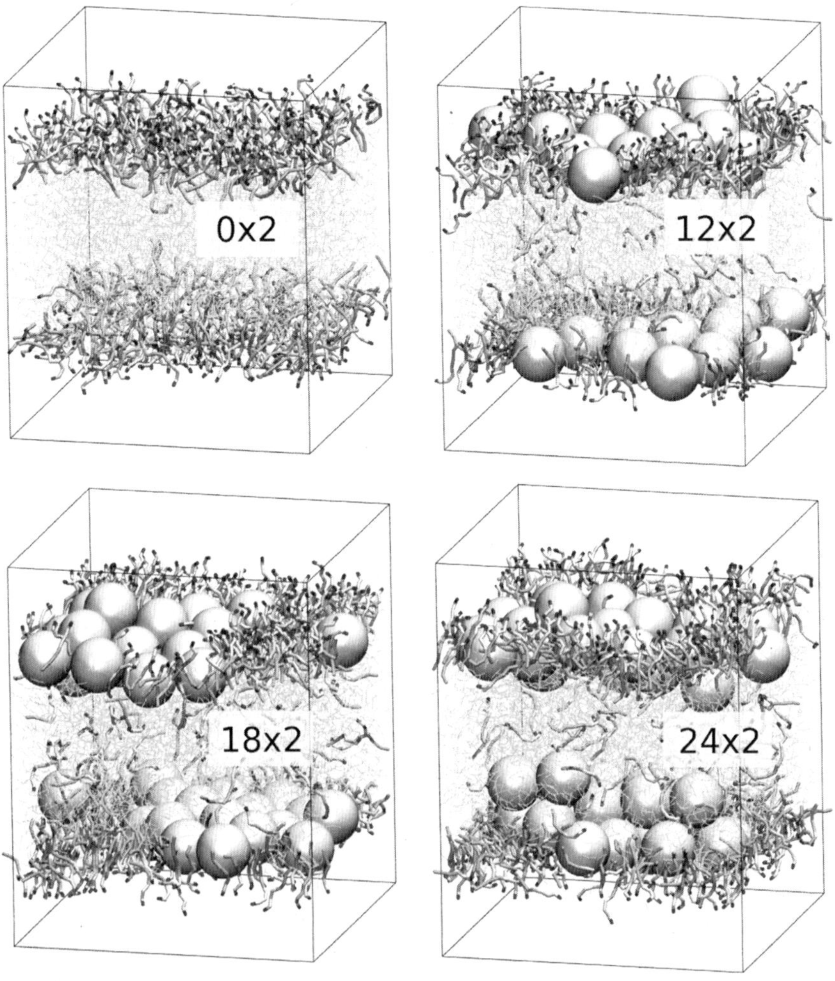

Figure 6. Snapshots of oil/water interfaces with varying numbers (labeled in each panel) of radius 10 Å NPs and 240 $C_{12}E_3$ surfactants. The ×2 is used to indicate the presence of two interfaces of identical composition. The coloring scheme is identical to Figure 1.

These results can be rationalized by acknowledging that the localization of NPs at the interface is driven by the last term of Equation 4, $-\gamma_{ow}\pi(R^2 - \eta^2)$. Surfactants act to lower the value of γ_{ow}, decreasing the thermodynamic driving force for particle adsorption to the interface. Past a critical value of the oil/water surface tension, the excess interfacial free energy of the oil/water interface, $A_{ow}\gamma_{ow}$, is no longer the dominant factor and further localization of particles is not favorable. This is seen in Figure 6 where the equilibrium position of the NPs shifts towards the oil phase, and particles are also pushed into the oil phase.

If the NPs used in this study instead preferred water over oil (see Figure 2) we would expect the opposite behavior, namely that the NPs would be displaced towards the water phase upon addition of surfactants.

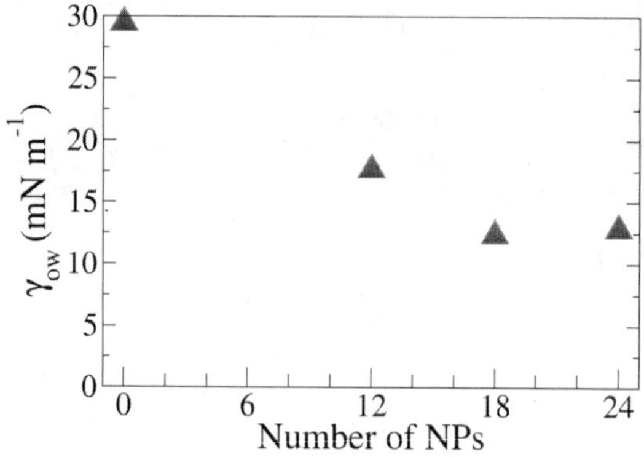

Figure 7. Oil/water interfacial tension for the systems shown in Figure 6. Initial addition of NPs lowers the γ_{ow} value although the synergistic effect diminishes with increasing numbers of NPs.

Surfactant Functionalized NPs

We now turn our attention to NPs functionalized with chemically absorbed surfactants. Organic ligands are often used to functionalize NPs to improve their colloidal stability, chemical stability, and interfacial activity (*38*). An example is the use of (mercaptoundecyl)tetra(ethylene glycol) ether functionalized gold nanoparticles to stabilize oil in water emulsions (*39*). It should be noted that this is a commercially available product, and is identical to the C_xE_y (with x = 12, y = 4) class of surfactants, except for an additional thiol moiety at the alkane terminus.

Ongoing MD simulation studies have shown the dramatic changes such chemical functionalization has on the interfacial activity and the dimerization free energy profile of NPs. For example, consider the 5.6 Å radius NP shown in Figure 2, which shows a free energy minimum of ~10.0 kcal mol^{-1}. Attachment of 20 $C_{12}E_2$ surfactants to this particle lowers the interfacial binding free energy to ~60 kcal mol^{-1}. Furthermore, preliminary results indicate that the dimerization free energy of NPs is also changed significantly. For the 10 Å radius particle, addition of even 10 $C_{12}E_3$ surfactants removes the attractive well, and the free energy of dimerization becomes completely repulsive.

The focus of this section is on simulations carried out on a NP of core radius 5.6 Å and ε_{NP-W} = 0.378 kcal mol^{-1}, with 20 bound $C_{12}E_2$ ligands. Emphasis is given to the role the surfactants play in shape deformations of the NP in the vicinity of an oil/water interface. A parameter for quantifying the tendency to deform is introduced as δ, the system deformability.

In the beginning of the chapter a simple theoretical model of the energetics of NPs at oil/water interfaces was presented. Some deficiencies of this theory are known, but a critical assumption made in its derivation is the ideal spherical shape of the NP itself. The flexibility of the surface functionalizing ligands violates this assumption (here we consider the NP to be the composite of the core and the functionalizing ligands). For larger cores (e.g. micron sized particles) the deformable layer is negligible and the assumption of spherical geometry may be accurate. However the importance of the deformable layer increases with smaller

NP core sizes, and could play an important role in some cases in determining the behavior of NPs at fluid interfaces (*32*).

Surfactant-Solvent Interactions and Preferential Orientation of Surfactant Ligands Functionalizing a NP Core

For the 5.6 Å particle functionalized with 20 $C_{12}E_2$ ligands considered here, the core is of ~1.1 nm diameter and the surfactant chains are of fully extended length ~1.5 nm, i.e., the two components are of similar length scales. Simulations of the particle at different positions relative to an oil/water interface were carried out. Representative images from these simulations are shown in Figure 8 from which it is clear that the surfactants strongly interact with the interface. The preferential orientations of the surfactants cause shape deformations of the NP as seen in Figure 8.

The preferred orientation of the surfactants is quantified by the angle distributions shown in Figure 9. At η values where the NP is in a bulk solvent environment corresponding to the first and last panels in Figure 8, there is no preferred orientation of the surfactant chains and hence uniform distributions are seen in Figure 9. Pronounced peaks are visible at η positions where the surfactants can interact with the interface (bottom left panel of Figure 9). It should be noted that Equation 4 predicts changes in ΔG only between $-R \leq \eta \leq R$ and at larger η positions the particle is completely solvated in bulk solvent. However, dispersive interactions of the deformable coating of the NP with the interface broadens the η range over which the free energy varies.

All the observed results can be explained through the preferred solvation environments of the PEG and alkane segments of the surfactants. When the NP is deep within the heptane phase, the surfactants do not need to protect the surface of the core (since Figure 2 established that the core prefers oil over water) and hence show extended conformations, swelling the ligand corona of the NP. At η values within dispersive interaction range of the interface, the PEG units are attracted towards the water phase resulting in preferred orientations of the surfactant chains at the interface. At the interface the surfactants create an annular ring with the PEG segment curling into the water phase (see the middle panel of Figure 8). Similar behaviors can be observed in the water phase ($\eta > 0$), but here the surfactants protect the high-energy NP/water surface by collapsing onto the NP core.

The extent of swelling of the surfactant layer in oil versus water is important because it is indicative of the conformational freedom of the surfactants with respect to the NP core and hence the susceptibility of the NP to deform in shape. To measure the extension of the ligands from the NP core the radius of gyration, R_g, can be used. In this context R_g can be taken as,

$$R_g^2 = \frac{1}{N}\left\langle \sum_{i=1}^{N} (r_i - r_{NP})^2 \right\rangle \tag{6}$$

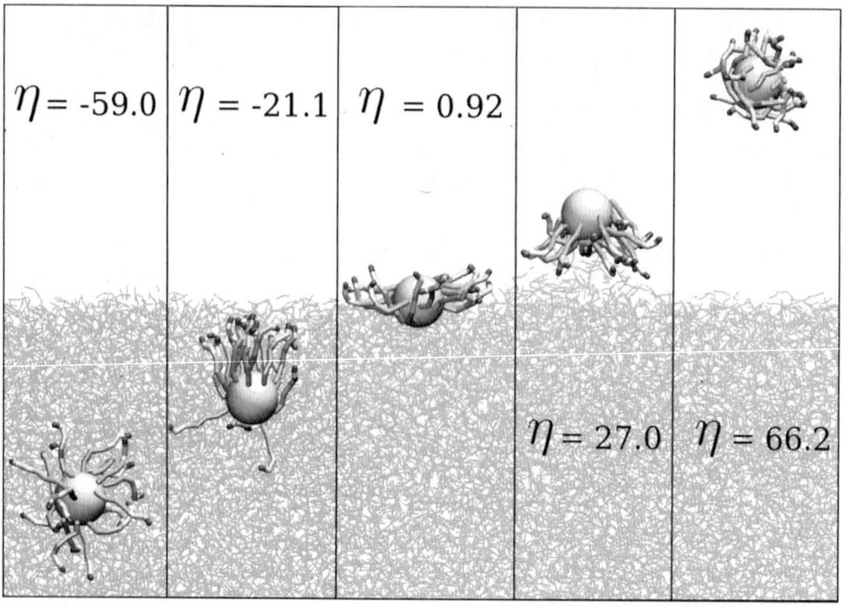

Figure 8. Snapshots of a core radius 5.6 Å NP functionalized with 20 $C_{12}E_2$ ligands in the vicinity of a planar oil/water interface. The surfactants adopt conformations in which the PEG block is close to water and the alkane block is close to heptane. These preferential orientations and conformations cause a position dependent deformation of the NP shape. η values are given in units of Å.

where r_i is the position of the ith surfactant bead, r_{NP} is the nanoparticle core position and N the total number of surfactant beads.

The radius of gyration for the functionalized NP in oil, $R_g(o)$, and water, $R_g(w)$, are 16.00 Å and 13.50 Å respectively. The values for $R_g(o)$ and $R_g(w)$ for other selected NP systems are given in Table 4. It is clear from the table that in all cases the NP is swollen to a greater extent in oil than in water.

Deformability

As stated previously, the difference of the radius of gyration values in the two bulk solvent phases is related to the susceptibility of the particle to deform according to its environment. Using this information, a dimensionless parameter, the system deformability (δ) can be defined,

$$\delta = \frac{C|R_g(o) - R_g(w)|}{R_g(w) + R_g(o)} \tag{7}$$

where C is a scaling factor taken as 20 (to bring the value to a convenient scale). The δ value for the NP studied in the previous section is 1.69. Table 4 also includes δ values for other NP systems, in order to see some trends in deformability. The deformability increases with the ability of the particle to change shape, based

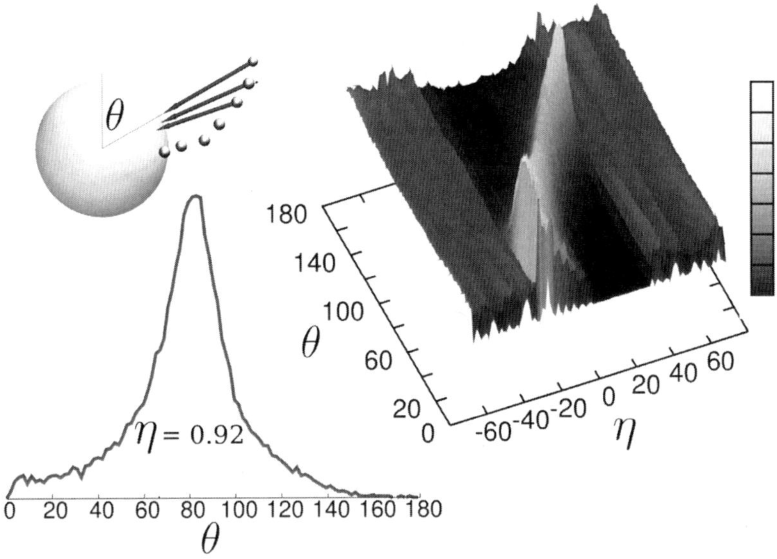

Figure 9. Angle distribution plots of 20 $C_{12}E_2$ surfactant ligands chemisorbed to a 5.6 Å radius NP core as a function of the interfacial position η of the NP (in units of Å). The angle between the oil/water interfacial normal vector and each NP center to surfactant bead vector (shown in the top left panel) was tabulated to obtain orientational distributions for specific values of η (bottom left). Plotting these distributions versus η produces the multidimensional graph shown in the top right panel.

on the ratio of the change in the radius of gyration values (numerator) to the average size of the particle (denominator). This parameter captures the effect of (1) increased density of surfactant coverage, which tends to lock the surfactants and hence lower the numerator, (2) changes in surfactant composition, which affects the R_g values, and (3) the ratio of the average extension of the surfactants to the NP core size. A graphical illustration of different NP types and their corresponding deformability parameters is given in Figure 10.

Measures of the 'softness' of particles are important in a number of fields. It has been shown that the deformability of particles has an effect on their rheological (*40*) properties as well as their self-assembly properties (*41*). Other measures of softness exist, an example being the χ factor used by Whetten *et al.* (*42*). This quantity is simply the ratio of the fully extended length of the functionalizing molecule to the size of the rigid core, which is used to predict the three-dimensional packing patterns of particle assemblies. The importance of characterizing the softness of particles was asserted by Tsai (*43*) who suggested that for any medicinal application of NPs rigorous physical and chemical characterization of particle size and structure is necessary. In the same article the author articulates the need for 'complementary, orthogonal methods' for this characterization. The δ parameter calculated through molecular dynamics simulations as introduced above is a prime example as a valuable characteristic property of functionalized NPs.

Table 4. Radius of gyration values for surfactant functionalized NPs

System	NP Core Size* (Å)	Surfactant Ligand	Number of Ligands	$R_g(o)$ (Å)	$R_g(w)$ (Å)	δ
1	5.6	$C_{12}E_2$	20	16.00	13.50	1.69
2	5.6	$C_{12}E_2$	40	16.93	15.50	1.77
3	5.6	$C_{12}E_3$	20	16.97	15.52	1.79
4	5.6	$C_{12}E_3$	40	18.05	16.30	2.03
5	10.0	$C_{12}E_3$	46	21.05	20.08	0.94
6	10.0	$C_{12}E_3$	91	22.42	21.09	1.22
7	10.0	$C_{12}E_3$	183	23.92	23.70	0.18
8	20.0	C_6E_1	286	27.15	26.94	0.16
9	20.0	$C_{12}E_3$	286	32.60	31.76	0.52
10	20.0	$C_{18}E_5$	286	38.18	35.71	1.34

* other than system 1, for which $\varepsilon_{NP-W} = 0.378$ kcal mol^{-1}, all other NP cores (systems 2-10) considered in the table have an $\varepsilon_{NP-W} = 0.596$ kcal mol^{-1}

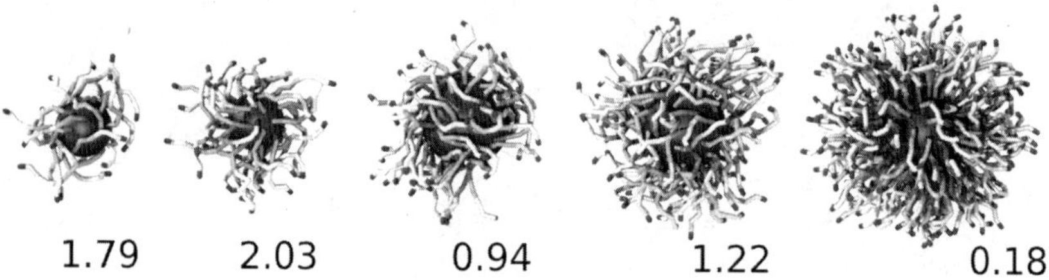

Figure 10. Selected NPs with different deformability (δ) values, shown below each NP image. From left to right the particles correspond to systems 3-7 of Table 4.

Conclusions

We have studied systems containing NPs and surfactants at oil/water interfaces using molecular dynamics computer simulations. We separately considered situations where the surfactants were free versus chemically absorbed to the NP core surface. It was found that at low surfactant and NP concentration, these species showed a synergistic effect in lowering the oil/water surface tension. However, as the effective area per surfactant is decreased the energetic drive for particle localization to the interface is severely attenuated. Studies of surfactant functionalized NPs showed that the interaction of the surfactants with the oil/water interface causes deformations in the NP shape. The ability of a particle to deform due to its environment was quantified using a deformability parameter δ. We hope to use this parameter in future studies of surfactant functionalized NPs to relate the surfactant layer properties to the properties of NP interfacial assemblies.

References

1. Pickering, S. U. *J. Chem. Soc.* **1907**, *91*, 2001–2021.
2. Ramsden, W. *Proc. R. Soc. London, Ser. A* **1903**, *72*, 156–164.
3. *Colloidal Particles at Liquid Interfaces*; Binks, B. P., Ed.; Cambridge University Press: New York, 2006.
4. *Colloids and Colloid Assemblies*; Caruso, F., Ed.; Wiley-VCH: Weinheim, 2004.
5. Day, L. In *An Encyclopaedia of the History of Technology*; McNeil, I., Ed.;Routledge: New York, 1990; pp 203–204.
6. Hunter, T. N.; Pugh, R. J.; Franks, G. V.; Jameson, G. J. *Adv. Colloid Interface Sci.* **2008**, *137*, 57–81.
7. Binks, B. P. *Curr. Opin. Colloid Interface Sci.* **2002**, *7*, 21–41.
8. Daniel, M. C.; Astruc, D. *Chem. Rev.* **2004**, *104*, 293–346.
9. Murray, C. B.; Kagan, C. R.; Bawendi, M. G. *Annu. Rev. Mater. Sci.* **2000**, *30*, 545–610.
10. Gupta, A. K.; Gupta, M. *Biomaterials* **2005**, *26*, 3995–4021.
11. Russel, W.; Saville, D.; Schowalter, W. *Colloidal Dispersions*; Cambridge University Press: Cambridge, 1989.
12. Shevchenko, E. V.; Talapin, D. V.; Kotov, N. A.; O'Brien, S.; Murray, C. B. *Nature* **2006**, *439*, 55–59.
13. Nagarajan, R.; Ruckenstein, E. *Langmuir* **1991**, *7*, 2934–2965.
14. Shinoda, W.; Mikami, M. *J. Comput. Chem.* **2003**, *24*, 920–930.
15. Hoover, W. G. *Phys. Rev.* **1985**, *31*, 1695–1697.
16. Andersen, H. C. *J. Chem. Phys.* **1980**, *72*, 2384–2393.
17. Tuckerman, M. E.; Berne, B. J.; Martyna, G. J. *J. Chem. Phys.* **1991**, *94*, 6811–6815.
18. Wang, J.; Wang, D.; Sobal, N. S.; Giersig, M.; Jiang, M.; Mohwald, H. *Angew. Chem., Int. Ed.* **2006**, *45*, 7963–7966.
19. Frenkel, D.; Smit, B. In *Understanding Molecular Simulation*; McNeil, I., Ed.;Academic Press: San Diego, 2002; p 34.
20. Müller, M.; Katsov, K.; Schick, M. *Phys. Rep.* **2006**, *434*, 113–176.
21. Shinoda, W.; DeVane, R.; Klein, M. L. *Soft Matter* **2008**, *4*, 2454–2462.
22. Kalescky, R. J. B.; Shinoda, W.; Moore, P. B.; Nielsen, S. O. *Langmuir* **2009**, *25*, 1352–1359.
23. Hamaker, H. C. *Physica (Amsterdam)* **1937**, *4*, 1058–1072.
24. Vázquez, U. O. M.; Shinoda, W.; Moore, P. B.; Nielsen, S. O. *J. Math. Chem.* **2009**, *45*, 161–174.
25. Chiu, C.-c.; Moore, P. B.; Shinoda, W.; Nielsen, S. O. *J. Chem. Phys.* **2009**, *131*, 244706.
26. Rosen, M. J. *Surfactants and Interfacial Phenomena*; John Wiley & Sons, Inc.: New York, 2004.
27. Lange, R. K. *Surfactants: A Practical Handbook*; Hanser Publishers: Munich, 1999.
28. Bresme, F.; Oettel, M. *J. Phys.: Condens. Matter* **2007**, *19*, 413101–413133.
29. Murai, J.; Marukawa, T.; Mima, T.; Arai, S.; Sasaki, K.; Saka, H. *J. Mater. Sci.* **2006**, *41*, 2723–2727.

30. Chandler, D. *Introduction to Modern Statistical Mechanics*; Oxford University Press: New York, 1987.
31. Frenkel, D; Smit, B. *Understanding Molecular Simulation*; Academic Press: San Diego, 2002.
32. Ranatunga, R. J. K. U.; Kalescky, R. J. B.; Chiu, C.-c.; Nielsen, S. O. *J. Phys. Chem. C* **2010**, *114*, 12151–12157.
33. Lin, Y.; Boker, A.; Skaff, H.; Cookson, D.; Dinsmore, A. D.; Emrick, T.; Russell, T. P. *Langmuir* **2005**, *21*, 191–194.
34. Liang, Y.; Hilal, N.; Langston, P.; Starov, V. *Adv. Colloid Interface Sci.* **2007**, *134–135*, 151–166.
35. Bresme, F.; Lehle, H.; Oettel, M. *J. Chem. Phys.* **2009**, *130*, 214711.
36. Eskander, N. G.; Simovic, S.; Prestidge, C. A. *Phys. Chem. Chem. Phys.* **2007**, *9*, 6426–6434.
37. Luo, M.; Song, Y.; Dai, L. L. *Mol. Simul.* **2009**, *35*, 773–784.
38. Sperling, R. A.; Parak, W. J. *Philos. Trans. R. Soc., A* **2010**, *368*, 1333–1383.
39. Glogowski, E.; Tangirala, R.; He, J.; Russell, T. P.; Emrick, T. *Nano Lett.* **2007**, *7*, 389–393.
40. Zackrisson, M.; Stradner, P.; Schurtenberger, A.; Bergenholtz, H. *Phys. Rev. E* **2006**, *73*, 011408.
41. Landman, U.; Luedtke, W. D. *Faraday Discuss.* **2003**, *125*, 1–22.
42. Whetten, R. L.; Shafigullin, M. N.; Khoury, J. T.; Schaaf, T. G.; Verzmar, I.; Wilkinson, A. *Acc. Chem. Res.* **1999**, *32*, 397–406.
43. Tsai, D.-H.; Zangmeister, R. A.; Pease, L. F., III; Tarlov, M. J.; Zachariah, M. R. *Langmuir* **2008**, *24*, 8483–8490.

Editor's Biography

Ramanathan Nagarajan

Dr. Nagarajan is Emeritus Professor of Chemical Engineering at The Pennsylvania State University and is currently a research scientist at the Natick Soldier Research, Development & Engineering Center. His research interests span the fundamentals and applications of soft nanoparticles, particularly those generated by self-assembly of surfactant and block copolymer molecules. He has authored approximately 100 publications which have been cited over 3000 times with an H index of 30. Nagarajan is the Program Chair of ACS Division of Colloid and Surface Chemistry and has served as the Guest Editor of "Colloids and Surfaces" and "Advances in Colloid and Interface Science" and as editorial board member of "Journal of Colloid and Interface Science", “Encyclopedia of Surface and Colloid Science” and “Journal of Macromolecular Science Part A. Pure and Applied Chemistry”. Nagarajan is co-editor of the ACS Symposium Series books, “Nanoparticles: Synthesis, Stabilization, Passivation and Functionalization” published in 2008 and “Nanoscience and Nanotechnology for Chemical and Biological Defense” published in 2009.

© 2011 American Chemical Society

Indexes

Author Index

Subject Index

C

D

E

F

M

N

O

P

S

T

U

V

W

X

Z